Volume 23

CRM MONOGRAPH SERIES

Centre de Recherches Mathématiques
Université de Montréal

Mathematical Techniques for Analyzing Concurrent and Probabilistic Systems

J. J. M. M. Rutten
Marta Kwiatkowska
Gethin Norman
David Parker

Prakash Panangaden
Franck van Breugel
Editors

The Centre de Recherches Mathématiques (CRM) of the Université de Montréal was created in 1968 to promote research in pure and applied mathematics and related disciplines. Among its activities are special theme years, summer schools, workshops, postdoctoral programs, and publishing. The CRM is supported by the Université de Montréal, the Province of Québec (FCAR), and the Natural Sciences and Engineering Research Council of Canada. It is affiliated with the Institut des Sciences Mathématiques (ISM) of Montréal, whose constituent members are Concordia University, McGill University, the Université de Montréal, the Université du Québec à Montréal, and the Ecole Polytechnique. The CRM may be reached on the Web at www.crm.umontreal.ca.

American Mathematical Society
Providence, Rhode Island USA

The production of this volume was supported in part by the Fonds pour la Formation de Chercheurs et l'Aide à la Recherche (Fonds FCAR) and the Natural Sciences and Engineering Research Council of Canada (NSERC).

2000 *Mathematics Subject Classification.* Primary 68Q60; Secondary 68Q85.

For additional information and updates on this book, visit
www.ams.org/bookpages/crmm-23

Library of Congress Cataloging-in-Publication Data

Mathematical techniques for analyzing concurrent and probabilistic systems / J.J.M.M. Rutten... [et al.]; Prakash Panangaden, Franck van Breugel, editors.
p. cm. — (CRM monograph series, ISSN 1065-8599; v. 23)
Includes bibliographical references.
ISBN 0-8218-3571-8 (alk. paper)
1. System analysis. I. Rutten, J.J.M.M. II. Panangaden, P. (Prakash). III. Breugel, Franck van, 1966– IV. Series.
QA402.M3594 2004
003–dc22 2004041069

∞ The paper used in this book is acid-free and falls within the guidelines
established to ensure permanence and durability.
This volume was submitted to the American Mathematical Society
in camera ready form by the Centre de Recherches Mathématiques.
Visit the AMS home page at http://www.ams.org/

10 9 8 7 6 5 4 3 2 1 09 08 07 06 05 04

Contents

Preface

The impact of mathematics and logic on theoretical computer science has been great from its very birth. It is perhaps less widely known among general mathematicians that mathematics has had a profound impact on computer systems. It has been about 40 years since pioneers like McCarthy, Strachey, Landin, Dijkstra, Scott and deBakker, among others, provided a firm mathematical underpinning to the study of programming languages. It is also about 30 years since Robin Milner, Tony Hoare, Carl Petri and others began looking at concurrent systems from a mathematical point of view.

At this point the theory of concurrency is a very rich subject with annual conferences and journals devoted to research in concurrency theory. The first set of lectures presents a new way of thinking about concurrency; the **coalgebraic** approach. Though it was realized implicitly from the start that coalgebra plays a key role in concurrent systems, it was only made very explicit quite recently in large part because of the work of Prof. Rutten. Roughly speaking, one can say that algebra gives the syntactic structure while a coalgebra describes the behaviour or dynamics of the system. Rutten presents an elegant "calculus of streams" for reasoning coalgebraically and shows some striking formal analogies between calculus as understood in the sense of analysis and the stream calculus.

A relatively young offshoot of concurrency theory is the theory of probabilistic, concurrent systems. The second set of lectures gives an overview of modelling and analysis of such systems. In particular the focus is on **automated tools** for reasoning about the behaviour and correctness of such systems. Prof. Kwiatkowska is one of the pioneers of this area and has been the driving force behind making the transition from theory to practice with a large active group developing implementations of the reasoning tools, carrying out case studies and, of course, enriching the theory further.

Both sets of lectures are suitable for beginning graduate students and even some advanced undergraduates. This helps fill a gap which should make it easier for graduate students to gain entry into these exciting research areas. The notes were reviewed by the editors. We wish to thank the authors for being willing to take time out of their busy research schedules to write these detailed expository notes. We hope that people will enjoy reading them as much as we enjoyed listening to the lectures in Montréal.

The two sets of notes in this volume were based on lectures delivered by Prof. Jan Rutten and Prof. Marta Kwiatkowska at a workshop on "Mathematical Techniques for Analyzing Systems" held from September 30th to October 4th 2002. This workshop was part of the Année Thématique of the Centre de Recherche Mathématiques devoted to theoretical computer science and we are very grateful to the CRM and its director, Jacques Hurtubise, whose generous support made the workshop possible. We would also like to thank the staff at CRM: Jean Le-Tourneux, André Montpetit, Louise Letendre, Louis Pelletier and Josée Laferriére who were very helpful in organizing the workshop and producing the book.

The workshop was attended by about 30 participants, including nearly a dozen students, and included several other individual lectures on the theme of the workshop. We would like to thank all the participants for their stimulating comments and the exciting atmosphere at the workshop.

Prakash Panangaden
McGill University
Montréal, Québec

Franck van Breugel
York University
Toronto, Ontario

On Streams and Coinduction

J. J. M. M. Rutten

Preface

The natural numbers are the most famous instance of an initial algebra and constitute the typical example for the explanation of induction. Similarly and dually, streams are the best known example of a final coalgebra and offer a perfect playground for the use of coinduction, both for definitions and for proofs. In these notes, we explain the basics of streams over an arbitrary alphabet in Chapter 1. Next we study the special case of streams over the reals in Chapter 2, and show how the (field) structure of the reals induces an equally rich structure on the set of streams. This leads to a calculus of streams, in close analogy to classical calculus in analysis, which is similar but more general than the calculus of generating functions. We apply the calculus of streams to the problem of solving differential equations in Chapter 3 and to various counting problems from enumerative combinatorics in Chapter 4. In Chapter 5, we apply streams and coinduction to a problem stemming from the world of computer science, more specifically, software engineering, by constructing a model for (software) component connectors. In Chapter A, the definitions of the most important operations on streams are summarized.

Chapter 1 is the basis for all subsequent chapters. Chapter 2 is needed for both Chapter 3 and Chapter 4 but not for Chapter 5, which depends solely on Chapter 1. In addition to the work by many other authors, which is discussed at the end of each chapter, the material of Chapters 1 through Chapter 4 is based on the following papers: [**35–40**]. Chapter 5 is based on very recent joint work with Farhad Arbab (CWI, Amsterdam), which appeared as: F. Arbab and J. J. M. M. Rutten, *A coinductive calculus of component connectors*, Report SEN-R0216, CWI, 2002.

Amsterdam, September 2002
Jan Rutten

Acknowledgments

I am very grateful to Prakash Panangaden for many years of inspiring discussions and meetings (starting in a discotheque in Tampere, Finland, in 1988), for organising the meeting 'Mathematical Models and Techniques for Analysing Systems' in the fall of 2002, and for asking me to present my work on streams and coinduction there. Many thanks are due to the following persons, for all what I have learned from them over the past years, through their papers, discussions, comments, and suggestions: Jiri Adamek, Farhad Arbab, Jaco de Bakker, Alexandru Baltag, Falk Bartels, Frank de Boer, Marcello Bonsangue,Franck van Breugel, Andrea Corradini, Philippe Darondeau, Jan van Eijk, Martin Escardó, H. Peter Gumm, Jan Heering, Furio Honsell, Bart Jacobs, Jan Willem Klop, Joost Kok, Jan Komenda, Clemens Kupke, Alexander Kurz, Elena Marchiori, Doug McIlroy, Wilfried Meyer Viol, Mike Mislove, Ugo Montanari, Larry Moss, Ataru Nakagawa, Maurice Nivat, Prakash Panangaden, Dusko Pavlović, Andy Pitts, Gordon Plotkin, John Power, Horst Reichel, Martin Rößiger, Jan van Schuppen, Nico Temme, Hendrik Tews, Daniele Turi, Erik de Vink, and Fer-Jan de Vries. I am very grateful to Franck van Breugel for many detailed comments on an earlier version of these notes.

CHAPTER 1

Streams and Coinduction

We introduce the set of streams (over a given set) and explain, using some elementary notions from universal coalgebra, how to define streams and operations on streams by coinduction, and how to prove facts about streams by coinduction.

1.1. Streams

Let A be any set. We define the set A^ω of all streams over A as

$$A^\omega = \{\sigma \mid \sigma\colon \{0,1,2,\ldots\} \to A\}.$$

In this chapter, no assumptions on A are made but later we shall look at specific choices for A. In particular, A will be the set $\mathbb{R}$ of real numbers in Chapter 2. For a stream σ, we call $\sigma(0)$ the *initial value* of σ. We define the *derivative* σ' of a stream σ, for all $n \geq 0$, by

$$\sigma'(n) = \sigma(n+1).$$

Initial value and derivative are usually called *head* and *tail* but the present terminology helps us, as we shall see, to develop a *calculus* of streams in close analogy to classical calculus in analysis.

Although streams will be viewed and handled as single mathematical entities, it will at various moments be convenient to refer to the individual elements of which they are made. For this, we shall use the following notation:

$$\sigma = \big(\sigma(0), \sigma(1), \sigma(2), \ldots\big).$$

(Similarly, we shall write $\tau = \big(\tau(0), \tau(1), \tau(2), \ldots\big)$ and the like.) With this notation, the derivative of σ is given by

$$\sigma' = \big(\sigma(1), \sigma(2), \sigma(3), \ldots\big).$$

For any $n \geq 0$, $\sigma(n)$ is called the nth *element* of σ. It can also be expressed in terms of higher-order stream derivatives, defined, for all $k \geq 0$, by

$$\sigma^{(0)} = \sigma, \quad \sigma^{(k+1)} = (\sigma^{(k)})'.$$

Now the nth element of σ is also given by

$$\sigma(n) = \sigma^{(n)}(0). \tag{1}$$

1.2. Proofs by Coinduction

In order to conclude that two streams σ and τ are equal, it is both necessary and sufficient to prove

$$\forall n \geq 0, \quad \sigma(n) = \tau(n). \tag{2}$$

What general methods for establishing the validity of (2) are there? A straightforward *induction* on the natural number n (prove $\sigma(0) = \tau(0)$ and show that

$\sigma(n) = \tau(n)$ implies $\sigma(n+1) = \tau(n+1)$) may seem the obvious answer, but very often it is not. There will be numerous occasions where we will have no workable description or formula for $\sigma(n)$ and $\tau(n)$, and where, consequently, induction simply cannot be applied.

Instead, we shall take a coalgebraic perspective on A^ω, and use almost exclusively the proof principle of *coinduction*, which is formulated in terms of the following notion.

DEFINITION 1.2.1. A *bisimulation* on A^ω is a relation $R \subseteq A^\omega \times A^\omega$ such that, for all σ and τ in A^ω, if $\langle \sigma, \tau \rangle \in R$ then

(i) $\sigma(0) = \tau(0)$ and
(ii) $\langle \sigma', \tau' \rangle \in R$.

(We shall sometimes write $\sigma \, R \, \tau$ for $\langle \sigma, \tau \rangle \in R$.)

One easily checks that unions and (relational) compositions of bisimulations are bisimulations again. If there exists a bisimulation relation R with $\sigma \, R \, \tau$ then we write $\sigma \sim \tau$ and say that σ and τ are *bisimilar*. This bisimilarity relation $\sim$ is the union of all bisimulations and, therewith, the greatest bisimulation.

THEOREM 1.2.2 (Coinduction). *For all* $\sigma, \tau \in A^\omega$,

$$\sigma \sim \tau \implies \sigma = \tau.$$

PROOF. Consider two streams σ and τ and let $R \subseteq A^\omega \times A^\omega$ be a bisimulation on A^ω containing the pair $\langle \sigma, \tau \rangle$. It follows by induction on n that $\langle \sigma^{(n)}, \tau^{(n)} \rangle \in R$, for all $n \geq 0$, because R is a bisimulation. This implies, again because R is a bisimulation, that $\sigma^{(n)}(0) = \tau^{(n)}(0)$, for all $n \geq 0$. By identity (1), $\sigma(n) = \tau(n)$, for all $n \geq 0$. Now $\sigma = \tau$ follows. □

The converse of the theorem above holds trivially, since $\{\langle \sigma, \sigma \rangle \mid \sigma \in A^\omega\}$ is a bisimulation relation on A^ω. The theorem can be used as a proof principle: in order to prove the equality of two streams σ and τ, it is sufficient to establish the existence of a bisimulation relation $R \subseteq A^\omega \times A^\omega$ with $\langle \sigma, \tau \rangle \in R$.

The above coinduction proof principle can be seen as a systematic way of strengthening the statement one is trying to prove: instead of proving only the single identity $\sigma = \tau$, one extends the set $\{\langle \sigma, \tau \rangle\}$ to a generally much larger set $R \subseteq A^\omega \times A^\omega$, with $\langle \sigma, \tau \rangle \in R$, such that R is a bisimulation relation. (As we shall see, the way such a larger set is constructed is always the same, and amounts to the computation of the closure of the original set under the taking of derivatives.) For such bisimulation relations, Theorem 1.2.2 states, once and for all, that for *all* pairs $\langle \alpha, \beta \rangle \in R$, we have $\alpha = \beta$. Since the proof of Theorem 1.2.2 is by induction, the theorem can thus be viewed as a systematic way of what sometimes is called *induction loading.*

The fact that the coinduction proof principle is proved by induction on the natural numbers, suggests that induction is a more fundamental principle than coinduction. We do not really address this issue here, but only remark that it is possible to develop a theory of so-called coinductive natural numbers. These consist of the ordinary natural numbers plus one extra element called infinity, and satisfy a basic coinduction proof principle from which the usual induction proof principle is derivable.

We shall see many examples of proofs by coinduction: one of the main reasons why we have become interested in A^ω at all, is that it offers a perfect playground for demonstrating the use and usefulness of coinduction. First, however, we discuss a corresponding definition principle.

1.3. Definitions by Coinduction

We use the word coinduction also as a term for certain definitions, called behavioural differential equations. They are based on a universal property of A^ω, which we introduce next, using again some elementary universal coalgebra.

A *stream coalgebra* or *stream automaton* is a pair $Q = (Q, \langle o_Q, t_Q \rangle)$ consisting of a set Q of states, together with an output function $o_Q \colon Q \to A$, and a transition function $t_Q \colon Q \to Q$. For a state q, the output value $o_Q(q)$ can be viewed as its *observable behaviour*, and the transition value $t_Q(q)$ (the next state) can be viewed as its *dynamical behaviour.*

A *homomorphism*

$$f \colon (P, \langle o_P, t_P \rangle) \to (Q, \langle o_Q, t_Q \rangle)$$

between two stream automata is a behaviour preserving function $f \colon P \to Q$: for all p in P, $o_P(p) = o_Q\big(f(p)\big)$ and $f\big(t_P(p)\big) = t_Q\big(f(p)\big)$. In other words, a function $f \colon P \to Q$ is a homomorphism if the following diagram commutes:

$$\begin{array}{ccc} P & \xrightarrow{\quad f \quad} & Q \\ {\scriptstyle \langle o_P, t_P \rangle}\big\downarrow & & \big\downarrow{\scriptstyle \langle o_Q, t_Q \rangle} \\ A \times P & \xrightarrow[1_A \times f]{} & A \times Q \end{array} \; .$$

Here $(1_A \times f)(\langle a, p \rangle) = \langle a, f(p) \rangle$, for all $\langle a, p \rangle \in A \times P$. The set A^ω of all streams can itself be turned into a stream automaton as follows. Defining $o \colon A^\omega \to A$ by $o(\sigma) = \sigma(0)$ and $t \colon A^\omega \to A^\omega$ by $t(\sigma) = \sigma'$, we obtain a stream automaton $(A^\omega, \langle o, t \rangle)$. It has the following universal property.

THEOREM 1.3.1. *The automaton $(A^\omega, \langle o, t \rangle)$ is* final *among the family of all stream automata. That is, for any automaton $(Q, \langle o_Q, t_Q \rangle)$ there exists a unique homomorphism $l \colon (Q, \langle o_Q, t_Q \rangle) \to (A^\omega, \langle o, t \rangle)$:*

$$\begin{array}{ccc} Q & \overset{\exists ! l}{- - - \to} & A^\omega \\ {\scriptstyle \langle o_Q, t_Q \rangle}\big\downarrow & & \big\downarrow{\scriptstyle \langle o, t \rangle} \\ A \times Q & \underset{1_A \times l}{- - \to} & A \times A^\omega \end{array} \; .$$

PROOF. Let $(Q, \langle o_Q, t_Q \rangle)$ be a stream automaton and define $l \colon Q \to A^\omega$ as

$$l(q) = \big(o(q), o\big(t(q)\big), o\big(t\big(t(q)\big)\big), \ldots\big)$$

for q in Q. It is straightforward to show that l is a homomorphism from $(Q, \langle o_Q, t_Q \rangle)$ to $(A^\omega, \langle o, t \rangle)$. For uniqueness, suppose f and g are homomorphisms from Q to A^ω. The equality of f and g follows by coinduction from the fact that $R = \{\langle f(q), g(q) \rangle \mid q \in Q\}$ is a bisimulation on A^ω, which is proved next. Consider $\langle f(q), g(q) \rangle \in R$. Because f and g are homomorphisms, $o\big(f(q)\big) = o_Q(q) = o\big(g(q)\big)$. Furthermore, $t\big(f(q)\big) = f\big(t_Q(q)\big)$ and $t\big(g(q)\big) = g\big(t_Q(q)\big)$. Because $\big\langle f\big(t_Q(q)\big), g\big(t_Q(q)\big)\big\rangle \in R$, this shows that R is a bisimulation. Now $f = g$ follows by the coinduction proof principle Theorem 1.2.2. $\square$

The finality of $(A^\omega, \langle o, t\rangle)$ provides the formal basis for *coinductive definitions* of (operators on) streams, which we call *behavioural differential equations.* Similarly to differential equations in analysis (such as $f(0) = 1$ and $f' = f$), such equations define streams by specifying their initial value and derivative, that is, their "behaviour". We present a few examples and shall become more general in later chapters. Consider the following behavioural differential equation:

derivative	initial value
$\sigma' = \sigma$	$\sigma(0) = a$

(We see that a behavioural differential equation consists of two equalities: one specifying the derivative and one specifying the initial value.) It is clear that the stream $\sigma = (a, a, a, \ldots)$ satisfies the equalities above, and that it is the only stream with this property. More formally, the unique existence of a solution of the behavioural differential equation can be established with the help of the finality of the set of streams. To this end, consider a stream automaton $(Q, \langle o_Q, t_Q\rangle)$ consisting of a singleton set $Q = \{q\}$ with $o_Q(q) = a$ and $t_Q(q) = q$. By Theorem 1.3.1, there exists a unique homomorphism

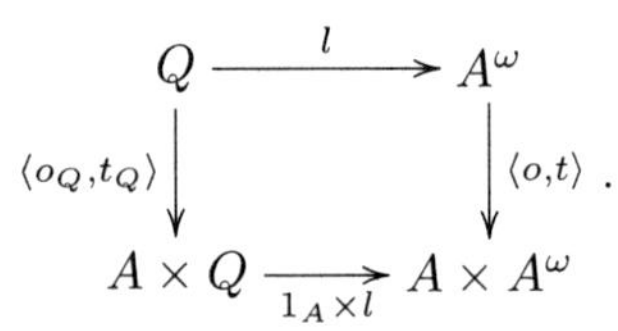

We now define $\sigma = l(q)$. Because l is a homomorphism, it follows that σ is a solution of the differential equation: $\sigma' = t(\sigma) = t\big(l(q)\big) = l\big(t_Q(q)\big) = l(q) = \sigma$ and $\sigma(0) = o(\sigma) = o\big(l(q)\big) = o_Q(q) = a$. If τ is a stream with $\tau' = \tau$ and $\tau(0) = a$, then $\sigma = \tau$ follows, by the coinduction proof principle Theorem 1.2.2, from the fact that $\{\langle \sigma, \tau\rangle\}$ is a bisimulation relation on A^ω. Thus σ is the only solution of the differential equation.

As a second example, we consider the following system of behavioural differential equations (one for each σ and τ in A^ω):

derivative	initial value
$\text{zip}(\sigma, \tau)' = \text{zip}(\tau, \sigma')$	$\text{zip}(\sigma, \tau)(0) = \sigma(0)$

In order to show that the above system of differential equations uniquely determines a binary operation zip: $A^\omega \times A^\omega$, we consider a stream automaton $(Q, \langle o_Q, t_Q\rangle)$, consisting of $Q = A^\omega \times A^\omega$ with, for all $\langle \sigma, \tau\rangle \in Q$,

$$t_Q(\langle \sigma, \tau\rangle) = \langle \tau, \sigma'\rangle, \quad o_Q(\langle \sigma, \tau\rangle) = \sigma(0).$$

As before, there exists by Theorem 1.3.1, a unique homomorphism $l \colon Q \to A^\omega$. We now define $zip(\sigma, \tau) = l(<\sigma, \tau>)$. Similar to the first example, it is not difficult to prove that *zip* is the unique function satisfying the above system of behavioural differential equations. Note that zip could have been, equivalently, defined by the following "global" formula, for every $n \geq 0$:

$$\text{zip}(\sigma, \tau)(2n) = \sigma(n), \quad \text{zip}(\sigma, \tau)(2n + 1) = \tau(n).$$

Although it may seem more complicated in this particular example, we prefer to work with behavioural differential equations for two reasons. Firstly, they are more general. We shall see many examples of streams defined by behavioural differential equations for which there is no easy "global" formula such as the one for zip above. Secondly, we can use the coinduction proof principle to reason about streams that have been defined coinductively, that is, by means of behavioural differential equations. Again, many convincing examples will follow.

The first example above involved a stream automaton with only one state; in the second example, an infinite automaton was used. One can, more generally, solve more complicated systems of behavioural differential equations by using more complicated stream automata. Much more can be said about the classification of families of differential equations and their solutions. In these notes, however, we shall only work with examples, and we shall refer to the literature for general results.

1.4. Examples of Coinductive Proofs

In order to present a few simple examples of coinductive reasoning, we introduce two more operators on streams:

derivative	initial value
$\text{even}(\sigma)' = \text{even}(\sigma'')$	$\text{even}(\sigma)(0) = \sigma(0)$
$\text{odd}(\sigma)' = \text{odd}(\sigma'')$	$\text{odd}(\sigma)(0) = \sigma'(0)$

The well-definedness of these operators can be proved as before. The reader should have no trouble convincing himself that these operations satisfy the following identities:

$$\begin{aligned}\text{even}(\sigma) &= (\sigma(0), \sigma(2), \sigma(4), \ldots),\\ \text{odd}(\sigma) &= (\sigma(1), \sigma(3), \sigma(5), \ldots).\end{aligned}$$

As an example, we prove the following identity by coinduction: for all $\sigma, \tau \in A^\omega$,

$$\text{even}(\text{zip}(\sigma, \tau)) = \sigma. \tag{3}$$

In order to apply the coinduction proof principle Theorem 1.2.2, we define

$$R = \{\langle \text{even}(\text{zip}(\sigma, \tau)), \sigma\rangle \mid \sigma, \tau \in A^\omega\}$$

and prove that R is a bisimulation relation. We have to show that every pair $\langle \text{even}(\text{zip}(\sigma, \tau)), \sigma\rangle \in R$ satisfies the two bisimulation conditions of Definition 1.2.1:

(i) $\text{even}(\text{zip}(\sigma, \tau))(0) = \sigma(0)$ and

(ii) $\langle \text{even}(\text{zip}(\sigma, \tau))', \sigma'\rangle \in R$

For (i), note that $\text{even}(\text{zip}(\sigma, \tau))(0) = \text{zip}(\sigma, \tau)(0) = \sigma(0)$. For (ii),

$$\begin{aligned}\langle \text{even}(\text{zip}(\sigma, \tau))', \sigma'\rangle &= \langle \text{even}(\text{zip}(\sigma, \tau)''), \sigma'\rangle\\ &= \langle \text{even}(\text{zip}(\tau, \sigma')'), \sigma'\rangle\\ &= \langle \text{even}(\text{zip}(\sigma', \tau')), \sigma'\rangle \in R.\end{aligned}$$

This concludes the proof that R is a bisimulation. Identity (3) then follows from the proof principle Theorem 1.2.2.

That was easy. We just defined our R as the set of all pairs that we wanted to prove equal. As it turned out, R was a bisimulation and the identity followed by coinduction. This is the right way to start coinductive proofs in general. Often, however, the relation has to be extended further before it satisfies condition (ii) of the definition of a bisimulation. In other words, the relation has to be closed under the taking of derivatives. At every such extension, condition (i), on initial values, has to be checked again. The proof of the following identity, for all $\sigma \in A^\omega$, clearly illustrates what we mean:

$$\text{zip}\big(\text{even}(\sigma), \text{odd}(\sigma)\big) = \sigma. \tag{4}$$

In order to apply coinduction, we begin again by defining

$$R = \big\{\big\langle \text{zip}\big(\text{even}(\sigma), \text{odd}(\sigma)\big), \sigma\big\rangle \mid \sigma \in A^\omega\big\}.$$

Next we check whether the pairs in R satisfy the bisimulation conditions (i) and (ii) of Definition 1.2.1. For (i), we have

$$\text{zip}\big(\text{even}(\sigma), \text{odd}(\sigma)\big)(0) = \text{even}(\sigma)(0) = \sigma(0).$$

For (ii), we compute

$$\begin{aligned}\big\langle \text{zip}\big(\text{even}(\sigma), \text{odd}(\sigma)\big)', \sigma'\big\rangle &= \big\langle \text{zip}\big(\text{odd}(\sigma), \text{even}(\sigma)'\big), \sigma'\big\rangle \\ &= \big\langle \text{zip}\big(\text{odd}(\sigma), \text{even}(\sigma'')\big), \sigma'\big\rangle\end{aligned}$$

which is *not* in R. The way to overcome this problem is as simple as it is effective: we extend R by including these latter pairs as well. In other words, our second proposal for R now looks like

$$\begin{aligned}R = \big\{\big\langle \text{zip}\big(\text{even}(\sigma), \text{odd}(\sigma)\big), \sigma\big\rangle \mid \sigma \in A^\omega\big\}& \\ \cup \big\{\big\langle \text{zip}\big(\text{odd}(\sigma), \text{even}(\sigma'')\big), \sigma'\big\rangle \mid \sigma \in A^\omega\big\}&.\end{aligned}$$

Next, the two bisimulation conditions should be checked for all the newly added pairs $\big\langle \text{zip}\big(\text{odd}(\sigma), \text{even}(\sigma'')\big), \sigma'\big\rangle \in R$. For (i),

$$\text{zip}\big(\text{odd}(\sigma), \text{even}(\sigma'')\big)(0) = \text{odd}(\sigma)(0) = \sigma'(0).$$

Computing derivatives, for (ii), gives

$$\begin{aligned}\big\langle \text{zip}\big(\text{odd}(\sigma), \text{even}(\sigma'')\big)', \sigma''\big\rangle &= \big\langle \text{zip}\big(\text{even}(\sigma''), \text{odd}(\sigma)'\big), \sigma''\big\rangle \\ &= \big\langle \text{zip}\big(\text{even}(\sigma''), \text{odd}(\sigma'')\big), \sigma''\big\rangle\end{aligned}$$

which *is* an element of R. This concludes both the construction of the relation R and, at the same time, the proof that it is in fact a bisimulation. Identity (4) now follows from the coinduction proof principle.

The above proof principle (and related principles) is sometimes referred to as the *bisimulation proof method*. It can prove identities in the way we saw above. But it can also be used to *disprove* an equality: one might encounter, at some stage of the construction of a bisimulation relation, a pair that has to be added in order to satisfy bisimulation condition (ii), but which does not satisfy condition (i). Then the conclusion is that all the pairs that have been considered so far, are *not* equal.

1.5. Coinduction and Greatest Fixed Points

There is yet another, in fact more classical, way of understanding bisimulations and the coinduction proof principle. Consider the set

$$\mathcal{P}(A^\omega \times A^\omega) = \{R \mid R \subseteq A^\omega \times A^\omega\},$$

of binary relations on A^ω, and the function

$$\Phi\colon \mathcal{P}(A^\omega \times A^\omega) \to \mathcal{P}(A^\omega \times A^\omega),$$

defined, for any $R \subseteq A^\omega \times A^\omega$, by

$$\Phi(R) = \{\langle \alpha, \beta\rangle \mid \alpha(0) = \beta(0) \wedge \langle \alpha', \beta'\rangle \in R\}.$$

As an immediate consequence of the definition of bisimulation, we have

$$R \text{ is a bisimulation} \iff R \subseteq \Phi(R).$$

Bisimulation relations are, in other words, *post-fixed points* of Φ. Consequently, the coinduction proof principle (Theorem 1.2.2) is equivalent to the following equality, where $\mathrm{id}_{A^\omega} = \{\langle \alpha, \alpha \rangle \mid \alpha \in A^\omega\}$:

$$\mathrm{id}_{A^\omega} = \bigcup\{R \mid R \subseteq \Phi(R)\}.$$

Since id_{A^ω} is itself a (bisimulation and thus a) post-fixed point, it is in fact the greatest fixed point of Φ. Therefore the above equality is an instance of the following well-known greatest fixed point theorem [**45**]. Let X be any set and let $\mathcal{P}(X) = \{V \mid V \subseteq X\}$ be the set of all its subsets. If $\Psi\colon \mathcal{P}(X) \to \mathcal{P}(X)$ is a monotone operator, that is, $R \subseteq S$ implies $\Psi(R) \subseteq \Psi(S)$ for all $R \subseteq X$ and $S \subseteq X$, then Ψ has a greatest fixed point $P = \Psi(P)$ satisfying

$$P = \bigcup\{R \mid R \subseteq \Psi(R)\}. \tag{5}$$

This equality can be used as a proof principle in the same way as Theorem 1.2.2: In order to prove that $R \subseteq P$, for any $R \subseteq X$, it suffices to show that R is a post-fixed point of Ψ, that is, $R \subseteq \Psi(R)$. We shall see further instances of this theorem (and related proof principles) in Chapter 5.

1.6. Other Coalgebras

In the language of category theory, a *coalgebra* of a functor F : Set $\to$ Set is a pair $\langle S, f\colon S \to F(S)\rangle$ consisting of a set S, called the carrier or state set of the coalgebra, and a function $f\colon S \to F(S)$ called the structure map of the coalgebra. Thus a stream automaton

$$Q \xrightarrow{\langle o,t\rangle} A \times Q$$

is a coalgebra of the functor $A \times (-)\colon$ Set $\to$ Set (which maps a set V to the Cartesian product $A \times V$ and a function f to $1_A \times f$). This functor can be viewed as a special instance of the functor $A \times (-)^B\colon$ Set $\to$ Set, which in addition to the set A has a set B as a second parameter. This functor maps a set V to the set $A \times V^B$, where $V^B = \{f \mid f\colon B \to V\}$ is the set of all functions from B to V. A coalgebra of this functor looks like

$$Q \xrightarrow{\langle o,t\rangle} A \times Q^B$$

with $o\colon Q \to A$, as before, but now with $t\colon Q \to Q^B$. Such a coalgebra is an automaton with state space Q, with outputs in A, given for any $q \in Q$ by $o(q)$, and with inputs from B, leading, for any input $b \in B$, to a next state $t(q)(b) \in Q$. Also the functor $A \times (-)^B$ has an interesting final coalgebra, consisting of the set

$$A^{B^*} = \{f \mid f\colon B^* \to A\}$$

of all functions from B^* to A (B^* is the set of finite sequences of elements of B). This set becomes particularly interesting if the set of outputs A is well-structured. Notably, if A is a *semi-ring* (which is, roughly speaking, a ring without an operation of subtraction) then A^{B^*} is the set of *formal power series* in many non-commutative variables from B and coefficients in A. Examples of such semi-rings are the set of real numbers (which is actually a ring) and the set $2 = \{0, 1\}$ of Boolean values.

The corresponding final coalgebras are the set $\mathbb{R}^{B^*}$ of formal power series over B with coefficients in $\mathbb{R}$, and the set

$$\begin{aligned} 2^{B^*} &= \{f \mid f : B^* \to 2\} \\ &\cong \{L \mid L \subseteq B^*\} \end{aligned}$$

of all languages over B. The following diagram summarises all the functors that we have mentioned so far. In the diagram, arrows should be read as "specialises to":

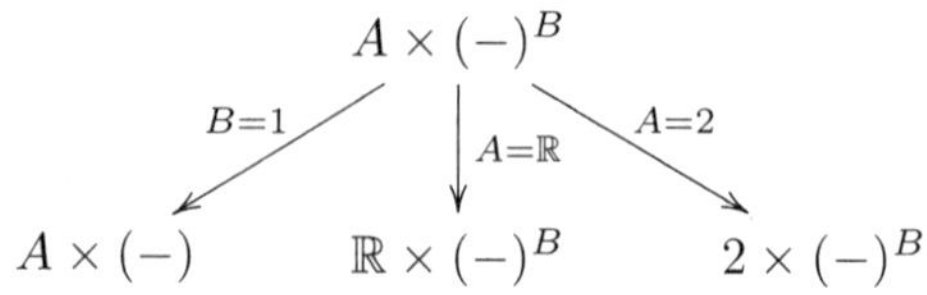

The common specialisation of the left two functors to $\mathbb{R} \times (-)$:

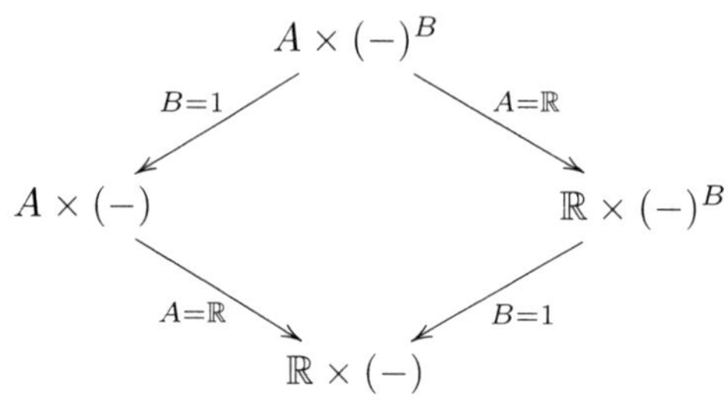

turns out to have (almost) all the interesting properties of the general case. The corresponding final coalgebra, which is the set $\mathbb{R}^\omega$ of all streams of real numbers, will be the subject of Chapter 2.

1.7. Discussion

Streams are just one out of many (final) coalgebras. We saw already some examples in Section 1.6. For many more examples and some of the basic elements of the theory of "universal" coalgebra, see [**23, 38**]. The notion of bisimulation is due to Park and Milner [**31, 33**], who designed it as a notion of equivalence for a theory of concurrent processes. (Earlier, it already arose, under the name of p-relation, in the world of Kripke frames and modal logic [**46**].) Final coalgebras have been used as models for certain dynamical systems at least since [**6, 27**]. It was not until a categorical generalisation of the notion of bisimulation was introduced, by Aczel and Mendler in [**2**], that coinduction (both as a definition and as a proof principle) was taken more seriously. Notably, Aczel used it as the basis for a theory of non-well-founded sets in [**1**]; see also earlier related models by Forti and Honsell [**18**]. By now, there is a host of literature on many aspects of both theory and applications of coalgebra. Rather than trying to give an overview of the relevant literature here, we refer the reader to the proceedings of the annual international workshops on *Coalgebraic Methods in Computer Science* (CMCS), which started in 1998. The proceedings have been published in Elsevier's Electronic lecture Notes in Theoretical Computer Science (ENTCS) series, as volumes 11, 19, 33, 44, and 65. Formal power series, automata, and languages, have been extensively studied in the literature (see for instance [**11, 26**], to mention just two out of many references). A coalgebraic treatment of these structures can be found in [**35, 36**].

CHAPTER 2

Stream Calculus

We study the set $\mathbb{R}^\omega$ of streams of real numbers. Definitions will be given in terms of behavioural differential equations, and proofs by coinduction. We introduce a number of constants and operators, and develop a calculus of streams that can be viewed as a more general alternative for so-called generating functions. As a first application, we use the calculus to solve difference equations. More applications of this calculus will follow in subsequent chapters, on differential equations and counting problems.

2.1. Behavioural Differential Equations

We already saw some simple examples of behavioural differential equations in Section 1.3, which were shown to have unique solutions on the basis of the finality of the set of all streams. In what follows, we shall encounter many other, more complicated (systems of) equations. For two of them, we shall prove in the present section the existence of a unique solution in some detail. All other equations occurring in this chapter can be solved in precisely the same way.

We shall prove that for any two streams σ and τ, there exist unique streams $\sigma+\tau$ and $\sigma\otimes\tau$ that satisfy the following system of behavioural differential equations:

derivative	initial value
$(\sigma+\tau)' = \sigma' + \tau'$	$(\sigma+\tau)(0) = \sigma(0) + \tau(0)$
$(\sigma\otimes\tau)' = (\sigma'\otimes\tau) + (\sigma\otimes\tau')$	$(\sigma\otimes\tau)(0) = \sigma(0)\times\tau(0)$

.

(We use $+$ both as an operation on streams and as the operation of addition of real numbers.) The rationale behind these equations will be given later, for $+$ in Section 2.2 and for $\otimes$ in Chapter 3. Here they merely serve as an example. Technically, the above system of equations is more complicated than the ones we saw earlier, in Chapter 1, because the derivative of $\sigma\otimes\tau$ is expressed in terms of not only $\otimes$ but also $+$. As a consequence, it is now not so easy to see in one glance how the solutions should look like. Still, the equations can be rather straightforwardly interpreted as recipes for the construction of the respective elements of the solutions. For instance,

$$\begin{aligned}
(\sigma\otimes\tau)(0) &= \sigma(0)\times\tau(0), \\
(\sigma\otimes\tau)(1) &= (\sigma\otimes\tau)'(0) \\
&= \big((\sigma'\otimes\tau) + (\sigma\otimes\tau')\big)(0) \\
&= \big(\sigma'(0)\times\tau(0)\big) + \big(\sigma(0)\times\tau'(0)\big) \\
&= \big(\sigma(1)\times\tau(0)\big) + \big(\sigma(0)\times\tau(1)\big)
\end{aligned}$$

$$
\begin{aligned}
(\sigma \otimes \tau)(2) &= (\sigma \otimes \tau)''(0) \\
&= \big((\sigma' \otimes \tau) + (\sigma \otimes \tau')\big)'(0) \\
&= \cdots \\
&= \big(\sigma(2) \times \tau(0)\big) + \big(2 \times \sigma(1) \times \tau(1)\big) + \big(\sigma(0) \times \tau(2)\big)
\end{aligned}
$$

and so on. This illustrates the algorithmic aspect of the behavioural differential equations. A formal proof of the unique existence of their solutions can again be based on the finality of $\mathbb{R}^\omega$, as follows.

THEOREM 2.1.1. *For all $\sigma, \tau \in \mathbb{R}^\omega$, there exist unique streams $\sigma + \tau$ and $\sigma \otimes \tau$ such that*

derivative	initial value
$(\sigma + \tau)' = \sigma' + \tau'$	$(\sigma + \tau)(0) = \sigma(0) + \tau(0)$
$(\sigma \otimes \tau)' = (\sigma' \otimes \tau) + (\sigma \otimes \tau')$	$(\sigma \otimes \tau)(0) = \sigma(0) \times \tau(0)$

.

PROOF. We construct what could be called a syntactic stream automaton, whose states are given by expressions that include all the possible shapes that occur on the right side of the behavioural differential equations above. The solutions are then given by the unique homomorphism into $\mathbb{R}^\omega$. More precisely, let the set $\mathcal{E}$ of expressions be given by the following syntax:

$$E ::= \underline{\sigma} \mid E \,\underline{+}\, F \mid E \,\underline{\otimes}\, F.$$

The set $\mathcal{E}$ contains for every stream σ in $\mathbb{R}^\omega$ a constant symbol $\underline{\sigma}$. The operators that we are in the process of defining, are represented in $\mathcal{E}$ by a syntactic counterpart, denoted by $\underline{+}$ and $\underline{\otimes}$. The set $\mathcal{E}$ is next supplied with an automaton structure $(\mathcal{E}, \langle o_\mathcal{E}, t_\mathcal{E}\rangle)$ by defining functions $o_\mathcal{E} \colon \mathcal{E} \to \mathbb{R}$ and $t_\mathcal{E} \colon \mathcal{E} \to \mathcal{E}$ by induction on the syntactic structure of expressions. For the definition of $o_\mathcal{E}$ and $t_\mathcal{E}$ on the constant symbols $\underline{\sigma}$, the automaton structure $(\mathbb{R}^\omega, \langle o, t\rangle)$ on $\mathbb{R}^\omega$ is used:

$$t_\mathcal{E}(\underline{\sigma}) = \underline{t(\sigma)}, \quad o_\mathcal{E}(\underline{\sigma}) = o(\sigma).$$

(Recall from Section 1.3 that $o(\sigma) = \sigma(0)$ and $t(\sigma) = \sigma'$ for $\sigma \in \mathbb{R}^\omega$.) Thus the constant $\underline{\sigma}$ behaves in the automaton $\mathcal{E}$ precisely as the stream σ behaves in the automaton $\mathbb{R}^\omega$. (This includes $\mathbb{R}^\omega$ as a subautomaton in $\mathcal{E}$.) For composite expressions, the definitions of $o_\mathcal{E}$ and $t_\mathcal{E}$ literally follow the definition of the corresponding behavioural differential equations. Writing $E(0)$ for $o_\mathcal{E}(E)$ and E' for $t_\mathcal{E}(E)$, for any expression E in $\mathcal{E}$, we put

definition of $t_\mathcal{E}$	definition of $o_\mathcal{E}$
$(E \,\underline{+}\, F)' = E' \,\underline{+}\, F'$	$(E \,\underline{+}\, F)(0) = E(0) + F(0)$
$(E \,\underline{\otimes}\, F)' = (E' \,\underline{\otimes}\, F) + (E \,\underline{\otimes}\, F')$	$(E \,\underline{\otimes}\, F)(0) = E(0) \times F(0)$

.

The above defines the functions $o_\mathcal{E}$ and $t_\mathcal{E}$ on all of $\mathcal{E}$. Because $\mathbb{R}^\omega$ is a final automaton, there exists, by Theorem 1.3.1, a unique homomorphism

$$
\begin{array}{ccc}
\mathcal{E} & \xrightarrow{\;\; l \;\;} & \mathbb{R}^\omega \\
{\scriptstyle \langle o_\mathcal{E}, t_\mathcal{E}\rangle} \downarrow & & \downarrow {\scriptstyle \langle o, t\rangle} \\
A \times \mathcal{E} & \xrightarrow[1_A \times l]{} & A \times \mathbb{R}^\omega
\end{array}
$$

which assigns to each expression E the stream $l(E)$ it represents. With the help of l, we now define operators

$$+ \colon \mathbb{R}^\omega \times \mathbb{R}^\omega \to \mathbb{R}^\omega, \quad \otimes \colon \mathbb{R}^\omega \times \mathbb{R}^\omega \to \mathbb{R}^\omega$$

by putting, for all $\sigma, \tau \in \mathbb{R}^\omega$,

$$\sigma + \tau = l(\underline{\sigma} \mathbin{\underline{+}} \underline{\tau}), \quad \sigma \otimes \tau = l(\underline{\sigma} \mathbin{\underline{\otimes}} \underline{\tau}).$$

In order to show that these operators satisfy the behavioural differential equations above, one first notes the following facts:

(i) For all $E \in \mathcal{E} : l(E)(0) = E(0)$ and $\big(l(E)\big)' = l(E')$.
(ii) For all $\sigma \in \mathbb{R}^\omega : l(\underline{\sigma}) = \sigma$.
(iii) The homomorphism $l \colon \mathcal{E} \to \mathbb{R}^\omega$ is compositional. That is, for all expressions E and F in $\mathcal{E}$,
  (a) $l(E \mathbin{\underline{+}} F) = l(E) + l(F)$
  (b) $l(E \mathbin{\underline{\otimes}} F) = l(E) \otimes l(F)$.
  Note that the operators on the left are syntactic constructors of expressions, while on the right there are the operators on streams.

Fact (i) is equivalent to the fact that l is a homomorphism (recall that by definition $\sigma(0) = o(\sigma)$, $\sigma' = t(\sigma)$, for $\sigma \in \mathbb{R}^\omega$, and $E(0) = o_{\mathcal{E}}(E)$, $E' = t_{\mathcal{E}}(E)$, for $E \in \mathcal{E}$). Facts (ii) and (iii) can be proved by coinduction.

Using these three facts, it follows, for all $\sigma, \tau \in \mathbb{R}^\omega$, that

$(\sigma + \tau)' = \sigma' + \tau'$	$(\sigma + \tau)(0) = \sigma(0) + \tau(0)$
$(\sigma \otimes \tau)' = (\sigma' \otimes \tau) + (\sigma \otimes \tau')$	$(\sigma \otimes \tau)(0) = \sigma(0) \times \tau(0)$

.

For instance, we have

$$\begin{aligned}
(\sigma \otimes \tau)' &= \big(l(\underline{\sigma} \mathbin{\underline{\otimes}} \underline{\tau})\big)' \\
&= l\big((\underline{\sigma} \mathbin{\underline{\otimes}} \underline{\tau})'\big) && [\text{(i): } l \text{ is a homomorphism}] \\
&= l\left(\left((\underline{\sigma})' \mathbin{\underline{\otimes}} \underline{\tau}\right) \mathbin{\underline{+}} \left(\underline{\sigma} \mathbin{\underline{\otimes}} (\underline{\tau})'\right)\right) && [\text{by the definition of } t_{\mathcal{E}}] \\
&= l\big((\underline{\sigma'} \mathbin{\underline{\otimes}} \underline{\tau}) \mathbin{\underline{+}} (\underline{\sigma} \mathbin{\underline{\otimes}} \underline{\tau'})\big) && [\text{by the definition of } t_{\mathcal{E}}] \\
&= l(\underline{\sigma'} \mathbin{\underline{\otimes}} \underline{\tau}) + l(\underline{\sigma} \mathbin{\underline{\otimes}} \underline{\tau'}) && [\text{(iii): } l \text{ is compositional}] \\
&= \big(l(\underline{\sigma'}) \otimes l(\underline{\tau})\big) + \big(l(\underline{\sigma}) \otimes l(\underline{\tau'})\big) && [\text{(iii): } l \text{ is compositional}] \\
&= (\sigma' \otimes \tau) + (\sigma \otimes \tau') && [\text{(ii): } l \text{ is the identity on stream constants}]
\end{aligned}$$

and similarly for the other cases. This proves that the operators $+$ and $\otimes$ satisfy the behavioural differential equations above. Using coinduction, one can easily prove that they are the unique operators with this property. (The few details that have been omitted in the above proof can be found in [**37**, Section 13].) □

From now on, we shall without further ado introduce all kinds of streams and operators by means of behavioural differential equations. All of them can be justified in the same manner as above, by coinduction and finality of $\mathbb{R}^\omega$.

2.2. Basic Stream Calculus

Taking the existence of solutions of behavioural differential equations from now on for granted, much energy will be spent on the computation of closed formulas for such solutions, expressing them in terms of a basic set of stream constants and stream operators, which we introduce next.

First of all, we want to be able to view any real number as a constant stream. For $r \in \mathbb{R}$, let $[r]$ be the unique stream satisfying the following behavioural differential equation:

derivative	initial value
$[r]' = [0]$	$[r](0) = r$

The following definition is clearly equivalent:

$$[r] = (r, 0, 0, 0, \dots).$$

Having once formally introduced the inclusion of the reals into the set of streams by means of this operator $[\ ]\colon \mathbb{R} \to \mathbb{R}^\omega$, we immediately observe that in stream calculus there will hardly be any chance of confusing the stream $[r]$ with the real number r it represents. Therefore we shall usually simply write r for $[r]$.

We introduce one more constant: Let X be the unique stream satisfying

derivative	initial value
$X' = [1]$	$\mathrm{X}(0) = 0$

The constant X plays a crucial role in stream calculus, and may be thought of as a formal variable. Again, there is a more explicit equivalent definition:

$$X = (0, 1, 0, 0, 0, \dots).$$

Next we repeat the definition of the operation of sum that we have already seen as an example in Section 2.1. The sum $\sigma + \tau$ of two streams σ and τ is defined as the unique stream satisfying:

derivative	initial value
$(\sigma + \tau)' = \sigma' + \tau'$	$(\sigma + \tau)(0) = \sigma(0) + \tau(0)$

(Note that we are using the same symbol $+$ both for the sum of streams and the sum of real numbers.) Alternatively and equivalently, the sum of $\sigma = (s_0, s_1, s_2, \dots)$ and $\tau = (t_0, t_1, t_2, \dots)$ is given by

$$\sigma + \tau = (s_0 + t_0, s_1 + t_1, s_2 + t_2, \dots).$$

For the motivation of the next operator, let us briefly look at the world of (real-valued) functions, which at more than one occasion, will be a source of inspiration for stream calculus. Consider a function $f\colon \mathbb{R} \to \mathbb{R}$ and a stream (of coefficients) $\sigma = (s_0, s_1, s_2, \dots)$ such that $f(x) = s_0 + s_1 x + s_2 x^2 + \cdots$. Writing $\bar{f}(x) = s_1 + s_2 x + s_3 x^2 + \cdots$, we have

$$f(x) = s_0 + \big(x \times \bar{f}(x)\big).$$

Let $g : \mathbb{R} \to \mathbb{R}$ and $\tau = (t_0, t_1, t_2, \dots)$ be another such function and stream, with $g(x) = t_0 + t_1 x + t_2 x^2 + \cdots$ and write, similarly, $g(x) = t_0 + \big(x \times \bar{g}(x)\big)$. Computing now the (elementwise) function product $f(x) \times g(x)$, one finds

$$\begin{aligned} f(x) \times g(x) &= \big(s_0 + \big(x \times \bar{f}(x)\big)\big) \times \big(t_0 + \big(x \times \bar{g}(x)\big)\big) \\ &= (s_0 \times t_0) + x \times \big(\bar{f}(x) \times g(x) + s_0 \times \bar{g}(x)\big). \end{aligned}$$

This is one way of motivating the following definition. Let the *(convolution) product* $\sigma \times \tau$ of two streams σ and τ be the unique stream satisfying:

derivative	initial value
$(\sigma \times \tau)' = (\sigma' \times \tau) + \big(\sigma(0) \times \tau'\big)$	$(\sigma \times \tau)(0) = \sigma(0) \times \tau(0)$

(Note that we are using the same symbol $\times$ both for the product of streams and the product of real numbers. Further note that in the above definition, $\sigma(0) \times \tau'$ is a shorthand for $[\sigma(0)] \times \tau'$.) We shall use the following standard conventions: for all $n \geq 0$,

$$-\sigma \equiv [-1] \times \sigma, \quad \sigma\tau \equiv \sigma \times \tau, \quad \sigma^0 \equiv 1, \quad \sigma^{n+1} \equiv \sigma \times \sigma^n.$$

On the basis of the analogy between function multiplication and stream multiplication, one might have expected

$$(\sigma \times \tau)' = (\sigma' \times \tau) + (\sigma \times \tau')$$

which generally is *not* valid. Apparently, stream differentiation does not quite correspond to function derivation in analysis. Rather, it corresponds to the above transformation of a function f into $\bar{f}$, an operation which is not usually present in analysis. Later we shall see variations on both the notion of stream derivative and the operator of multiplication, which have more familiar properties. Let it be noted, however, that in stream calculus, both the operations of stream derivative and of convolution product, with their non-standard properties, are of central importance.

There is also the following formula for the nth element of $\sigma \times \tau$, for any $n \geq 0$,

$$(\sigma \times \tau)(n) = \sum_{k=0}^{n} \sigma(n-k) \times \tau(k) \tag{6}$$

which could have been (and traditionally is) taken as an alternative definition of $\sigma \times \tau$. (Since it will play no role in what follows, a proof of this identity, which would not be too difficult, is omitted.) Note that stream product shares this property with the function product considered above, since

$$f(x) \times g(x) = s_0t_0 + (s_1t_0 + s_0t_1) \times x + (s_2t_0 + s_1t_1 + s_0t_2) \times x^2 + \cdots.$$

As we shall see time and again, coinductive reasoning directly in terms of the above behavioural differential equation is quite a bit simpler than the use of (6), because of the summation over the indices k and $(n-k)$ in the latter. Equally importantly, we shall see examples of other definitions, such as that of inverse below, where no simple formula for the nth element is known.

We introduce an operator on streams that is inverse to multiplication in the following sense. Given a stream σ, we look for a stream σ^{-1} such that $\sigma \times \sigma^{-1} = 1$. Constant streams $[r] = (r, 0, 0, 0, \ldots)$ with $r \neq 0$, clearly have $[r]^{-1} = (r^{-1}, 0, 0, 0, \ldots) = [r^{-1}]$ as their inverse, because

$$(r, 0, 0, 0, \ldots) \times (r^{-1}, 0, 0, 0, \ldots) = (1, 0, 0, 0, \ldots).$$

Considering now an arbitrary stream σ with $\sigma(0) \neq 0$, and assuming for a second that σ^{-1} indeed existed, it would have to satisfy

$$0 = 1' = (\sigma \times \sigma^{-1})' = (\sigma' \times \sigma^{-1}) + (\sigma(0) \times (\sigma^{-1})').$$

This implies

$$\sigma(0) \times (\sigma^{-1})' = -1 \times \sigma' \times \sigma^{-1}.$$

Multiplying (on the left) with the inverse of (the stream) $\sigma(0)$, one obtains

$$(\sigma^{-1})' = -\sigma(0)^{-1} \times (\sigma' \times \sigma^{-1}).$$

Thus the equation $1 = \sigma \times \sigma^{-1}$ determines what $(\sigma^{-1})'$ should be. It also determines the initial value $(\sigma^{-1})(0)$: taking the initial value on both sides of the equation (recall that 1 stands for the stream $[1] = (1, 0, 0, 0, \dots)$) gives

$$1 = \sigma(0) \times \sigma^{-1}(0)$$

whence $(\sigma^{-1})(0) = \sigma(0)^{-1}$. What has happened is that we have derived from our "specification" $\sigma \times \sigma^{-1} = 1$ a behavioural differential equation, which we can now simply take as the definition of inverse. For all $\sigma \in \mathbb{R}^\omega$ with $\sigma(0) \neq 0$, let the (*multiplicative*) *inverse* σ^{-1} of σ be the unique stream satisfying the following equation:

derivative	initial value
$(\sigma^{-1})' = -\sigma(0)^{-1} \times (\sigma' \times \sigma^{-1})$	$(\sigma^{-1})(0) = \sigma(0)^{-1}$

Whenever we write σ^{-1}, we shall silently assume that $\sigma(0) \neq 0$. As usual, we write

$$\sigma^{-n} \equiv (\sigma^{-1})^n, \quad \frac{\sigma}{\tau} \equiv \sigma \times \tau^{-1}.$$

With this convention, we can write $(\sigma^{-1})'$ as

$$\left(\frac{1}{\sigma}\right)' = \frac{-\sigma'}{\sigma(0) \times \sigma}$$

which is a bit easier to remember. For reasons similar to the case of stream multiplication, stream derivation of inverse behaves differently from what we are used to in analysis. In particular, $(\sigma^{-1})' = -\sigma' \times \sigma^{-2}$ generally does *not* hold.

Was our preference to use behavioural differential equations for the definition of the operators so far based on either obsession (in the case of the constants) or computational convenience (in the case of the product), in the case of inverse, it turns out to be essential. For inverse, no explicit formula such as (6) for product, is known (not in the literature, not to us):

$$\sigma^{-1}(n) = ?.$$

The best one can do is to define inverse by means of a recurrence relation as follows. The elements of σ^{-1} are given by $\sigma^{-1}(0) = \sigma(0)^{-1}$ and

$$\sigma^{-1}(n) = \sigma(0)^{-1} \sum_{k=0}^{n-1} \sigma(n-k) \times \sigma^{-1}(k)$$

for $n \geq 1$. This is an immediate consequence of the requirement that $\sigma \times \sigma^{-1} = 1$ and identity (6)) above. Note that this recurrence is of so-called unbounded order: the nth element $\sigma^{-1}(n)$ depends on all of the n preceding values $\sigma^{-1}(0)$ through $\sigma^{-1}(n-1)$. Reasoning in terms of such recurrences is not only extremely tedious but even next to impossible.

The good news is that in stream calculus, there is no need for a closed formula for $\sigma^{-1}(n)$. All our reasoning about inverse, in fact about all streams and stream operators, will be coinductive, calculating with stream derivatives as specified by behavioural differential equations.

The definition of the following operation on streams is again clearly inspired by analysis. For all streams σ and τ with $\tau(0) = 0$, the *composition* $\sigma \circ \tau$ is defined as the unique stream satisfying the following behavioural differential equation:

derivative	initial value
$(\sigma \circ \tau)' = \tau' \times (\sigma' \circ \tau)$	$(\sigma \circ \tau)(0) = \sigma(0)$

We shall often write

$$\sigma \circ \tau \equiv \sigma(\tau).$$

The condition that $\tau(0) = 0$ is needed in the proof of Theorem 2.5.3 in Section 2.5, which shows that composition behaves as usual.

2.3. Coinduction-Up-To

Before proving a number of basic properties of the operators introduced above, we present another proof principle, called coinduction-up-to, which is a mild generalisation of the coinduction proof principle of Theorem 1.2.2.

DEFINITION 2.3.1. A *bisimulation-up-to* is a relation $R \subseteq \mathbb{R}^\omega \times \mathbb{R}^\omega$ such that, for all σ and τ in $\mathbb{R}^\omega$, if $\sigma\ R\ \tau$ then

(i) $\sigma(0) = \tau(0)$ and
(ii) there exist $n \geq 0$, $\alpha_0, \ldots, \alpha_n$, $\beta_0, \ldots, \beta_n \in \mathbb{R}^\omega$, such that $\sigma' = \alpha_0 + \cdots + \alpha_n$ and $\tau' = \beta_0 + \cdots + \beta_n$ and, for all $0 \leq i \leq n$: either $\alpha_i = \beta_i$ or $\alpha_i\ R\ \beta_i$. The latter condition will usually be denoted by

$$\alpha_0 + \cdots + \alpha_n\ (\Sigma R)\ \beta_0 + \cdots + \beta_n.$$

THEOREM 2.3.2. *If $\sigma\ R\ \tau$, for streams σ and τ and a bisimulation-up-to relation $R \subseteq \mathbb{R}^\omega \times \mathbb{R}^\omega$, then $\sigma = \tau$.*

PROOF. Let $R \subseteq \mathbb{R}^\omega \times \mathbb{R}^\omega$ be a bisimulation-up-to. Let $\overline{R}$ be the smallest relation satisfying

(1) $R \subseteq \overline{R}$,
(2) $\{\langle \sigma, \sigma \rangle \mid \sigma \in \mathbb{R}^\omega\} \subset \overline{R}$,
(3) for all $\alpha_0, \alpha_1, \beta_0, \beta_1 \in \mathbb{R}^\omega$, if $\alpha_0\ \overline{R}\ \beta_0$ and $\alpha_1\ R\ \beta_1$ then $(\alpha_0 + \alpha_1)\ \overline{R}\ (\beta_0 + \beta_1)$.

Using the fact that R is a bisimulation-up-to, one can easily prove by induction on the definition of $\overline{R}$ that the latter is an (ordinary) bisimulation relation on $\mathbb{R}^\omega$. Since $R \subseteq \overline{R}$, the theorem follows by (ordinary) coinduction, Theorem 1.2.2. □

For a simple but typical example, consider streams σ, τ and ρ such that

derivative	initial value
$\sigma' = \sigma + \sigma$	$\sigma(0) = 1$
$\tau' = \tau + \rho$	$\tau(0) = 1$
$\rho' = \tau + \rho$	$\rho(0) = 1$

Then $\{\langle \sigma, \tau \rangle, \langle \sigma, \rho \rangle\}$ is a bisimulation-up-to and $\sigma = \tau = \rho$ follows by coinduction-up-to. Other examples can be found in the proof of Theorem 2.4.1.

2.4. Basic Properties

We prove a number of elementary properties of stream operators.

THEOREM 2.4.1. *For all* $r, s \in \mathbb{R}$ *and* $\sigma, \tau, \rho \in \mathbb{R}^\omega$,

$$\sigma + 0 = \sigma,$$

$$\sigma + \tau = \tau + \sigma, \tag{7}$$

$$\sigma + (\tau + \rho) = (\sigma + \tau) + \rho,$$

$$0 \times \sigma = 0,$$

$$1 \times \sigma = \sigma,$$

$$[r] \times [s] = [r \times s],$$

$$\sigma \times \tau = \tau \times \sigma, \tag{8}$$

$$\sigma \times (\tau + \rho) = (\sigma \times \tau) + (\sigma \times \rho), \tag{9}$$

$$\sigma \times (\tau \times \rho) = (\sigma \times \tau) \times \rho,$$

$$\sigma \times \sigma^{-1} = 1, \tag{10}$$

$$(\sigma^{-1})^{-1} = \sigma, \tag{11}$$

$$(\sigma \times \tau)^{-1} = \tau^{-1} \times \sigma^{-1}, \tag{12}$$

$$(X \times \sigma)' = \sigma. \tag{13}$$

PROOF. We shall prove only some of the above identities (for the others see [**39**]). For (7) let $R = \{\langle \sigma + \tau, \tau + \sigma \rangle \mid \sigma, \tau \in \mathbb{R}^\omega\}$. Since $(\sigma + \tau)(0) = \sigma(0) + \tau(0) = (\tau + \sigma)(0)$ and

$$\begin{aligned} (\sigma + \tau)' &= (\sigma' + \tau') \\ &\ R\ (\tau' + \sigma') \\ &= (\tau + \sigma)' \end{aligned}$$

for any σ and τ, R is a bisimulation relation on $\mathbb{R}^\omega$. Identity (7) now follows by coinduction (Theorem 1.2.2). Identity (8) is, on the basis of the non-symmetrical definition of stream product, somewhat surprising. It is an immediate consequence of identity (6). A proof by coinduction can be given too, but is slightly complicated (see [**39**]). For (9), let $R = \{\langle \sigma \times (\tau + \rho), (\sigma \times \tau) + (\sigma \times \rho) \rangle \mid \sigma, \rho, \tau \in \mathbb{R}^\omega\}$. We prove that R is a bisimulation-up-to. Initial values match and computing derivatives, we find

$$\begin{aligned} (\sigma \times (\tau + \rho))' &= (\sigma' \times (\tau + \rho)) + (\sigma(0) \times (\tau' + \rho')) \\ &\ (\Sigma R)\ ((\sigma' \times \tau) + (\sigma' \times \rho)) + ((\sigma(0) \times \tau') + (\sigma(0) \times \rho')) \\ &= ((\sigma' \times \tau) + (\sigma(0) \times \tau')) + ((\sigma' \times \rho) + (\sigma(0) \times \rho')) \\ &= ((\sigma \times \tau) + (\sigma \times \rho))' \end{aligned}$$

with (ΣR) as defined in Definition 2.3.1. Thus, R is a bisimulation-up-to and identity (9) follows by coinduction-up-to (Theorem 2.3.2). For identity (10) note that

$$\begin{aligned}(\sigma \times \sigma^{-1})' &= (\sigma' \times \sigma^{-1}) + \big(\sigma(0) \times (-\sigma(0)^{-1} \times \sigma' \times \sigma^{-1})\big) \\ &= (\sigma' \times \sigma^{-1}) + (-\sigma' \times \sigma^{-1}) \\ &= 0\end{aligned}$$

and coinduction does the rest. For (11), note that

$$\begin{aligned}\big((\sigma^{-1})^{-1}\big)' &= -\big(\sigma^{-1}(0)\big)^{-1} \times (\sigma^{-1})' \times (\sigma^{-1})^{-1} \\ &= -\sigma(0) \times \big(-\sigma(0)^{-1} \times \sigma' \times \sigma^{-1}\big) \times (\sigma^{-1})^{-1} \\ &= \sigma' \times \big(\sigma^{-1} \times (\sigma^{-1})^{-1}\big) \\ &= \sigma' && \text{[by identity (10)]}\end{aligned}$$

and use coinduction. For (12), note that

$$\begin{aligned}(\sigma \times \tau) \times (\tau^{-1} \times \sigma^{-1}) &= \sigma \times (\tau \times \tau^{-1}) \times \sigma^{-1} \\ &= \sigma \times \sigma^{-1} \\ &= 1.\end{aligned}$$

Multiplying both sides of the resulting equality with $(\sigma \times \tau)^{-1}$ and using identity (10) implies (12). □

2.5. Infinite Sum

As was already suggested by the comparison between function product and stream product, in Section 2.2, streams can be viewed as power series in one formal variable (namely, the constant stream X). In order to make this more precise, we need to introduce (countably) infinite sums of streams.

Let $\sigma_0, \sigma_1, \sigma_2, \ldots$ be a sequence of streams such that, for all $k \geq 0$, $\sigma_0(k) + \sigma_1(k) + \sigma_2(k) + \cdots < \infty$. Such a sequence is called *summable.* We define:

derivative	initial value
$\left(\sum_{n=0}^{\infty} \sigma_n\right)' = \sum_{n=0}^{\infty} (\sigma_n')$	$\left(\sum_{n=0}^{\infty} \sigma_n\right)(0) = \sum_{n=0}^{\infty} \sigma_n(0)$

Often we shall also write

$$\sum_{n=0}^{\infty} \sigma_n = \sigma_0 + \sigma_1 + \sigma_2 + \cdots .$$

Infinite sum is as well-behaved as binary sum. For instance, we have

$$\left(\sum_{n=0}^{\infty} \sigma_n\right) \times \tau = \sum_{n=0}^{\infty} (\sigma_n \times \tau)$$

and similarly for the other basic properties of sum. For the proof of the theorem below, we shall need the following basic property, which easily follows from identity (13) by induction: for all $n \geq 0$,

$$(X^{n+1})' = X^n.$$

THEOREM 2.5.1. *For all streams* $\sigma = (s_0, s_1, s_2, \ldots)$,

$$s_0, s_1 \times X^1, \quad s_2 \times X^2, \ldots$$

is summable, and satisfies

$$\sigma = s_0 + (s_1 \times X^1) + (s_2 \times X^2) + \cdots .$$

PROOF. Using the defining equation for infinite sum; the fact that if a stream σ is summable, then also σ' is summable; and identity (2.5), the proof is straightforward by coinduction. □

This theorem is to be contrasted with the following way to represent streams $\sigma = (s_0, s_1, s_2, \ldots)$, which is very common in mathematics: if $f\colon \mathbb{R} \to \mathbb{R}$ is a function and r is a positive real number such that for all $|x| < r$,

$$f(x) = s_0 + s_1 x + s_2 x^2 + \cdots$$

(that is, the infinite sum on the right exists and is equal to $f(x)$), then f is called a *generating function* for the stream σ. The characterisation of streams given in Theorem 2.5.1 above, is different in two respects. The infinite sum is not merely an *encoding* of σ, but is itself a stream (built from the constant streams $s_i = (s_i, 0, 0, 0, \ldots)$, for all $i \geq 0$, the constant stream $X = (0, 1, 0, 0, 0, \ldots)$, and the operators of sum and product), which is *equal* to σ. Moreover, this equality is not subject to a condition of summability, but simply holds for any σ. In particular, this also holds for streams such as $(0!, 1!, 2!, \ldots)$, for which no generating function exists (because $0! + 1!x + 2!x^2 + \cdots$ diverges for every $x \neq 0$).

Here are some identities involving infinite sums.

THEOREM 2.5.2. *For all streams* $\sigma = (s_0, s_1, s_2, \ldots)$,

$$\frac{1}{1-X} = 1 + X + X^2 + X^3 + \cdots , \tag{14}$$

$$X \times \sigma = (0, s_0, s_1, s_2, \ldots), \tag{15}$$

$$\frac{1}{1+X} = 1 - X + X^2 - X^3 + \cdots , \tag{16}$$

$$\frac{1}{1-X^2} = 1 + X^2 + X^4 + \cdots , \tag{17}$$

$$\frac{1}{(1-X)^2} = 1 + 2X + 3X^2 + 4X^3 + \cdots , \tag{18}$$

$$\frac{1}{1-rX} = 1 + rX + r^2X^2 + r^3X^3 + \cdots , \tag{19}$$

$$\frac{X}{1+X^2} = X - X^3 + X^5 - X^7 + \cdots . \tag{20}$$

PROOF. All of these identities can be easily proved by coinduction. □

The following theorem shows that stream composition $\sigma \circ \tau = \sigma(\tau)$, introduced at the end of Section 2.2, behaves as it should.

THEOREM 2.5.3. *For all streams* $\sigma = (s_0, s_1, s_2, \ldots)$, *and* τ *with* $\tau(0) = 0$,

$$\sigma(\tau) = s_0 + (s_1 \times \tau^1) + (s_2 \times \tau^2) + \cdots . \tag{21}$$

PROOF. Because $\tau(0) = 0$, we have $(\tau^{n+1})' = \tau' \times \tau^n$, for all $n \geq 0$. As a consequence,

$$\begin{aligned}\big(s_0 + (s_1 \times \tau^1) + (s_2 \times \tau^2) + \cdots\big)' &= (s_1 \times \tau') + (s_2 \times \tau' \times \tau) + (s_3 \times \tau' \times \tau^2) + \cdots \\ &= \tau' \times \big(s_1 + (s_2 \times \tau) + (s_3 \times \tau^2) + \cdots\big).\end{aligned}$$

For $\sigma = (s_0, s_1, s_2, \ldots)$, let

$$R = \{\langle \rho \times \sigma(\tau), \rho \times (s_0 + (s_1 \times \tau^1) + (s_2 \times \tau^2) + \cdots)\rangle \mid \sigma, \tau, \rho \in \mathbb{R}^\omega\}.$$

Because

$$\begin{aligned}
(\rho \times \sigma(\tau))' &= (\rho' \times \sigma(\tau)) + (\rho(0) \times (\tau' \times \sigma'(\tau))) \\
&= (\rho' \times \sigma(\tau)) + (\rho(0) \times \tau') \times \sigma'(\tau) \\
&\Sigma R\, \rho' \times (s_0 + (s_1 \times \tau^1) + (s_2 \times \tau^2) + \cdots) + \\
&\qquad (\rho(0) \times \tau') \times (s_1 + (s_2 \times \tau^1) + (s_3 \times \tau^2) + \cdots) \\
&= \rho' \times (s_0 + (s_1 \times \tau^1) + (s_2 \times \tau^2) + \cdots) + \\
&\qquad \rho(0) \times (\tau' \times (s_1 + (s_2 \times \tau^1) + (s_3 \times \tau^2) + \cdots)) \\
&= (\rho \times (s_0 +, (s_1 \times \tau^1) + (s_2 \times \tau^2) + \cdots))'.
\end{aligned}$$

R is a bisimulation-up-to. The theorem follows by coinduction-up-to. □

Here are a few useful instances and applications of Theorem 2.5.3:

$$\begin{aligned}
\sigma(X) &= \sigma, \\
\sigma(-X) &= s_0 - s_1 X + s_2 X^2 - s_3 X^3 + \cdots \\
&= (s_0, -s_1, s_2, -s_3, \ldots), \\
\tfrac{1}{2}(\sigma(X) + \sigma(-X)) &= (s_0, 0, s_3, 0, \ldots) \\
\sigma(X^2) &= s_0 + s_1 X^2 + s_2 X^4 + s_3 X^6 + \cdots. \qquad (22)
\end{aligned}$$

2.6. Solving Behavioural Differential Equations

We introduce the "Fundamental Theorem" of stream calculus, named after the corresponding theorem from analysis. As we discussed in Section 2.1, the *existence* of a (unique) solution of many behavioural differential equations, can be proved using the finality of $\mathbb{R}^\omega$. With the help of the Fundamental Theorem, we can actually compute an *expression* for such solutions in terms of the basic operators on streams, namely, constants, sum, product, and inverse. In particular, we show how to obtain such expressions for the solutions of (homogeneous or non-homogeneous) linear equations, possibly with non-constant coefficients. Later, these solutions will be used to solve difference equations and analytical differential equations of a similar kind.

Here is the Fundamental Theorem of stream calculus.

THEOREM 2.6.1. *For all $\sigma \in \mathbb{R}^\omega$: $\sigma = \sigma(0) + (X \times \sigma')$.*

PROOF. We first note that $(X \times \sigma)' = \sigma$, for all $\sigma \in \mathbb{R}^\omega$. Next, we define $R = \{\langle \sigma, \sigma\rangle \mid \sigma \in \mathbb{R}^\omega\} \cup \{\langle \sigma, \sigma(0) + (X \times \sigma')\rangle \mid \sigma \in \mathbb{R}^\omega\}$. Because

$$\begin{aligned}
&\sigma'\, R\, \sigma' \\
&= (\sigma(0) + (X \times \sigma'))'.
\end{aligned}$$

R is a bisimulation (all other conditions that are to be checked are trivial). The theorem follows by coinduction. □

Theorem 2.6.1 is named after the fundamental theorem from analysis: for all (well-behaved) functions $f\colon \mathbb{R} \to \mathbb{R}$,

$$f(x) = f(0) + \int_0^x f'.$$

The exact correspondence between both theorems will be made precise later, after we have introduced Taylor series, in Definition 3.4.1 below. Theorem 2.6.1 tells us that we can obtain σ out of σ' by multiplication with the constant stream X (and addition with the initial value $\sigma(0)$). Multiplication with the constant stream X can therefore be viewed as a kind of stream integration. We observe that stream calculus has the pleasant property that every stream is both differentiable and integrable.

Next, we show how the Fundamental Theorem can be used to compute the solution of behavioural differential equations in stream calculus in an algebraic manner. We start with linear behavioural differential equations with constant coefficients:

$\sigma^{(n)} + r_{n-1}\sigma^{(n-1)} + \cdots + r_1\sigma^{(1)} + r_0\sigma = \tau$	derivative
$\sigma^{(k)}(0) = c_k, \quad 0 \leq k \leq n-1$	initial value

where $n \geq 1$, $\sigma^{(k)}$ denotes the kth stream derivative of σ, $r_0, \ldots, r_{n-1}$ are real numbers (interpreted as constant streams $[r_k]$), $c_0, \ldots, c_{n-1}$ are also real numbers, and τ is an arbitrary stream. If $\tau = 0$ the equation is called homogeneous, otherwise it is called non-homogeneous. The claim is that

(1) there exists a unique stream σ satisfying the equation;
(2) σ can be expressed in terms of τ, the coefficients $r_0, \ldots, r_{n-1}$, the initial values $c_0, \ldots, c_{n-1}$, the constant stream X, and the operators of plus, product and inverse.

Rather than giving a proof of this claim in its full generality, we prefer to treat a few special cases, being confident that they will bring the main idea across. Let us start with a simple example of a homogeneous equation:

derivative	initial value
$\sigma'' - \sigma' - \sigma = 0$	$\sigma(0) = 0, \sigma'(0) = 1$

.

In order to solve it we multiply the left and right side of the equation with X^2 (where 2 is the coefficient of the highest derivative):

$$(X^2 \times \sigma'') - (X^2 \times \sigma') - (X^2 \times \sigma) = 0.$$

Applying the Fundamental Theorem 2.6.1 to both σ and σ', we obtain, using $\sigma(0) = 0$ and $\sigma'(0) = 1$,

$$\sigma = X \times \sigma', \quad \sigma' = 1 + (X \times \sigma'')$$

implying $X^2 \times \sigma'' = -X + \sigma$ and $X^2 \times \sigma' = X \times \sigma$. Substituting this above gives

$$\begin{aligned} 0 &= (X^2 \times \sigma'') - (X^2 \times \sigma') - (X^2 \times \sigma) \\ &= -X + \sigma - (X \times \sigma) - (X^2 \times \sigma) \\ &= -X + (1 - X - X^2) \times \sigma \end{aligned}$$

yielding as the final outcome the following expression for the solution of the differential equation:

$$\sigma = \frac{X}{1 - X - X^2}.$$

One could continue by manipulating this expression still further in order to obtain a so-called *closed formula* for the nth element of σ, expressing $\sigma(n)$ in terms of a formula on the natural numbers depending on (the variable) n. We shall come back to this in Section 2.7.

Here is another example, this time non-homogeneous:

derivative	initial value
$\sigma' - \sigma = (1 - X)^{-1}$	$\sigma(0) = 1$

.

By Theorem 2.6.1, we obtain $X \times \sigma' = \sigma - 1$, whence

$$\begin{aligned} X \times (1 - X)^{-1} &= (X \times \sigma') - (X \times \sigma) \\ &= \sigma - 1 - (X \times \sigma) \\ &= -1 + (1 - X) \times \sigma. \end{aligned}$$

As a consequence, the following expression is obtained:

$$\sigma = \frac{1}{(1 - X)^2}. \tag{23}$$

The following table contains some further examples which the reader may wish to use to test his or her stream-calculus abilities:

derivative	initial value
$\sigma' - 5\sigma = -2(1 - 2X)^{-1}$	$\sigma(0) = 3$
$\tau'' + r^2\tau = 1$	$\tau(0) = 1,\ \tau'(0) = 0$
$\rho'' - \rho' = 2 + 6X$	$\rho(0) = 1,\ \rho'(0) = 0$

.

The solutions of these equations are

$$\sigma = \frac{3 \quad 8X}{1 - 7X + 10X^2}, \quad \tau = \frac{1 + X + X^2}{1 + r^2X^2}, \quad \rho = \frac{1 - X + 2X^2 + 6X^3}{1 - X}.$$

The fact that the coefficients of the equations so far have been constants is by no means crucial. Here is the more general formulation:

$\sigma^{(n)} + (\rho_{n-1} \times \sigma^{(n-1)}) + \cdots + (\rho_0 \times \sigma) = \tau$	derivative
$\sigma^{(k)}(0) = c_k, 0 \leq k \leq n - 1$	initial value

where now not only τ but also $\rho_0, \ldots, \rho_{n-1}$ are arbitrary streams. The claim is again that

(1) there exists a unique stream σ satisfying the equation;
(2) σ can be expressed in terms of the streams τ and $\rho_0, \ldots, \rho_{n-1}$, the initial values $c_0, \ldots, c_{n-1}$, the constant stream X, and the operators of plus, product and inverse.

The proof for arbitrary $n \geq 1$ is not more difficult (just more writing) than the proof for $n = 1$, which is presented next. Consider the equation

derivative	initial value
$\sigma' + (\rho \times \sigma) = \tau$	$\sigma(0)$

where ρ and τ are arbitrary streams. Calculating as before (multiplying both sides of the equation by X, invoking Theorem 2.6.1), we obtain for the solution of this equation the following expression:

$$\sigma = \frac{\sigma(0) + (X \times \tau)}{1 + (X \times \rho)}.$$

2.7. Solving Difference Equations

As an application of the techniques of Section 2.6, we next show how they can be used to solve linear difference equations (also called linear recurrence relations) with constant coefficients. The main idea is to transform difference equations, which can be seen as inductive definitions of streams, in a canonical fashion into behavioural differential equations. The heart of the matter is thus a systematic transformation of inductive stream definitions into coinductive ones.

More precisely, we consider non-homogeneous linear difference equations of order k with constant coefficients:

difference equation	initial values
$s_{n+k} + r_{k-1}s_{n+k-1} + \cdots + r_1 s_{n+1} + r_0 s_n = t_n$	$s_0, \ldots, s_{k-1}$

where $k \geq 1$ (the order of the equation), $n \geq 0$, the coefficients $r_0, \ldots, r_{n-1}$ are real numbers (the coefficient for s_{n+k} has been taken identical to 1 for convenience), and $t_0, t_1, t_2, \ldots$ is an arbitrary sequence of real numbers. Defining

$$\sigma = (s_0, s_1, s_2, \ldots), \quad \tau = (t_0, t_1, t_2, \ldots)$$

we set out to transform the above difference equation into a behavioural differential equation in the variable σ, by multiplying both sides of the equation with the constant stream X^n:

$$s_{n+k}X^n + r_{k-1}s_{n+k-1}X^n + \cdots + r_1 s_{n+1}X^n + r_0 s_n X^n = t_n X^n.$$

Next we take on both sides the infinite sum over all $n \geq 0$:

$$\sum_{n=0}^{\infty} s_{n+k}X^n + r_{k-1}\sum_{n=0}^{\infty} s_{n+k-1}X^n + \cdots + r_1 \sum_{n=0}^{\infty} s_{n+1}X^n + r_0 \sum_{n=0}^{\infty} s_n X^n = \sum_{n=0}^{\infty} t_n X^n.$$

Since, for any $i \geq 0$,

$$\sigma^{(i)} = s_i + s_{i+1}X^1 + s_{i+2}X^2 + \cdots$$

which is an immediate corollary of Theorem 2.5.1, we obtain

$$\sigma^{(k)} + r_{k-1}\sigma^{(k-1)} + \cdots + r_1\sigma^{(1)} + r_0\sigma = \tau$$

with initial values

$$\sigma(0) = s_0, \quad \sigma^{(1)}(0) = s_1, \ldots, \sigma^{(k-1)}(0) = s_{k-1}.$$

And so we are done, since we have learned in Section 2.6 how to solve this type of behavioural differential equation.

Let us look at a few examples. Here is maybe the most famous difference equation of all:

difference equation	initial value
$s_{n+2} - s_{n+1} - s_n = 0$	$s_0 = 0, \ s_1 = 1$

defining the Fibonacci numbers. The method above transforms it into

derivative	initial values
$\sigma'' - \sigma' - \sigma = 0$	$\sigma(0) = 0, \sigma'(0) = 1$

which we recognise from Section 2.6, where the solution was shown to be

$$\sigma = \frac{X}{1 - X - X^2}.$$

This in principle answers the question as to what stream is defined by the difference equation we started with. A further question that is often raised is how the nth element $\sigma(n)$ looks like, ideally as a function of n. The method of partial fractions, well-known from the theory of generating functions, also works in stream calculus. Defining

$$r_+ = \frac{1+\sqrt{5}}{2}, \quad r_- = \frac{1-\sqrt{5}}{2}$$

which are the roots of $1 - X - X^2 = (1 - r_+X)(1 - r_-X)$, we have

$$\begin{aligned}
\sigma &= \frac{X}{1-X-X^2}, \\
&= \frac{1}{r_+ - r_-}\left(\frac{1}{1-r_+X} - \frac{1}{1-r_-X}\right) \\
&= \frac{1}{\sqrt{5}}\left(\sum_{n=0}^{\infty} r_+^n X^n - \sum_{n=0}^{\infty} r_-^n X^n\right) && \text{[by identity (19)]} \\
&= \sum_{n=0}^{\infty} \frac{1}{\sqrt{5}}(r_+^n - r_-^n)X^n
\end{aligned}$$

giving the familiar answer

$$\sigma(n) = \frac{1}{\sqrt{5}}(r_+^n - r_-^n).$$

Note that the entire game is played here inside the world of stream calculus, without any reference to (generating) functions. For a second example, non-homogeneous this time, consider

difference equation	initial values
$s_{n+1} - s_n = 1$	$s_0 = 1$

.

Our method transforms it into

derivative	initial value
$\sigma' - \sigma = (1-X)^{-1}$	$\sigma(0) = 1$

(using the fact that $(1-X)^{-1} = 1 + X + X^2 + \cdots$, identity (14)). For the solution we look again at Section 2.6:

$$\begin{aligned}
\sigma &= (1-X)^{-2} && \text{[identity (23)]} \\
&= 1 + 2X + 3X^2 + 4X^3 + \cdots && \text{[by identity (18)]} \\
&= (1, 2, 3, 4, \ldots)
\end{aligned}$$

which comes as no surprise. Here is yet another example of a non-homogeneous equation:

difference equation	initial value
$s_{n+1} - 5s_n = -2^{n+1}$	$s_0 = 3$

.

Writing $-2^{n+1} = -2 \times 2^n$ and using identity (19), the following differential equation is obtained:

derivative	initial value
$\sigma' - 5\sigma = -2(1-2X)^{-1}$	$\sigma(0) = 3$

.

Using the method of partial fractions again, the solution found in Section 2.6 can be rewritten as

$$\begin{aligned}\sigma &= \frac{3-8X}{1-7X+10X^2} \\ &= \frac{2}{3}(1-2X)^{-1} + \frac{7}{3}(1-5X)^{-1}\end{aligned}$$

yielding, again by identity (19),

$$\sigma(n) = \frac{2}{3}2^n + \frac{7}{3}5^n.$$

2.8. Streams and Weighted Automata

In this section, we introduce the notion of weighted stream automaton, the transition diagrams of which can serve as pleasant graphical representations of streams and their successive derivatives.

An $\mathbb{R}$-*weighted stream automaton*, or *weighted automaton* for short, is a pair $Q = (Q, \langle o, t\rangle)$ consisting of a set Q of states, together with an output function $o\colon Q \to \mathbb{R}$, and a transition function $t\colon Q \to (Q \to_f \mathbb{R})$, where the latter set contains only functions of finite *support*:

$$Q \to_f \mathbb{R} = \{\phi\colon Q \to \mathbb{R} \mid \text{the set } \{q \in Q \mid \phi(q) \neq 0\} \text{ is finite}\}.$$

The output function o assigns to each state q in Q a real number $o(q)$ in $\mathbb{R}$. The transition function t assigns to a state q in Q a function $t(q)\colon Q \to_f \mathbb{R}$, which specifies for any state u in Q a real number $t(q)(u)$ in $\mathbb{R}$. This number can be thought of as the weight (cost, multiplicity, duration, etc.) with which the transition from q to u occurs. The following notation will be used: $q \xrightarrow{r} u$ denotes $t(q)(u) = r$ and $\underline{q}_r$ denotes $o(q) = r$. Moreover, for $t(q)(u) = 1$ we shall often simply write $q \to u$, and we write $\underline{q}$ when $o(q) = 1$, in which case q is called an *output* state. In pictures, we usually include only non-zero output values and only arrows with a non-zero label.

The *stream behaviour* $S(q) \in \mathbb{R}^\omega$ of a state q in a weighted automaton $(Q, \langle o, t\rangle)$, can be defined in two equivalent ways. First, there is the following formula, for any $k \geq 0$:

$$(24)\quad S(q)(k) = \sum\{l_1 \times \cdots \times l_k \times l \mid \Sigma q_0, \ldots, q_k\colon \; q = q_0 \xrightarrow{l_1} \cdots \xrightarrow{l_k} q_k \text{ and } o(q_k) = l\}.$$

This formula computes the kth element $S(q)(k)$ of the stream $S(q)$ as a weighted count of the number of paths (transition sequences) of length k that start in the state q (and end in a state with a non-zero output). It is precisely this aspect of counting paths that makes weighted automata very suitable for the representation of counting problems in general. This will be the subject of Chapter 4.

At the same time, formula (24) does not yield any compact representations for $S(q)$ and is thereby not very suited for actual reasoning. Here we can benefit from stream calculus as follows. Consider again a state q in a weighted automaton $(Q, \langle o, t\rangle)$, and let $\{q_1, \ldots, q_n\}$ be the set of all states u for which $t(q)(u) \neq 0$. Then the stream behaviour $S(q) \in \mathbb{R}^\omega$ of the state q can also be defined by the following system of behavioural differential equations:

derivative	initial value
$S(q)' = t(q)(q_1) \times S(q_1) + \cdots + t(q)(q_n) \times S(q_n)$	$S(q)(0) = o(q)$

.

We saw in Section 2.6 how to solve such a system of behavioural differential equations. It follows from the Fundamental Theorem 2.6.1 that it is equivalent to the following system of equations:

$$\boxed{S(q) = o(q) + \big(t(q)(q_1) \times X \times S(q_1)\big) + \cdots + \big(t(q)(q_n) \times X \times S(q_n)\big)}\;.$$

If the automaton (that is, the state space Q) is finite, this is a finite system of equations, which can be solved in the familiar algebraic way.

It is not very difficult to prove the equivalence of the definition of $S(q)$ by formula (24), on the one hand, and the definition of $S(q)$ by means of behavioural differential equations, on the other.

THEOREM 2.8.1. *For a weighted automaton $(Q, \langle o, t\rangle)$ and $q \in Q$, let $S(q)$ be the stream defined by the above system of behavioural differential equations. Then $S(q)$ satisfies identity* (24) *above. That is, for all $k \geq 0$,*

$$S(q)(k) = \Sigma\{l_1 \times \cdots \times l_k \times l \mid \exists q_0, \ldots, q_k \colon q = q_0 \xrightarrow{l_1} \cdots \xrightarrow{l_k} q_k \text{ and } o(q_k) = l\}.$$

PROOF. Using the differential equation for $S(q)$ and the observation that

$$\begin{aligned} S(q)(k+1) &= S(q)^{(k+1)}(0) \\ &= \big(S(q)'\big)^{(k)}(0) \\ &= \big(t(q)(q_1) \times S(q_1) + \cdots + t(q)(q_n) \times S(q_n)\big)^{(k)}(0) \\ &= t(q)(q_1) \times S(q_1)^{(k)}(0) + \cdots + t(q)(q_n) \times S(q_n)^{(k)}(0) \\ &= t(q)(q_1) \times S(q_1)(k) + \cdots + t(q)(q_n) \times S(q_n)(k) \end{aligned}$$

the proof follows by induction on k. □

As an example of the stream behaviour of a weighted automaton, consider the automaton defined by the following picture

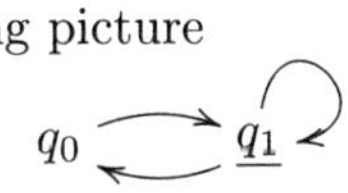

(which we shall consider in the counting example of Section 4.1). The stream $S(q_0)$ represented by the state q_0 can be computed in two ways. According to formula (24), one finds $S(q_0)(k)$ by simply counting (since all weights are 1) the number of paths of length k from q_0 to q_1 (which is the only state with a non-zero output), yielding $S(q_0) = (0, 1, 1, 2, 3, 5, 8, \ldots)$. By the second definition, one has $S(q_0) = X \times S(q_1)$, and $S(q_1) = 1 + \big(X \times S(q_0)\big) + \big(X \times S(q_1)\big)$, yielding $S(q_0) = X/1 - X - X^2$.

In the present chapter, we shall only deal with finite weighted automata, but infinite weighted automata, such as

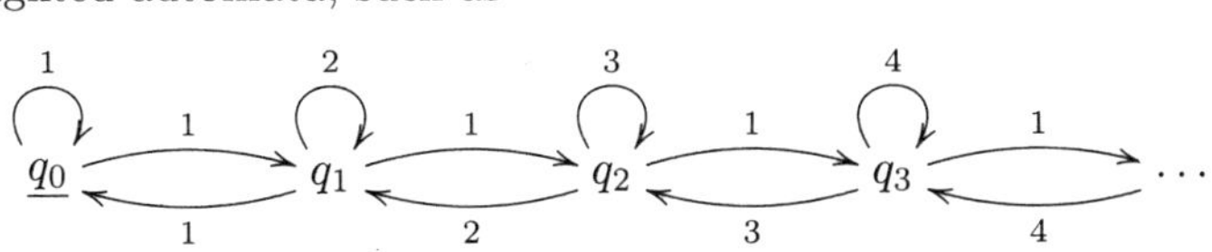

are interesting too. We shall see this and many similar examples of infinite weighted automata in Chapter 4.

2.9. Rational Streams

The streams that can be represented by finite weighted automata are precisely the *rational* streams, which are introduced and characterised in Theorem 2.9.1 below. First we present the following definition. A stream $\sigma \in \mathbb{R}^\omega$ is *polynomial* if there exist $n \geq 0$ and $s_0, \ldots, s_n \in \mathbb{R}$ such that

$$\sigma = s_0 + (s_1 \times X) + (s_2 \times X^2) + \cdots + (s_n \times X^n)$$

or, equivalently, $\sigma = (s_0, s_1, s_2, \ldots, s_n, 0, 0, 0, \ldots)$.

THEOREM 2.9.1. *The following three conditions are equivalent. Any stream σ satisfying them is called* rational:

(1) *There exist polynomial streams π and ρ with $\rho(0) \neq 0$ such that*

$$\sigma = \frac{\pi}{\rho}.$$

(2) *There exist $n \geq 1$, real numbers s_{ij}, for $1 \leq i \leq n$ and $1 \leq j \leq n$, real numbers $k_1, \ldots, k_n$, and streams $\sigma_1, \sigma_2, \ldots, \sigma_n$ with $\sigma = \sigma_1$, satisfying the following finite system of behavioural differential equations*:

derivative	initial value
$(\sigma_1)' = s_{11}\sigma_1 + s_{12}\sigma_2 + \cdots + s_{1n}\sigma_n$	$\sigma_1(0) = k_1$
$(\sigma_2)' = s_{21}\sigma_1 + s_{22}\sigma_2 + \cdots + s_{2n}\sigma_n$	$\sigma_2(0) = k_2$
$\ldots$	$\ldots$
$(\sigma_n)' = s_{n1}\sigma_1 + s_{n2}\sigma_2 + \cdots + s_{nn}\sigma_n$	$\sigma_n(0) = k_n$

.

(3) *σ has a finite representation (is recognizable): there exist a finite weighted automaton Q and a state $q \in Q$ with $\sigma = S(q)$.*

PROOF. A full proof can be found in [**37**], the main ingredients of which we mention next. For the proof that (1) implies (2), consider polynomials $\pi = p_0 + p_1X + \cdots + p_nX^n$ and $\rho = r_0 + r_1X + \cdots + r_mX^m$ and let $\sigma = \pi \times \rho^{-1}$. For notational convenience, we assume that $r_0 = 1$. We also assume that $0 < n < m$ (the case that $m \leq n$ can be dealt with similarly). Using the defining behavioural differential equations of sum, convolution product, and inverse, σ is readily seen to satisfy the following finite system of behavioural differential equations:

derivative	initial value
$(\pi \times \rho^{-1})' = v_1\rho^{-1} + v_2X\rho^{-1} + \cdots + v_mX^{m-1}\rho^{-1}$	$(\pi \times \rho^{-1})(0) = p_0$
$(\rho^{-1})' = -r_1\rho^{-1} - r_2X\rho^{-1} - \cdots - r_mX^{m-1}\rho^{-1}$	$(\rho^{-1})(0) = 1$
$(X\rho^{-1})' = \rho^{-1}$	$(X\rho^{-1})(0) = 0$
$\ldots$	$\ldots$
$(X^{m-1}\rho^{-1})' = X^{m-2}\rho^{-1}$	$(X^{m-1}\rho^{-1})(0) = 0$

where

$$v_i = \begin{cases} p_i - p_0r_i & \text{if } 1 \leq i \leq n, \\ -p_0r_i & \text{if } n < i \leq m. \end{cases}$$

The equivalence between (2) and (3) follows from the correspondence between this finite system of behavioural differential equations and the following finite weighted automaton for $\pi \times \rho^{-1}$:

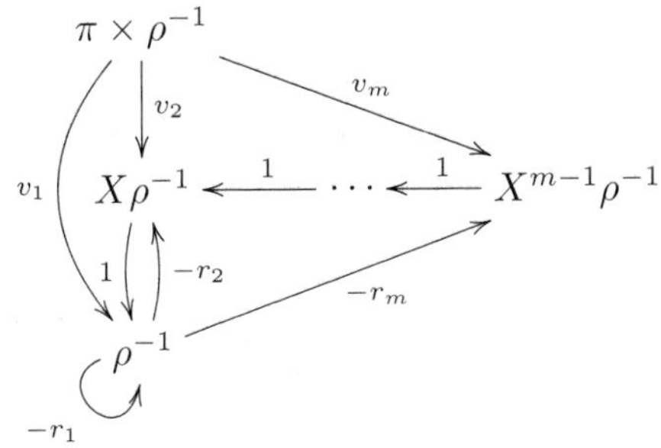

with $o(\pi \times \rho^{-1}) = p_0$ and $o(\rho^{-1}) = 1$. □

Note that in the construction of the weighted automaton for $\pi \times \rho^{-1}$ above, we have used the streams themselves ($\pi \times \rho^{-1}$, $X\rho^{-1}$, etc.) as states. The transitions of any stream (state) in this automaton are obtained by "splitting the derivative" of this stream into its "+" components. Here is a concrete instance of the above automaton, taking $\pi = 2 - X$ and $\rho = 1 - X + X^2$:

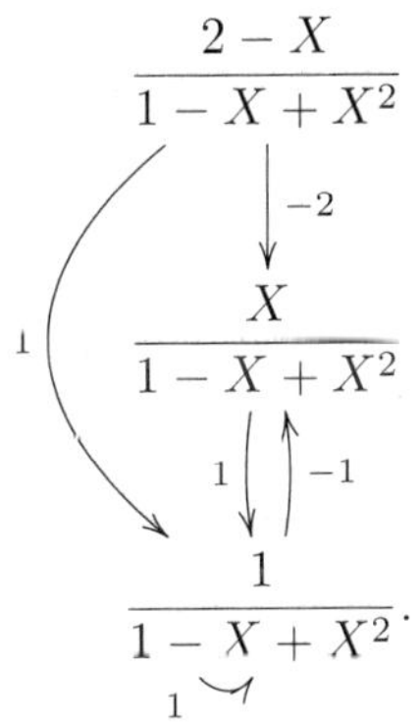

In this automaton, the transitions of for instance the lower state are determined by the fact that its derivative is a sum of two streams, as follows:

$$\left(\frac{1}{1-X+X^2}\right)' = -\frac{X}{1-X+X^2} + \frac{1}{1-X+X^2}.$$

2.10. Streams and Languages

Recall from Section 1.6 that stream automata are coalgebras of the functor $\mathbb{R} \times (-)$, which is a special instance of the more general functor $A \times (-)^B$. We also saw that $2 \times (-)^B$ is another special instance of the latter functor, with final coalgebra $2^{B^*} = \{L \mid L \subseteq B^*\}$, the set of all languages over B. Next, we briefly illustrate that much of the present chapter also applies to languages, by a straightforward generalisation as follows. First, we spell out the final coalgebra structure $(2^{B^*}, \langle o, t\rangle)$ on 2^{B^*}. It consists of functions $o\colon 2^{B^*} \to 2$ and $t\colon 2^{B^*} \to (2^{B^*})^B$, defined, for all $L \subseteq B^*$ and $b \in B$, as

$$o(L) = \begin{cases} 1 & \text{if } \varepsilon \in L, \\ 0 & \text{if } \varepsilon \notin L \end{cases}, \qquad t(L)(b) = L_b.$$

Here ε denotes the empty word, and L_b denotes the *input derivative* (or b-derivative) of L, which we define as

$$L_b = \{w \in B^* \mid b \cdot w \in L\}$$

($b \cdot w$ denotes the word obtained by prefixing w with b). The value $o(L)$, for which we shall write $L(0)$, corresponds to the initial value $\sigma(0)$ of a stream σ. The input derivative L_b generalises the stream derivative σ'. Using these operations, one can define (operators on) languages by means of behavioural differential equations, just as we did with streams. For instance, the following system of equations, which is a minor variation of the corresponding system for streams, defines the operations of sum (union) and product (concatenation) of languages $K, L \subseteq B^*$:

derivative	initial value
$(K + L)_b = K_b + L_b$	$(K + L)(0) = \max\{K(0), L(0)\}$
$(K \times L)_b = (K_b \times L) + \big(L(0) \times L_b\big)$	$(K \times L)(0) = \min\{K(0), L(0)\}$

.

(Note that $\big(L(0) \times L_b\big) = 0$, the empty set, if $L(0) = 0$; if $L(0) = 1$ then $\big(L(0) \times L_b\big) = L_b\big)$. These equations define the familiar operations of union and concatenation since, for instance,

$$K \times L = \{v \cdot w \mid v \in K, w \in L\}.$$

As with streams, the advantage of the use of behavioural differential equations is that they allow us to reason about languages by coinduction.

More generally still, all of this applies also to formal power series in many non-commutative variables and with coefficients in any semi-ring, which we also mentioned in Section 1.6. Details and many examples of definitions and proofs by coinduction for languages and formal power series, can be found in [**37**].

2.11. Discussion

The material of the present chapter is taken from [**39**]. Motivating sources of examples of streams and stream operators have been the books [**22, 47**], and the papers by McIlroy [**28, 29**] and Karczmarczuk [**24, 25**]. The solution of difference equations by means of stream calculus is conceptually very simple, since the entire game is played within the world of streams. In contrast to the classical technique of generating functions (as in [**22, 47**]), functions (from $\mathbb{R}$ to $\mathbb{R}$) are just not needed in stream calculus, and convergence issues thereby simply do not enter the picture. As we have seen, streams can be viewed as formal power series (in one variable), which are often used as a formal alternative to generating functions, precisely to avoid convergence considerations. We see some advantages of stream calculus also over the use of formal power series. First, there is the rigorous use of coinduction, both in definitions and in proofs, which makes stream calculus possibly more formal than the use of "formal" power series usually is. Notably this applies to the use of the operation of inverse, which is not always treated strictly formally within theories of power series. Secondly, stream calculus is very expressive, as we shall see in Chapter 3, where additional operators (including shuffle product and shuffle inverse) will be introduced. The notion of input derivative of a language, which we saw in Section 2.10 goes back to [**14**]. It plays a role also in [**16**], where the chapter "The differential calculus of events" already suggests a connection with classical calculus. The notion of bisimulation-up-to, in Definition 2.3.1, is a variation on a similar notion by Milner [**32**] (see also [**41**]). In [**9**], variations and coalgebraic generalisations of coinductive proof methods are given. The present notion of bisimulation-up-to (identity and sum), can be easily generalised along the lines of [**9**], to a version which would allow derivatives to be bisimilar up to arbitrary contexts (including product, inverse, and the other operators). For a general reference on (rational) formal power series, see [**11**]. Theorem 2.9.1 is a classical result. New about our use of weighted automata as representations for streams are: -the coinductive definition of their behaviour; -the way such automata are constructed by means of splitting derivatives (as in the proof of Theorem 2.9.1); -and our use of infinite weighted automata (in Chapter 4).

CHAPTER 3

Analytical Differential Equations

The power of stream calculus is further increased by the introduction of three new operators: shuffle product, shuffle inverse, and stream exponentiation. These operators and specifically their interplay with the ordinary (convolution) product and inverse, are interesting in themselves. They are also useful in applications, as we shall illustrate by the solution of analytical differential equations.

3.1. Shuffle Product and Shuffle Inverse

For streams σ and τ, let the *shuffle product* $\sigma \otimes \tau$ and the *shuffle inverse* $\sigma^{-\underline{1}}$ be the unique streams satisfying the following behavioural differential equations:

derivative	initial value
$(\sigma \otimes \tau)' = (\sigma' \otimes \tau) + (\sigma \otimes \tau')$	$(\sigma \otimes \tau)(0) = \sigma(0) \times \tau(0)$
$(\sigma^{-\underline{1}})' = -\sigma' \otimes (\sigma^{-\underline{1}} \otimes \sigma^{-\underline{1}})$	$\sigma^{-\underline{1}}(0) = \sigma(0)^{-1}$

(The shuffle inverse is defined only when $\sigma(0) \neq 0$.) We use an underlined symbol $\underline{1}$ to distinguish shuffle inverse from the inverse to the convolution product. As we observed earlier, stream differentiation of (convolution) product and inverse does not behave the way we are used to from analysis. A somewhat formalistic way of motivating the definition of shuffle product and inverse is that they constitute, in this respect, more familiarly behaved alternatives to product and inverse. Another motivation will follow in Section 3.4, where we shall see that the Taylor series of the (pointwise) product of two real-valued functions is equal to the shuffle product of their Taylor series. Finally, the *name* shuffle product is chosen because this operator is the stream variant of the shuffle (or merge) operation on languages.

Similar to formula (6) for the convolution product, there is the following formula for $(\sigma \otimes \tau)(n)$, for any $n \geq 0$,

$$(\sigma \otimes \tau)(n) = \sum_{k=0}^{n} \binom{n}{k} \times \sigma(n-k) \times \tau(k). \tag{25}$$

This formula could also have been taken as an alternative definition to the system of behavioural differential equations above. Reasoning in terms of (25), however, is again unnecessarily complicated, because of the use of the indices, and the occurrence of the binomial coefficient. And as with ordinary inverse, no similar such formula for shuffle inverse is known.

We shall use the following conventions, for all $n \geq 0$:

$$\sigma^{\underline{0}} \equiv 1, \quad \sigma^{\underline{n+1}} \equiv \sigma \otimes \sigma^{\underline{n}}, \quad \sigma^{-\underline{n}} \equiv (\sigma^{-\underline{1}})^{\underline{n}}.$$

The following theorem contains some basic properties of shuffle product and shuffle inverse.

THEOREM 3.1.1. *For all streams* $\sigma = (s_0, s_1, s_2, \dots)$, τ *and* ρ, *for all* $r \in \mathbb{R}$, *and for all* $n \geq 0$,

$$\sigma \otimes \tau = \tau \otimes \sigma, \tag{26}$$

$$r \otimes \sigma = r \times \sigma, \tag{27}$$

$$0 \otimes \sigma = 0, \tag{28}$$

$$1 \otimes \sigma = \sigma, \tag{29}$$

$$\sigma \otimes (\tau \otimes \rho) = (\sigma \otimes \tau) \otimes \rho, \tag{30}$$

$$\sigma \otimes (\tau + \rho) = (\sigma \otimes \tau) + (\sigma \otimes \rho), \tag{31}$$

$$(\sigma^{\underline{n+1}})' = (n+1) \otimes \sigma' \otimes \sigma^{\underline{n}}, \tag{32}$$

$$X^{\underline{n}} = n! \times X^n, \tag{33}$$

$$\sigma \otimes \sigma^{\underline{-1}} = 1, \tag{34}$$

$$(\sigma^{\underline{-1}})^{\underline{-1}} = \sigma, \tag{35}$$

$$(\sigma \otimes \tau)^{\underline{-1}} = \sigma^{\underline{-1}} \otimes \tau^{\underline{-1}}, \tag{36}$$

$$X \otimes \sigma = s_0 X^1 + 2s_1 X^2 + 3s_2 X^3 + \cdots = (0, s_0, 2s_1, 3s_2, \dots), \tag{37}$$

$$(X \otimes \sigma')' = s_1 + 2s_2 X^1 + 3s_3 X^2 + \cdots = (s_1, 2s_2, 3s_3, \dots), \tag{38}$$

$$(1 - X)^{\underline{-1}} = 1 + 1!X + 2!X^2 + 3!X^3 \cdots . \tag{39}$$

PROOF. We treat a few examples. For (26), define

$$R = \{\angle \sigma \otimes \tau, \tau \otimes \sigma\rangle \mid \sigma, \tau \in \mathbb{R}^\omega\}$$

and observe that

$$\begin{aligned}(\sigma \otimes \tau)' &= (\sigma' \otimes \tau) + (\sigma \otimes \tau') \\ &\Sigma R \; (\tau \otimes \sigma') + (\tau' \otimes \sigma) \\ &= (\tau' \otimes \sigma) + (\tau \otimes \sigma') \\ &= (\tau \otimes \sigma)'.\end{aligned}$$

Identity (26) now follows by coinduction-up-to (Theorem 2.3.2). Identity (32) can be proven by induction on n. For (34) define

$$Q = \{\langle \rho \otimes (\sigma \otimes \sigma^{\underline{-1}}), \rho\rangle \mid \rho, \sigma \in \mathbb{R}^\omega\}.$$

Computing derivatives gives

$$\begin{aligned}(\rho \otimes (\sigma \otimes \sigma^{\underline{-1}}))' &= \rho' \otimes (\sigma \otimes \sigma^{\underline{-1}}) + \rho \otimes ((\sigma' \otimes \sigma^{\underline{-1}}) + (\sigma \otimes (-\sigma' \otimes \sigma^{\underline{-1}} \otimes \sigma^{\underline{-1}}))) \\ &= \rho' \otimes ((\sigma \otimes \sigma^{\underline{-1}}) + \rho \otimes (\sigma' \otimes \sigma^{\underline{-1}}) - (\rho \otimes (\sigma' \otimes \sigma^{\underline{-1}})) \otimes (\sigma \otimes \sigma^{\underline{-1}})) \\ &\Sigma Q \; \rho' + \rho \otimes (\sigma' \otimes \sigma^{\underline{-1}}) - \rho \otimes (\sigma' \otimes \sigma^{\underline{-1}}) \\ &= \rho'\end{aligned}$$

which proves that Q is a bisimulation-up-to, allowing us again to apply coinduction-up-to. For identity (37), let

$$T = \{\langle X \otimes \sigma, s_0 X^1 + 2s_1 X^2 + 3s_2 X^3 + \cdots\rangle \mid \sigma = (s_0, s_1, s_2, \dots) \in \mathbb{R}^\omega\}$$

and note that

$$\begin{aligned}(X \otimes \sigma)' &= \sigma + (X \otimes \sigma') \\ &\Sigma T\ \sigma + (s_1 X^1 + 2s_2 X^2 + 3s_3 X^3 + \cdots) \\ &= (s_0 + s_1 X^1 + s_2 X^2 + \cdots) + (s_1 X^1 + 2s_2 X^2 + 3s_3 X^3 + \cdots) \\ &\qquad\qquad\qquad\qquad\qquad\qquad \text{[by Theorem 2.5.1]} \\ &= s_0 + 2s_1 X^1 + 3s_2 X^2 + 4s_3 X^3 + \cdots \\ &= (s_0 X + 2s_1 X^2 + 3s_2 X^3 + 4s_3 X^4 + \cdots)'\end{aligned}$$

showing that T is a bisimulation-up-to. Finally for (39), observe that

$$\{\langle n!(1-X)^{-\underline{(n+1)}}, n! + (n+1)!X + (n+2)!X^2 + \cdots\rangle \mid n \geq 0\}$$

is a bisimulation relation on $\mathbb{R}^\omega$. □

3.2. Stream Exponentiation

We next introduce the operation of *stream exponentiation.* For a stream σ let $\exp(\sigma)$ be the unique stream satisfying the following behavioural differential equation:

derivative	initial value
$\exp(\sigma)' = \sigma' \otimes \exp(\sigma)$	$\exp(\sigma)(0) = e^{\sigma(0)}$

where $e^{\sigma(0)}$ is the analytical function e^x applied to the real number $\sigma(0)$. The theorem below contains a number of more and less familiar identities for stream exponentiation.

THEOREM 3.2.1. *For all $r, s \in \mathbb{R}$, $\sigma, \tau \in \mathbb{R}^\omega$, and $n \geq 0$,*

(40) $$\exp(rX) = \frac{1}{1-rX},$$

(41) $$\exp(\sigma) = 1 + \frac{\sigma^{\underline{1}}}{1!} + \frac{\sigma^{\underline{2}}}{2!} + \cdots,$$

(42) $$\exp(\sigma) \otimes \exp(\tau) = \exp(\sigma + \tau),$$

(43) $$\exp(-\sigma) = \exp(\sigma)^{-\underline{1}},$$

(44) $$\frac{1}{1-rX} \otimes \frac{1}{1-sX} = \frac{1}{1-(r+s)X}.$$

PROOF. Again we treat only a few examples. For (40) consider $R = \{\langle s \times \exp(rX), s/1 - rX\rangle \mid r, s \in \mathbb{R}\}$. It is a bisimulation relation, since

$$\begin{aligned}(s \times \exp(rX))' &= (r \times s) \times \exp(rX) \\ &R\ \frac{r \times s}{1-rX} \\ &= \left(\frac{s}{1-rX}\right)'.\end{aligned}$$

For identity (43) let $Q = \{\langle \tau \otimes \exp(-\sigma), \tau \otimes \exp(\sigma)^{-\underline{1}} \rangle \mid \sigma, \tau \in \mathbb{R}^\omega\}$ and compute as follows:

$$\begin{aligned}
(\tau \otimes \exp(-\sigma))' &= (\tau' - (\tau \otimes \sigma')) \otimes \exp(-\sigma) \\
&\; Q \; (\tau' - (\tau \otimes \sigma')) \otimes \exp(\sigma)^{-\underline{1}} \\
&= (\tau' \otimes \exp(\sigma)^{-\underline{1}}) - (\tau \otimes \sigma' \otimes \exp(\sigma)^{-\underline{1}}) \\
&= (\tau' \otimes \exp(\sigma)^{-\underline{1}}) - (\tau \otimes \sigma' \otimes \exp(\sigma) \otimes \exp(\sigma)^{-\underline{1}} \otimes \exp(\sigma)^{-\underline{1}}) \\
&= (\tau' \otimes \exp(\sigma)^{-\underline{1}}) + (\tau \otimes (-\sigma' \otimes \exp(\sigma) \otimes \exp(\sigma)^{-\underline{1}} \otimes \exp(\sigma)^{-\underline{1}})) \\
&= (\tau' \otimes \exp(\sigma)^{-\underline{1}}) + (\tau \otimes (\exp(\sigma)^{-\underline{1}})') \\
&= (\tau \otimes \exp(\sigma)^{-\underline{1}})'
\end{aligned}$$

which proves that Q is a bisimulation. Identity (44) follows from identities (40) and (42). □

Identity (40) is somewhat surprising. Also identity (44) is a bit unusual, but note that it involves (ordinary) inverse combined with shuffle product.

3.3. Comparing Convolution and Shuffle Product

We investigate in some detail the relation between the two types of product and inverse, for which we shall make use of a new type of stream derivation.

Let the *analytical* stream derivative of a stream σ be defined by

(45) $$\frac{d}{dX}(\sigma) = (X \otimes \sigma')'.$$

As usual, we shall also write

$$\frac{d\sigma}{dX} \equiv \frac{d}{dX}(\sigma).$$

There is the following general formula for our new type of derivation:

(46) $$\frac{d}{dX}(s_0 + s_1 X^1 + s_2 X^2 + \cdots) = s_1 + 2s_2 X^1 + 3s_3 X^2 + \cdots$$

which we recognise as identity (38), and which explains the name of analytical stream derivation. Analytical stream derivation behaves for (convolution) product and inverse in the, from analysis, familiar way.

THEOREM 3.3.1. *For all* $\sigma, \tau \in \mathbb{R}^\omega$ *and* $n \geq 0$,

(47) $$\frac{d(\sigma \times \tau)}{dX} = \left(\frac{d\sigma}{dX} \times \tau\right) + \left(\sigma \times \frac{d\tau}{dX}\right),$$

(48) $$\frac{d\sigma^{-1}}{dX} = -\frac{d\sigma}{dX} \times \sigma^{-1} \times \sigma^{-1},$$

(49) $$\frac{d\sigma^{n+1}}{dX} = (n+1) \times \frac{d\sigma}{dX} \times \sigma^n.$$

PROOF. The proof of identity (47) uses the following equalities (the proofs of these are left to the reader). For all summable families $\{\sigma_n\}_{n=0}^{\infty}$ and all τ in $\mathbb{R}^\omega$, for all $k \geq 0$,

$$
\begin{aligned}
\frac{d(\sum_{n=0}^{\infty} \sigma_n)}{dX} &= \sum_{n=0}^{\infty} \frac{d\sigma_n}{dX}, \\
\frac{d(X^k \times \tau)}{dX} &= \left(\frac{dX^k}{dX} \times \tau\right) + \left(X^k \times \frac{d\tau}{dX}\right), \\
\frac{dX^{k+1}}{dX} &= (k+1) \times X^k.
\end{aligned}
$$

Next consider $\sigma = (s_0, s_1, s_2, \ldots)$ and τ in $\mathbb{R}^\omega$ and note that

$$
\begin{aligned}
\frac{d(\sigma \times \tau)}{dX} &= \frac{d((\sum_{n=0}^{\infty} s_n \times X^n) \times \tau)}{dX} \qquad \text{[Theorem 2.5.1]} \\
&= \frac{d(\sum_{n=0}^{\infty} s_n \times (X^n \times \tau))}{dX} \\
&= \sum_{n=0}^{\infty} \frac{d(s_n \times (X^n \times \tau))}{dX} \\
&= \sum_{n=0}^{\infty} s_n \times \frac{d(X^n \times \tau)}{dX} \\
&= \sum_{n=0}^{\infty} s_n \times \left(\frac{dX^n}{dX} \times \tau + X^n \times \frac{d\tau}{dX}\right) \\
&= \sum_{n=0}^{\infty} s_n \times \left(\frac{dX^n}{dX} \times \tau\right) + \sum_{n=0}^{\infty} s_n \times \left(X^n \times \frac{d\tau}{dX}\right) \\
&= \sum_{n=1}^{\infty} (n \times s_n \times X^{n-1} \times \tau) + \sum_{n=0}^{\infty} s_n \times \left(X^n \times \frac{d\tau}{dX}\right) \\
&= \left(\sum_{n=1}^{\infty} n \times s_n \times X^{n-1}\right) \times \tau + \left(\sum_{n=0}^{\infty} s_n \times X^n\right) \times \frac{d\tau}{dX} \\
&= \left(\frac{d\sigma}{dX} \times \tau\right) + \left(\sigma \times \frac{d\tau}{dX}\right) \qquad \text{[identity (46) and Theorem 2.5.1].}
\end{aligned}
$$

Identities (48) and (49) easily follow from (47). □

Note that properties (47)–(49) reflect precisely how ordinary stream derivation behaves on shuffle product and shuffle inverse, which satisfy

$$
\begin{aligned}
(\sigma \otimes \tau)' &= (\sigma' \otimes \tau) + (\sigma \otimes \tau'), \\
(\sigma^{-\underline{1}})' &= -\sigma' \otimes (\sigma^{-\underline{1}} \otimes \sigma^{-\underline{1}}), \\
(\sigma^{\underline{n+1}})' &= (n+1) \otimes \sigma' \otimes \sigma^{\underline{n}}.
\end{aligned}
$$

Analytical stream derivation is used in the following definition of a new operator on streams:

derivative	initial value
$\Lambda_C(\sigma)' = \Lambda_C\left(\frac{d\sigma}{dX}\right)$	$\Lambda_C(\sigma)(0) = \sigma(0)$

.

One can easily prove that Λ_C transforms a stream $(s_0, s_1, s_2, s_3, \dots)$ into $(0!s_0, 1!s_1, 2!s_2, 3!s_3, \dots)$, or, equivalently,

$$\Lambda_C\left(s_0 + \frac{s_1}{1!}X + \frac{s_2}{2!}X^2 + \frac{s_3}{3!}X^3 + \cdots\right) = s_0 + s_1 X + s_2 X^2 + s_3 X^3 + \cdots .$$

In combinatorics, this is referred to as the *Laplace–Carson* transform (cf. [**10**, p. 350] and [**15**, p. 48]), hence our notation. The following theorem shows that with Λ_C, we can relate the two types of product and inverse.

THEOREM 3.3.2. *For all $r \in \mathbb{R}$, $\sigma, \tau \in \mathbb{R}^\omega$,*

$$\Lambda_C\left(\frac{d\sigma}{dX}\right) = \Lambda_C(\sigma)',$$
$$\Lambda_C(r) = r,$$
$$\Lambda_C(X) = X,$$
$$\Lambda_C(\sigma + \tau) = \Lambda_C(\sigma) + \Lambda_C(\tau),$$
$$\Lambda_C(\sigma \times \tau) = \Lambda_C(\sigma) \otimes \Lambda_C(\tau),$$
$$\Lambda_C(\sigma^{-1}) = \Lambda_C(\sigma)^{-\underline{1}}.$$

PROOF. The first equality is by definition and the next two are trivial. For the latter three, let $R \subseteq \mathbb{R}^\omega \times \mathbb{R}^\omega$ be the smallest relation such that

(1) $\{\langle\sigma, \sigma\rangle \mid \sigma \in \mathbb{R}^\omega\} \subseteq R$,
(2) for all $\sigma_1, \sigma_2, \tau_1, \tau_2 \in \mathbb{R}^\omega$, if $\langle\Lambda_C(\sigma_1), \tau_1\rangle \in R$ and $\langle\Lambda_C(\sigma_2), \tau_2\rangle \in R$ then
   (a) $\langle\Lambda_C(\sigma_1 + \sigma_2), \tau_1 + \tau_2\rangle \in R$,
   (b) $\langle\Lambda_C(\sigma_1 \times \sigma_2), \tau_1 \otimes \tau_2\rangle \in R$,
   (c) $\langle\Lambda_C((\sigma_1)^{-1}), (\tau_1)^{-\underline{1}}\rangle \in R$.

Then R is a bisimulation on $\mathbb{R}^\omega$ and the result follows by coinduction. □

Thus the operator Λ_C allows one to switch between the two different ring structures on $\mathbb{R}^\omega$ that are determined by convolution product and shuffle product, each of which comes along with its own type of (partially defined) operation of inverse, and its own type of derivative:

$$\Lambda_C\colon \left\langle \mathbb{R}^\omega, +, \times, (-)^{-1}, \frac{d(-)}{dX} \right\rangle \to \langle \mathbb{R}^\omega, +, \otimes, (-)^{-\underline{1}}, (-)' \rangle.$$

Let us emphasize, however, that for the stream calculus we are developing, it is of crucial importance to have both structures present at the same time. Notably the interplay between the various operators from both worlds, turns out to constitute the most interesting part of the calculus. The following identities clearly illustrate this point, since they involve both the shuffle product and the (convolution) inverse. They all have in common that they provide a way of eliminating the occurrence of the shuffle product, a procedure we shall sometimes refer to as *shuffle elimination.*

THEOREM 3.3.3. *For all $r \in \mathbb{R}$ and $\sigma \in \mathbb{R}^\omega$:*

$$1 = \frac{1}{1+rX} \otimes \frac{1}{1-rX}, \tag{50}$$

$$X \otimes \sigma = \left(X^2 \times \frac{d\sigma}{dX}\right) + (X \times \sigma), \tag{51}$$

$$\left(\frac{1}{1-rX}\right) \otimes \sigma = \left(\frac{1}{1-rX}\right) \times \left(\sigma \circ \frac{X}{1-rX}\right) \tag{52}$$

$$= \frac{s_0}{1-rX} + \frac{s_1 X}{(1-rX)^2} + \frac{s_2 X^2}{(1-rX)^3} + \cdots. \tag{53}$$

PROOF. Identity (50) follows from (44). For (51), observe that

$$\begin{aligned}
\left(X^2 \times \frac{d\sigma}{dX}\right) + (X \times \sigma) &= \left(X^2 \times (X \otimes \sigma')'\right) + (X \times \sigma) \\
&= \left(X \times \left(X \times (X \otimes \sigma')'\right)\right) + (X \times \sigma) \\
&= \left(X \times (X \otimes \sigma')\right) + (X \times \sigma) \\
&\qquad \text{[Fundamental Theorem 2.6.1 } (X \otimes \sigma')(0) = 0] \\
&= X \times \left((X \otimes \sigma') + \sigma\right) \\
&= X \times \left((X \otimes \sigma)'\right) \\
&= X \otimes \sigma \quad \text{[Fundamental Theorem 2.6.1 } (X \otimes \sigma)(0) = 0].
\end{aligned}$$

And for (52), define

$$R = \left\{ \left\langle \left(\frac{s}{1-rX}\right) \otimes \sigma, \left(\frac{s}{1-rX}\right) \times \left(\sigma \circ \frac{X}{1-rX}\right) \right\rangle \;\middle|\; r, s \in \mathbb{R}, \sigma \in \mathbb{R}^\omega \right\}$$

and compute as follows to see that R is a bisimulation-up-to:

$$\begin{aligned}
\left(\left(\frac{s}{1-rX}\right) \otimes \sigma\right)' &= \left(\frac{r \times s}{1-rX}\right) \otimes \sigma + \left(\frac{s}{1-rX}\right) \otimes \sigma' \\
&\Sigma R \left(\frac{r \times s}{1-rX}\right) \times \left(\sigma \circ \frac{X}{1-rX}\right) + \left(\frac{s}{1-rX}\right) \times \left(\sigma' \circ \frac{X}{1-rX}\right) \\
&= \left(\frac{s}{1-rX}\right)' \times \left(\sigma \circ \frac{X}{1-rX}\right) + \left(s \times \left(\sigma \circ \frac{X}{1-rX}\right)'\right) \\
&= \left(\left(\frac{s}{1-rX}\right) \times \left(\sigma \circ \frac{X}{1-rX}\right)\right)'.
\end{aligned}$$

Identity (52) then follows by coinduction-up-to. Identity (53) follows from (52) and Theorem 2.5.3. □

To illustrate identity (51), we compute

$$\begin{aligned}
X \otimes \frac{1}{1+X^2} &= X^2 \times \frac{d}{dX}\left(\frac{1}{1+X^2}\right) + \left(X \times \frac{1}{1+X^2}\right) \\
&= X^2 \times \frac{-2X}{(1+X^2)^2} + \frac{X}{1+X^2} \qquad \text{[identity (48)]} \\
&= \frac{X - X^3}{(1+X^2)^2}.
\end{aligned}$$

Similar computations give

$$X \otimes \frac{X}{1+X^2} = \frac{2X^2}{(1+X^2)^2},$$

$$X^2 \otimes \frac{1}{1+X^2} = \frac{X^2 - 3X^4}{(1+X^2)^3},$$

$$X^2 \otimes \frac{X}{1+X^2} = \frac{-3X^3 + X^5}{(1+X^2)^3}.$$

The following two identities are useful special cases of (53):

$$X^n \otimes \frac{1}{1-rX} = \frac{X^n}{(1-rX)^{n+1}}, \tag{54}$$

$$(1+rX)^n \otimes \frac{1}{1-rX} = \frac{1}{(1-rX)^{n+1}}. \tag{55}$$

3.4. Solving Analytical Differential Equations

A classical technique in analysis for the solution of differential equations defining analytical functions, is the method of undetermined coefficients (cf. [**12**, p. 82]). The idea is quickly explained by means of an example. In order to solve the differential equation

derivative	initial value
$f'' + f = 0$	$f(0) = 0,\ f'(0) = 1$

one assumes the solution to be of the shape

$$f(x) = s_0 + \frac{s_1}{1!}x + \frac{s_2}{2!}x^2 + \cdots.$$

Computing f'' gives

$$f''(x) = s_2 + \frac{s_3}{1!}x + \frac{s_4}{2!}x^2 + \cdots.$$

Substituting the expressions for f and f'' in the differential equation, one obtains the following difference equation for the coefficients $(s_0, s_1, s_2, \dots)$ of f:

derivative	initial value
$s_{n+2} + s_n = 0$	$s_0 = 0,\ s_1 = 1$

.

Thus this method reduces the problem of solving a differential equation for f to the problem of solving a difference equation for the stream $(s_0, s_1, s_2, \dots)$ of Taylor coefficients of f. Though conceptually very simple, the method of undetermined coefficients has two major drawbacks. Firstly, more interesting differential equations quickly lead to very complicated difference equations. Secondly, there is no general technique for translating the solution of the difference equation (if found at all) back into a (closed) expression for f. Here we shall present a variant of the method above, which in many applications is free of these restrictions. Let

$$\mathcal{A} = \{f : \mathbb{R} \to \mathbb{R} \mid f \text{ is analytic in (a neighbourhood of) } 0\}.$$

The details of the definition of analytic function are not important here. It is for our present purposes sufficient to know that analytical functions are differentiable (in a neighbourhood of 0) and that their derivative is again analytic. These properties are used in the definition of the function $\mathcal{T} : \mathcal{A} \to \mathbb{R}^\omega$, which sends an analytic function f to its Taylor series $\mathcal{T}(f) = \big(f(0), f'(0), f''(0), \dots\big)$. The formal definition is given, as always, by means of behavioural differential equations.

DEFINITION 3.4.1. Let $\mathcal{T}(f)$ be defined by the following system of behavioural differential equations (one for each $f \in \mathcal{A}$):

derivative	initial value
$\mathcal{T}(f)' = \mathcal{T}(f')$	$\mathcal{T}(f)(0) = f(0)$

.

(Note that the first occurrence of $'$ in $\mathcal{T}(f)' = \mathcal{T}(f')$ stands for stream derivation, whereas the second denotes analytical function derivation.) We shall sometimes refer to the stream $\mathcal{T}(f)$ as the *Taylor transform* of f.

Our method is characterised by the following three steps:

(1) The function $\mathcal{T}$ is used to transform, in a systematic fashion, a differential equation for f into a behavioural differential equation for the stream $\mathcal{T}(f)$ of Taylor coefficients of f.
(2) The behavioural differential equation is solved in stream calculus by means of the techniques of Section 2.6.
(3) The resulting solution is translated back in a systematic manner into an expression for f.

First of all, the following theorem expresses that the function $\mathcal{T}$ transforms functions into their Taylor series in a systematic manner, indeed. For functions f and g, we use the following familiar definitions: for all $x \in \mathbb{R}$,

$$(f+g)(x) = f(x) + g(x), \quad (f \cdot g)(x) = f(x) \times g(x),$$
$$f^{-1}(x) = f(x)^{-1}, \quad e^f(x) = e^{f(x)}.$$

THEOREM 3.4.2. *For all analytic functions* $f, g \in \mathcal{A}$,

(56) $$\mathcal{T}(f)(0) = f(0),$$
(57) $$\mathcal{T}(f') = \mathcal{T}(f)',$$
(58) $$\mathcal{T}\Big(\int_0^x f(t)\,dt\Big) = X \times \mathcal{T}(f),$$
(59) $$\mathcal{T}(f+g) = \mathcal{T}(f) + \mathcal{T}(g),$$
(60) $$\mathcal{T}(f \cdot g) = \mathcal{T}(f) \otimes \mathcal{T}(g),$$
(61) $$\mathcal{T}(f^{-1}) = \mathcal{T}(f)^{-1},$$
(62) $$\mathcal{T}(e^f) = \exp\big(\mathcal{T}(f)\big).$$

PROOF. Identities (56) and (57) are immediate by the definition of $\mathcal{T}$. Identity (58) follows by coinduction, using the fundamental theorem from analysis $\big((\int_0^x f)' = f\big)$. For the others, define $R \subseteq \mathbb{R}^\omega \times \mathbb{R}^\omega$ as the smallest relation on streams such that $\langle \mathcal{T}(f), \mathcal{T}(f)\rangle \in R$, for all $f \in \mathcal{A}$, and such that if $\langle \mathcal{T}(f_1), \mathcal{T}(f_2)\rangle \in R$ and $\langle \mathcal{T}(g_1), \mathcal{T}(g_2)\rangle \in R$ then

(1) $\langle \mathcal{T}(f_1 + g_1), \mathcal{T}(f_2) + \mathcal{T}(g_2)\rangle \in R$,
(2) $\langle \mathcal{T}(f_1 \cdot g_1), \mathcal{T}(f_2) \otimes \mathcal{T}(g_2)\rangle \in R$,
(3) $\langle \mathcal{T}(f_1^{-1}), \mathcal{T}(f_2)^{-1}\rangle \in R$,
(4) $\big\langle \mathcal{T}(e^{f_1}), \exp\big(\mathcal{T}(f_2)\big)\big\rangle \in R$.

The relation R can be shown to be a bisimulation. For instance, consider $\langle \mathcal{T}(f \cdot g), \mathcal{T}(f) \otimes \mathcal{T}(g)\rangle \in R$. Both streams clearly have the same initial value.

And $\langle \mathcal{T}(f \cdot g)', (\mathcal{T}(f) \otimes \mathcal{T}(g))' \rangle \in R$, since

$$\begin{aligned}
\mathcal{T}(f \cdot g)' &= \mathcal{T}((f \cdot g)') \\
&= \mathcal{T}(f' \cdot g + f \cdot g') \\
&R\, \mathcal{T}(f') \otimes \mathcal{T}(g) + \mathcal{T}(f) \otimes \mathcal{T}(g') \\
&= \mathcal{T}(f)' \otimes \mathcal{T}(g) + \mathcal{T}(f) \otimes \mathcal{T}(g)' \\
&= \big(\mathcal{T}(f) \otimes \mathcal{T}(g)\big)'.
\end{aligned}$$

Similarly for the other elements of R. Now the theorem follows by coinduction. □

The following identities on the Taylor transforms of some well known functions will be useful when solving differential equations.

THEOREM 3.4.3. *For all* $r \in \mathbb{R}$, $n \geq 0$, $f \in \mathcal{A}$,

$$\mathcal{T}(r) = r, \tag{63}$$

$$\mathcal{T}(x^n) = n! X^n, \tag{64}$$

$$\mathcal{T}(e^{rx}) = \frac{1}{1 - rX}, \tag{65}$$

$$\mathcal{T}\big(\sin(rx)\big) = \frac{rX}{1 + r^2X^2}, \tag{66}$$

$$\mathcal{T}\big(\cos(rx)\big) = \frac{1}{1 + r^2X^2}, \tag{67}$$

$$\mathcal{T}(f \cdot e^{rx}) = \left(\frac{1}{1 - rX}\right) \times \left(\mathcal{T}(f) \circ \frac{X}{1 - rX}\right), \tag{68}$$

$$\mathcal{T}(f \cdot x) = \left(X^2 \times \frac{d\mathcal{T}(f)}{dX}\right) + (X \times \mathcal{T}(f)). \tag{69}$$

PROOF. The first two equalities are straightforward. For (65)–(67), define

$$R = \left\{ \left\langle \mathcal{T}(s \cdot e^{rx}), \frac{s}{1 - rX} \right\rangle \,\middle|\, r, s \in \mathbb{R} \right\} \cup \left\{ \left\langle \mathcal{T}(s \cdot \sin(rx)), \frac{(s \times r)X}{1 + r^2X^2} \right\rangle \,\middle|\, r, s \in \mathbb{R} \right\}$$
$$\cup \left\{ \left\langle \mathcal{T}(s \cdot \cos(rx)), \frac{s}{1 + r^2X^2} \right\rangle \,\middle|\, r, s \in \mathbb{R} \right\}$$

and use coinduction. For (68), note that

$$\begin{aligned}
\mathcal{T}(f \cdot e^{rx}) &= \mathcal{T}(f) \otimes \mathcal{T}(e^{rx}) && \text{[identity (60)]} \\
&= \mathcal{T}(f) \otimes \left(\frac{1}{1 - rX}\right) && \text{[identity (65)]} \\
&= \left(\frac{1}{1 - rX}\right) \times \left(\mathcal{T}(f) \circ \frac{X}{1 - rX}\right) && \text{[identity (52)].}
\end{aligned}$$

For (69), finally, we have:

$$\begin{aligned}
\mathcal{T}(f \cdot x) &= \mathcal{T}(f) \otimes \mathcal{T}(x) && \text{[identity (60)]} \\
&= \mathcal{T}(f) \otimes X && \text{[identity (64)]} \\
&= \left(X^2 \times \frac{d\mathcal{T}(f)}{dX}\right) + (X \times \mathcal{T}(f)). && \text{[identity (51)].}
\end{aligned}$$

□

By now, we are sufficiently prepared to tackle a variety of differential equations. As a warming up, consider the equation from the beginning of the present section:

differential equation	initial value
$f'' + f = 0$	$f(0) = 0,\ f'(0) = 1$

Applying $\mathcal{T}$ to both sides of the equation, and putting $\sigma = \mathcal{T}(f)$ gives

derivative	initial value
$\sigma'' + \sigma = 0$	$\sigma(0) = 0,\ \sigma'(0) = 1$

This constitutes step (1) of our method. In step (2), the resulting behavioural differential equation is solved according to the techniques of Section 2.6, yielding

$$\sigma = \frac{X}{1 + X^2}.$$

In order to translate this outcome back into the function f—step (3) of our method—we can apply identity (20) yielding

$$\sigma = X - X^3 + X^5 - X^7 + \cdots$$

which, in combination with identity (64) gives

$$f(x) = x - \frac{x^3}{3!} + \frac{x^5}{5!} - \frac{x^7}{7!} + \cdots.$$

Often, we can improve on the last step. In this particular example, it is sufficient to consult our set of basic identities on Taylor transforms, to find that identity (66) provides us immediately with an answer:

$$f(x) = sin(x).$$

For a second example, consider the following non-homogeneous equation:

differential equation	initial value
$f' - f = e^x$	$f(0) = 1$

Using identity (65) and writing $\sigma = \mathcal{T}(f)$, step (1) transforms this equation into

derivative	initial value
$\sigma' - \sigma = (1 - X)^{-1}$	$\sigma(0) = 1$

The solution of this equation, step (2), is identity (23). For step (3), we combine $\sigma = \mathcal{T}(f)$ with identities (18) and (64), yielding

$$f(x) = 1 + \frac{2}{1!}x + \frac{3}{2!}x^2 + \frac{4}{3!}x^3 \cdots.$$

Again, we can do better than this. Using some elementary stream calculus, we can rewrite σ as follows:

$$\begin{aligned}\sigma &= \frac{1}{(1 - X)^2} \\ &= (1 + X) \otimes \frac{1}{1 - X} \qquad \text{[by identity (55)]}.\end{aligned}$$

It follows that

$$f(x) = (1 + x) \cdot e^x.$$

Here is example number three:

differential equation	initial value
$f'' + r^2 f = 1$	$f(0) = 1,\ f'(0) = 1$

Putting $\sigma = \mathcal{T}(f)$, step (1) gives

derivative	initial value
$\sigma'' + r^2\sigma = 1$	$\sigma(0) = 1,\ \sigma'(0) = 1$

For step (2) we recall the solution of this equation from Section 2.6, and perform some elementary stream calculus on it:

$$\begin{aligned}\sigma &= \frac{1+X+X^2}{1+r^2X^2} \\ &= \frac{1}{1+r^2X^2} + \frac{X}{1+r^2X^2} + \frac{X^2}{1+r^2X^2} \\ &= \frac{1}{1+r^2X^2} + \frac{1}{r}\frac{rX}{1+r^2X^2} + \frac{1}{r^2}\left(1 - \frac{1}{1+r^2X^2}\right).\end{aligned}$$

The rewriting was done to make step (3) easy: applying identities (66) and (67) yields the final outcome:

$$f(x) = \cos(rx) + \frac{1}{r}\sin(rx) + \frac{1}{r^2}\big(1 - \cos(rx)\big).$$

As a last example, consider the following equation:

differential equation	initial value
$f'' - f' = 2 + 6x$	$f(0) = 1,\ f'(0) = 0$

Step (1) gives

derivative	initial value
$\sigma'' - \sigma' = 2 + 6X$	$\sigma(0) = 1,\ \sigma'(0) = 0$

As before, we recall the solution of this equation (step (2)) from Section 2.6, and perform some elementary stream calculus on it:

$$\begin{aligned}\sigma &= \frac{1 - X + 2X^2 + 6X^3}{1-X} \\ &= -7 - 8X - 6X^2 + 8\frac{1}{1-X}.\end{aligned}$$

For the last equality, we have used the following identity:

$$X^{n+1} \times \frac{1}{1-X} = -1 - X - \cdots - X^n + \frac{1}{1-X}.$$

Applying the by now familiar identities on the Taylor transforms, step (3) yields

$$f(x) = -7 - 8x - 3x^2 + 8e^x.$$

3.5. Discussion

Pavlović and Escardó's paper on calculus in coinductive form [**34**], emphasising the close connection between classical analysis and coinduction, has been an important source of inspiration for the work presented in this chapter. Definition 3.4.1 of $\mathcal{T}$ is a variation on a definition in [**34**], where $\mathcal{T}$ is characterised as a final coalgebra homomorphism, in order to give a coinductive characterization of the Laplace transform. Here $\mathcal{T}$ is studied in its own right, and serves rather as an alternative to Laplace transform.

Solving differential equations in stream calculus essentially amounts to the classical method of undetermined coefficients [**12**, p. 82]. The main difference is that the obtained difference equations for the Taylor coefficients of the analytical solution are solved within the world of streams. This approach is technically closely related to the use of Laplace transforms (cf. [**30, 42**]), because the operation of assigning the Taylor series $\mathcal{T}(f)$ to an analytical function f implicitly uses the Laplace–Carson transform (introduced in Section 3.3). Formulae such as (63)–(69) are similar to but different from the formulae of the corresponding Laplace transforms (cf. [**42**, p. 519]). Conceptually, the use of stream calculus is different and, again, a bit simpler than these traditional approaches. In particular, analytical integration plays no role, since the difference equations are solved by stream integration (applying the Fundamental Theorem of stream calculus, Theorem 2.6.1).

CHAPTER 4

Coinductive Counting

This chapter presents a further application of the stream calculus developed in Chapter 2, to counting problems from the world of enumerative combinatorics. The first section below introduces the kind of problems we shall be dealing with, and summarizes the approach.

4.1. Introduction

The following counting problem is taken from [**22**, p. 291]. Male bees are called drones and female bees are called queens. Drones are born out of a queen and have no father; a queen is born out of a father drone and a mother queen. The first few levels of the pedigree of a drone (drawn upside-down) look as follows:

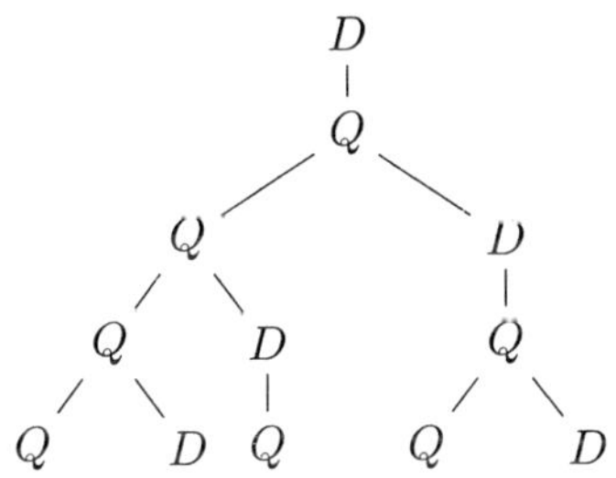

We see that each drone has one mother, one grandmother, two great-grandmothers, three great-great-grandmothers, and so on. What is, for any $k \geq 0$, the number s_k of female ancestors at level k? The key idea of coinductive counting is to use the very tree that enumerates all the (female) ancestors of a drone, as the basis for a representation of the infinite *stream* $\sigma = (s_0, s_1, s_2, \ldots)$ containing all the answers. To this end, the tree is turned into an automaton, in which the arrows indicate transitions and in which all the queen states are output states:

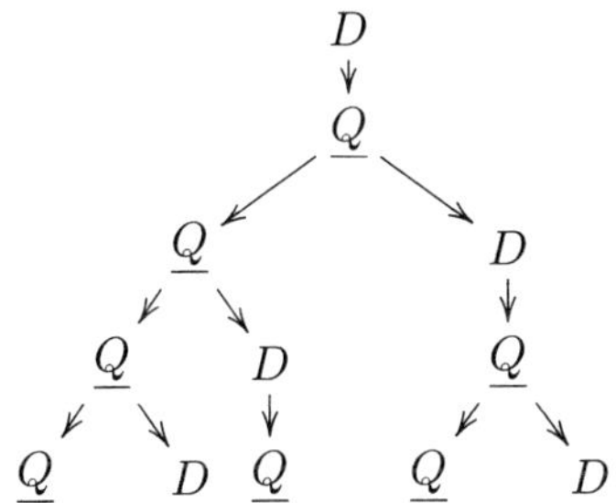

Formally, it is a weighted automaton, a notion that was introduced in Section 2.8. There we saw that the stream behaviour of such automata can be defined both by means of behavioural differential equations, and in terms of transition sequences.

In the present automaton, the number s_k of female ancestors at level k is encoded as the number of paths of length k leading from the initial (topmost) drone state to an output queen state, since there are as many such sequences as there are queens at level k. Thus we have translated the original counting problem into a question about streams and their representation by automata. Using stream bisimulations, the above automaton can be simplified by identifying all equivalent states as follows. Every drone state is bisimulation equivalent to the state q_0 and every queen state is equivalent to the state q_1 of the following two state automaton:

Intuitively, any queen state and the state q_1 have the same transition behaviour: both can take two transitions to states that are again, respectively, equivalent. Similarly for drone states and q_0, which have only one transition. As a consequence, (the state q_0 of) this new automaton is an equivalent representation of our stream of answers σ: s_k corresponds again to the number of paths of length k leading from q_0 to the (in this case only) output state q_1. Using stream calculus, we can easily compute a closed expression for the infinite stream σ of answers represented by the state q_0, yielding

$$\sigma = \frac{X}{1 - X - X^2}.$$

As with generating functions, this stream expression encodes the numbers s_k for all $k \geq 0$ (which turn out to be the Fibonacci numbers). We leave aside the computation of an explicit formula for s_k, and consider the above fraction, which is formulated in terms of stream constants and operators, as a satisfactory answer to our question.

Summarizing the above, we distinguish three phases in the procedure of coinductive counting:

(1) *Enumeration* of the objects to be counted in an infinite, tree-shaped weighted automaton.
(2) *Identification* of equivalent states using bisimulation.
(3) *Expression* of the resulting stream of counts in terms of stream constants and operators.

As we shall see shortly, the entire approach is essentially quantitative: both transitions and output states will generally be labelled with *weights* (here: real numbers), which are taken into account by the notion of stream bisimulation (and bisimulation-up-to).

4.2. Compositions of Natural Numbers

A *composition* of a natural number $k \geq 0$ is a sequence of natural numbers $n_1 \cdots n_l$ such that $k = n_1 + \cdots + n_l$. What is, for any $k \geq 0$, the number s_k of compositions of k? Or, equivalently, what is the stream of all counts $\sigma = (s_0, s_1, s_2, \ldots)$? The problem will be solved coinductively by going through the three phases mentioned in the introduction.

Enumeration. The following automaton enumerates all compositions for all natural numbers (here and in what follows, pictures show only the first few levels of what is understood to be an infinite automaton):

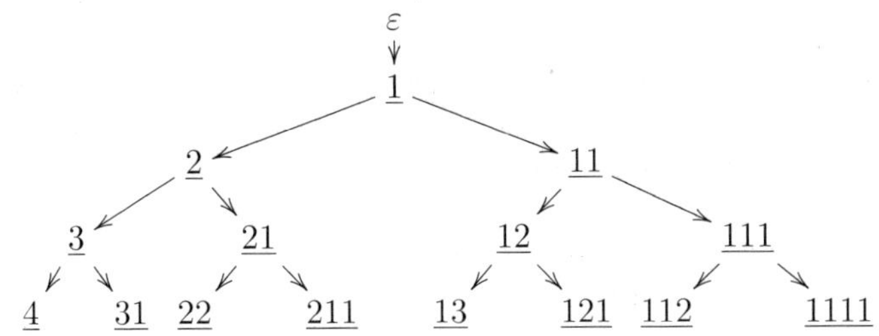

The automaton of this picture can be formally described by defining a state set $Q = \{w \mid w \in \mathbb{N}^*\}$; an output function $o \colon Q \to \mathbb{R}$ given by $o(\varepsilon) = 0$ and $o(w) = 1$ for all $w \neq \varepsilon$; and a transition function $t \colon Q \to (Q \to_f \mathbb{R})$, defined by $t(v)(w) = 1$ iff either $w = v \cdot 1$ or ($v = u \cdot k$ and $w = u \cdot (k+1)$, for some $u \in Q$ and $k \in \mathbb{N}$), and by $t(v)(w) = 0$ otherwise. (Here $u \cdot k$ denotes the concatenation of the word u and the one letter word k.) In what follows, however, we shall present automata usually by their pictures, becoming more formal only when strictly necessary.

Let $k \geq 0$ be any natural number. There is a one-to-one correspondence between paths of length k starting in the (topmost) initial state ε, on the one hand, and compositions of the natural number k (all of which are situated at the kth level of the automaton), on the other. As a consequence, the total number s_k of all compositions of k is equal to the total number of paths of length k starting in the initial state ε. We can now use identity (24), from Section 2.8, to conclude that the stream behaviour $S(\varepsilon)$ of the initial state ε equals the stream $\sigma = (s_0, s_1, s_2, \ldots)$ of answers we are after: $S(\varepsilon) = \sigma$.

Identification. Next we identify as many states as possible in our infinite weighted automaton, as follows:

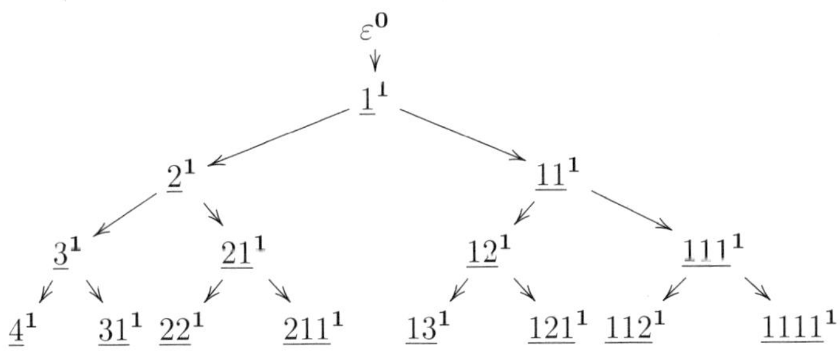

The superscripts that we have added to the states of our automaton, indicate to which state in the automaton below they are related:

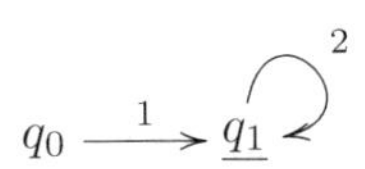

More explicitly and precisely, the above pictures suggests the definition of a relation $R \subseteq \mathbb{R}^\omega \times \mathbb{R}^\omega$ as follows:

$$R = \{\langle S(\varepsilon), S(q_0)\rangle\} \cup \{\langle S(w), S(q_1)\rangle \mid w \in \mathbb{N}^*, w \neq \varepsilon\}$$

relating the streams represented by (that is, the stream behaviour of) the states in the first automaton above, with the streams represented by the states q_0 and q_1 in the second. In order to show that R is a bisimulation-up-to, first note that the initial values of all related pairs match: $S(\varepsilon)(0) = 0 = S(q_0)(0)$ and, for all words $v \in \mathbb{N}^*$ with $v \neq \varepsilon$, $S(v)(0) = 1 = S(q_1)(0)$. Next we check derivatives: $S(\varepsilon)' = S(1)$, which is related to $S(q_1) = S(q_0)'$; and for all words $v \in \mathbb{N}^*$ and natural numbers n, we have $S(v \cdot n)' = S\big(v \cdot (n+1)\big) + S(v \cdot n \cdot 1)$, each component of which is related to $S(q_1)$, thus matching $S(q_1)' = 2 \times S(q_1) = S(q_1) + S(q_1)$. This proves that R is a bisimulation-up-to, indeed. Now it follows by coinduction-up-to that $S(\varepsilon) = S(q_0)$.

Expression. We can now easily compute a closed expression for $\sigma = S(q_0)$ by solving the system of equations, consisting of $S(q_0) = X \times S(q_1)$ and $S(q_1) = 1 + \big(2 \times X \times S(q_1)\big)$, yielding

$$\sigma = S(q_0) = \frac{X}{1 - 2X}.$$

It is worthwhile emphasizing the quantitative aspect of the notion of bisimulation (up-to): the fact that any state of the original weighted automaton labelled by a non-empty word w can take *two* transitions to similar such states, is reflected by a 2-*labelled* transition from q_1 to itself.

4.3. Surjections

What is, for any natural number $k \geq 0$, the number s_k of surjections from the set $\{1, \ldots, k\}$ onto the set $\{1, 2, 3\}$ (defining s_0 to be 0)? Below we shall see how the answer can be generalized to surjections onto the set $\{1, \ldots, n\}$, for a fixed but arbitrary $n \geq 1$.

Enumeration. Let us denote a function $f\colon \{1, \ldots, k\} \to \{1, 2, 3\}$ by means of the word $f(1) \cdots f(k)$. The following automaton enumerates at each level k all such functions:

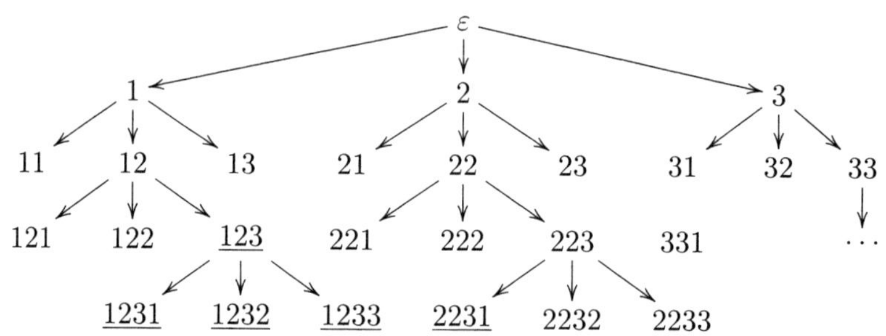

Note that all states labelled by a word representing a surjection (that is, containing at least one 1, one 2, and one 3), have been defined as output states. Also note that we have not only restricted the picture to the first five levels of the automaton, but that moreover not all transitions have been included, for lack of space. As before, it follows from (24) that the initial state ε represents the stream $\sigma = (s_0, s_1, s_2, \ldots)$ of answers we are interested in.

Identification. The automaton can be simplified by identifying all states (labelled by words) that have the same number of different symbols, as indicated by the superscripts below:

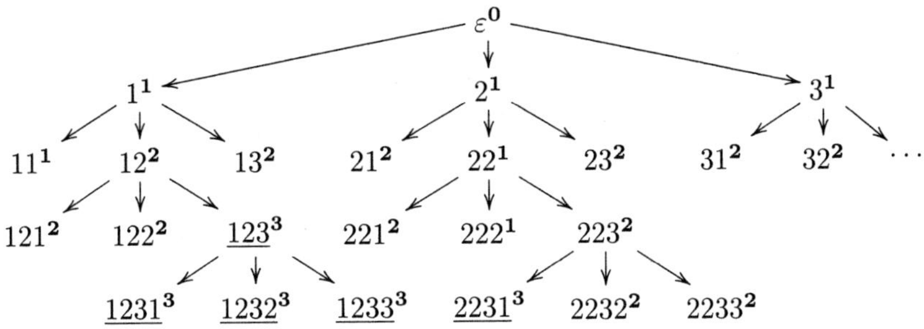

More precisely, (the streams represented by) the i-superscripted states above can be related with (the stream represented by) the state q_i in the automaton below,

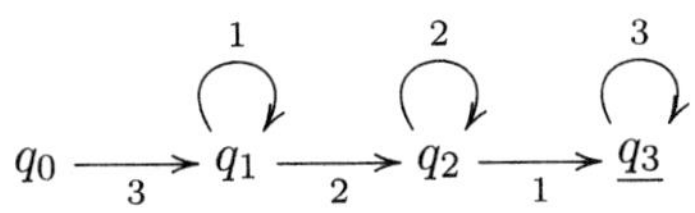

by means of a relation $R \subseteq \mathbb{R}^\omega \times \mathbb{R}^\omega$ that is defined as follows:

$$R = \{\langle S(w), S(q_i)\rangle \mid w \in \{1,2,3\}^*, i \in \{0,1,2,3\} : \sharp(w) = i\}$$

where, for this occasion, $\sharp(w)$ denotes the number of different symbols contained in w. It is straightforward to check that R is a bisimulation-up-to, from which $S(\varepsilon) = S(q_0)$ follows by coinduction-up-to.

Expression. The stream behaviour of our 4-state automaton is determined by the following system of equations:

$$\begin{aligned}
S(q_0) &= 3 \times X \times S(q_1), \\
S(q_1) &= \big(1 \times X \times S(q_1)\big) + \big(2 \times X \times S(q_2)\big), \\
S(q_2) &= \big(2 \times X \times S(q_2)\big) + \big(1 \times X \times S(q_3)\big), \\
S(q_3) &= 1 + \big(3 \times X \times S(q_3)\big).
\end{aligned}$$

Solving this system of equations, one finds:

$$\sigma = S(q_0) = \frac{3!X^3}{(1-X)(1-2X)(1-3X)}.$$

The above can be easily generalised. Let $n \geq 1$ be arbitrary but fixed, and let t_k, for any $k \geq 0$, be the number of all surjections from the set $\{1, \dots, k\}$ onto the set $\{1, \dots, n\}$. Then the stream of counts $\tau = (t_0, t_1, t_2, \dots)$ is given by

$$\begin{aligned}
\tau &= \frac{n!X^n}{(1-X)(1-2X)\cdots(1-nX)} \\
&= \left(\frac{1}{1-X} - 1\right)^n
\end{aligned}$$

where the latter equality can be proved using some basic stream calculus.

4.4. Counting with Probabilities

Consider a not necessarily fair coin with probability p of producing a head and probability $q = 1 - p$ of producing a tail. What is, for any $k \geq 0$, the probability s_k of getting, by flipping the coin k times, a sequence of heads and tails (of length k) without the occurrence of two consecutive heads but for the two very last outcomes, which are required to be heads? We hope the reader is by now already sufficiently experienced with coinductive counting, to be able to supply the details that have been left out in the succinct presentation of the solution below.

Enumeration. Here is a weighted automaton describing all possible scenarios:

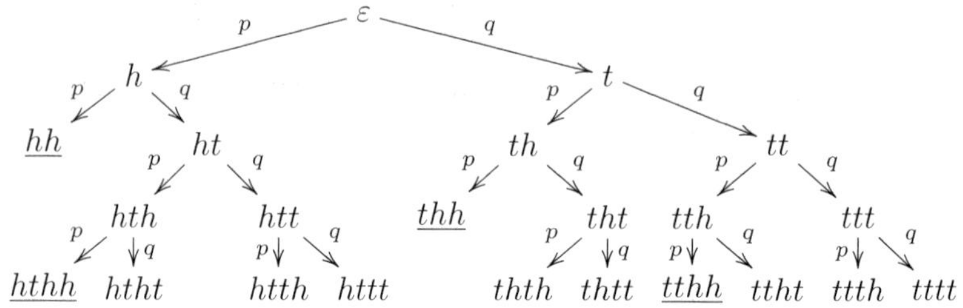

All states that are (labelled with a sequence) ending in two heads are output states, and have no further transitions. The stream $\sigma = (s_0, s_1, s_2, \ldots)$ of all counts is represented by the initial state ε, that is, $\sigma = S(\varepsilon)$.

Identification. States can be identified according to the number of final heads:

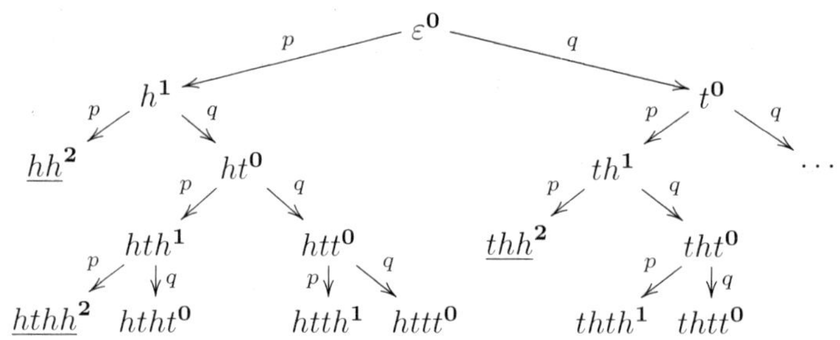

yielding the following automaton:

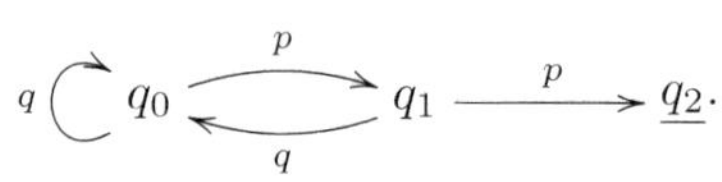

More precisely, one can prove by coinduction-up-to that $S(\varepsilon) = S(q_0)$.

Expression. The stream behaviour of $S(q_0)$ can be easily computed:

$$\sigma = S(q_0) = \frac{p^2 X^2}{1 - qX - pqX^2}.$$

4.5. Well-Bracketed Words

So far the identification phase of our coinductive counting exercises has always resulted in a finite automaton yielding a rational stream. Here is an example for which the stream of counts is no longer rational (but *algebraic*). Consider a two letter alphabet $\{(,)\}$ consisting of a left and a right bracket. What is, for any $k \geq 0$, the number s_k of well-bracketed words over this alphabet, of length k?

Enumeration. The output states at level k of the following automaton correspond precisely to all such words:

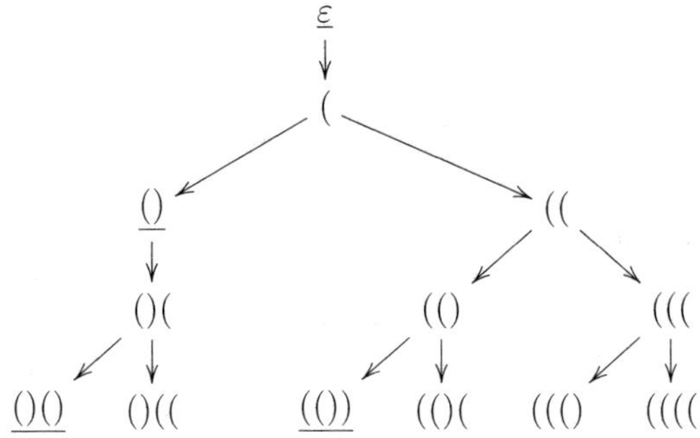

so the stream $\sigma = (s_0, s_1, s_2, \dots)$ of all counts is represented by the initial state: $\sigma = S(\varepsilon)$.

Identification. We identify states according to the number of left brackets they contain without a matching right bracket:

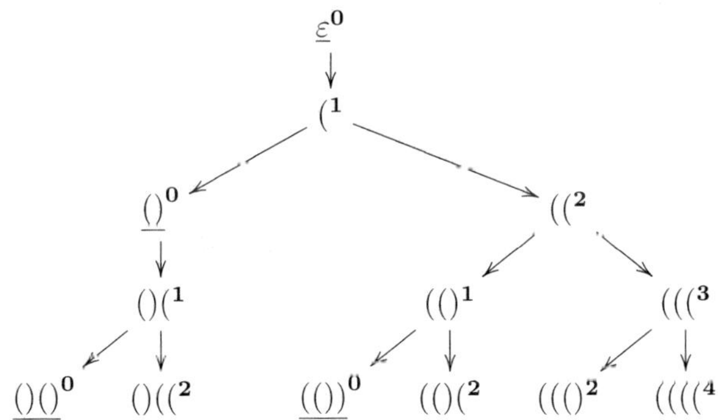

This yields the following well-structured, but still infinite automaton:

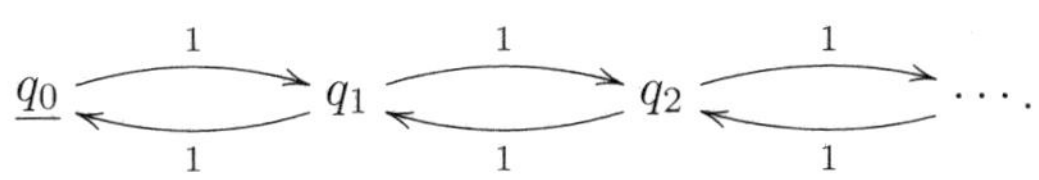

As before, one can prove $S(\varepsilon) = S(q_0)$ by coinduction-up-to.

Expression. The stream behaviour of the above automaton is given by the following infinite system of equations: $S(q_0) = 1 + \big(X \times S(q_1)\big)$ and

$$\boxed{S(q_i) = \big(X \times S(q_{i-1})\big) + \big(X \times S(q_{i+1})\big) \;\Big|\; i \geq 1}$$

but it is not immediately obvious how this infinite system of equations can be solved. Rather than trying to solve it, we shall in Section 4.6 develop some more general results directly on infinite automata of the type above. The discussion of the solution of the present problem is therefore postponed until Section 4.7.

4.6. Streams and Continued Fractions

In this section, we study the stream behaviour of infinite weighted automata of the following type:

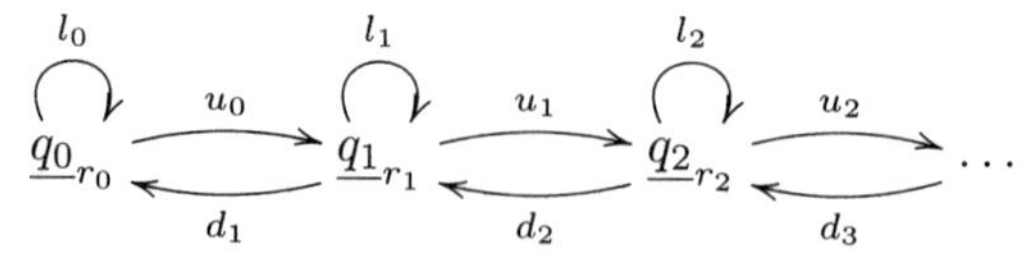

All of the labels and output values are arbitrary real numbers. The labels l_i and u_i, for $i \geq 0$, and d_i, for $i \geq 1$, could be pronounced as *level*, *up*, and *down*. The stream behaviour of this automaton will be expressed by means of a (generalised) continued fraction, but before we deal with it in its full generality, we first treat two special cases that are particularly relevant for many of the counting problems to come.

The first case concerns the situation that all output values r_i are 0 but for the output value r_0 of the first state q_0, which (may be arbitrary but for convenience) has been chosen to be 1:

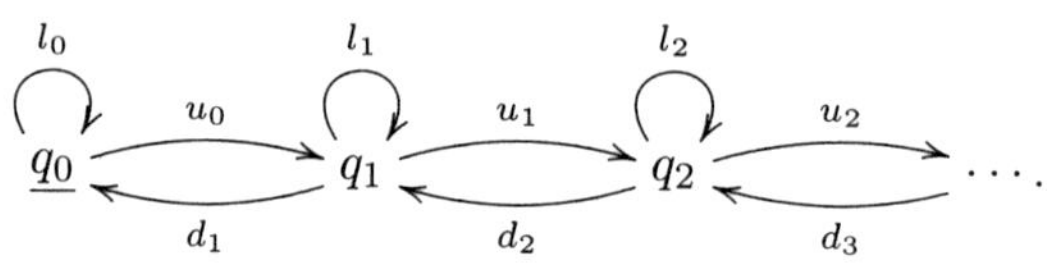

THEOREM 4.6.1. *The stream behaviour $S(q_0)$ of the state q_0 in the automaton above is given by the following continued fraction*:

$$\cfrac{1}{1 - (l_0 \times X) - \cfrac{(u_0 \times X) \times (d_1 \times X)}{1 - (l_1 \times X) - \cfrac{(u_1 \times X) \times (d_2 \times X)}{1 - (l_2 \times X) - \cfrac{(u_2 \times X) \times (d_3 \times X)}{\ddots}}}}$$

PROOF. Let us first make sense of the continued fraction, which is denoted by what seems to be an infinite expression. Formally, it can be defined as follows. Let the streams $\sigma_0, \sigma_1, \ldots$ be the unique solutions of the following system of behavioural differential equations:

differential equation	initial value
$\sigma_i' = (l_i \times \sigma_i) + (u_i \times \sigma_{i+1} \times d_i \times X \times \sigma_i)$	$\sigma_i(0) = 1$

By Theorem 2.6.1, it follows that

$$\sigma_i = \frac{1}{1 - (l_i \times X) - (u_i \times X) \times \sigma_{i+1} \times (d_{i+1} \times X)}$$

for all $i \geq 0$. The infinite continued fraction in the theorem can now simply be read as a suggestive notation for σ_0. Next we show that $S(q_0) = \sigma_0$. Consider the following relation on streams:

$$R = \{\langle r \times S(q_i), r \times \tau_i \rangle \mid r \in \mathbb{R}, i \geq 0\}$$

where the streams τ_i are defined by

$$\tau_i = \sigma_i \times (d_i \times X) \times \sigma_{i-1} \times (d_{i-1} \times X) \times \cdots \times \sigma_1 \times (d_1 \times X) \times \sigma_0.$$

Because R is a bisimulation relation on $\mathbb{R}^\omega$, $S(q_0) = \sigma_0$ follows by coinduction. □

For the second special case of the automaton presented at the beginning of this section, we assume that $d_i = 0$ for all $i \geq 1$, that is, there are no downward transitions. In contrast to the automaton of Theorem 4.6.1, however, all states may have again a non-trivial output value $r_i \in \mathbb{R}$:

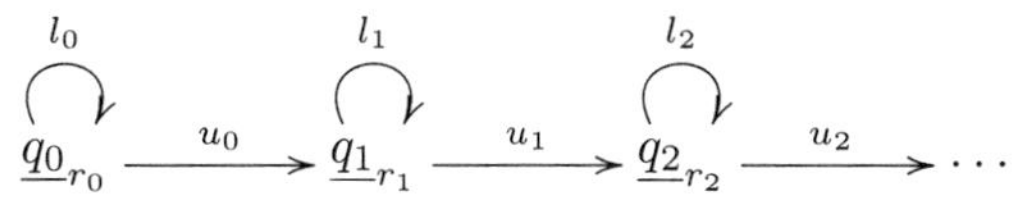

THEOREM 4.6.2. *The stream behaviour $S(q_0)$ of the state q_0 is given by the following "upward" continued fraction:*

$$\cfrac{r_0 + (u_0 \times X) \times \cfrac{r_1 + (u_1 \times X) \times \cfrac{r_2 + (u_2 \times X) \times \cfrac{\vdots}{1 - (l_3 \times X)}}{1 - (l_2 \times X)}}{1 - (l_1 \times X)}}{1 - (l_0 \times X)}$$

The proof is omitted. It can be given along the same lines as the proof of Theorem 4.6.1.

Finally, we return to the automaton at the beginning of this section, which is the most general of all:

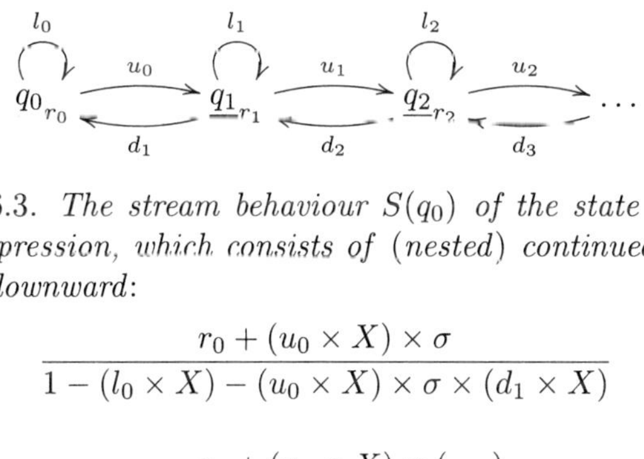

THEOREM 4.6.3. *The stream behaviour $S(q_0)$ of the state q_0 is given by the following crazy expression, which consists of (nested) continued fractions growing both upward and downward:*

$$\frac{r_0 + (u_0 \times X) \times \sigma}{1 - (l_0 \times X) - (u_0 \times X) \times \sigma \times (d_1 \times X)}$$

where

$$\sigma = \frac{r_1 + (u_1 \times X) \times (\cdots)}{1 - (l_1 \times X) - (u_1 \times X) \times (\cdots) \times (d_2 \times X)}.$$

PROOF. Also this theorem can be proved in the same manner as the previous two theorems. □

4.7. More on Well-Bracketed Words

We return to the problem of Section 4.5, and compute a closed expression for the stream $\sigma = (s_0, s_1, s_2, \ldots)$ of numbers s_k of well-bracketed words of length k (over the alphabet $\{(,)\}$).

At the end of Section 4.5, the following automaton representation for the stream σ of counts was found:

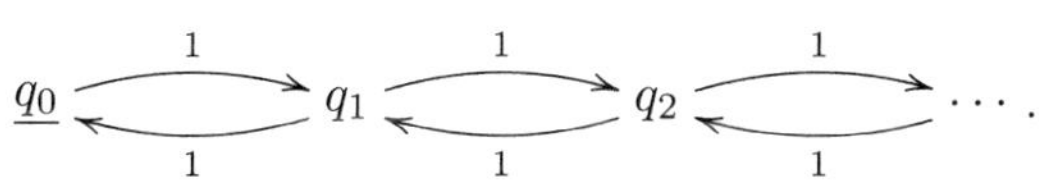

An application of Theorem 4.6.1 yields the following continued fraction for $\sigma = S(q_0)$:

$$\sigma = S(q_0) = \cfrac{1}{1 - \cfrac{X^2}{1 - \cfrac{X^2}{\ddots}}}$$

Because of the regular structure of this fraction, it is possible to come up with a more succinct expression for σ. It is easy to see that

$$\sigma = \frac{1}{1 - (X \times \sigma \times X)}$$

(cf. the proof of Theorem 4.6.1). This implies that σ is a solution of the following quadratic equation:

$$(X^2 \times \sigma^2) - \sigma + 1 = 0.$$

In order to solve it, we need a square root operator on streams. As usual in stream calculus, its definition can be given by means of a behavioural differential equation. Let $\sqrt{\sigma}$, for all streams σ with $\sigma(0) > 0$, be defined as the unique stream satisfying the following differential equation:

behavioural differential equation	initial value
$(\sqrt{\sigma})' = \frac{\sigma'}{\sqrt{\sigma(0)}+\sqrt{\sigma}}$	$\sqrt{\sigma}(0) = \sqrt{\sigma(0)}$

.

(Here $\sqrt{\sigma(0)}$ is the square root of the real number $\sigma(0)$.) With this operator, quadratic equations like the one above can be solved in much the same way we are used to in analysis (see [**39**] for details). This yields as the final outcome for our stream of counts of well-bracketed words:

$$\sigma = \frac{2}{1 + \sqrt{1 - 4X^2}}$$

(which equals the stream $(1, 0, 1, 0, 2, 0, 5, 0, 14, 0, \ldots)$ having the Catalan numbers at the even positions).

4.8. Permutations

A permutation (bijection) $p\colon \{1, \ldots, k\} \to \{1, \ldots, k\}$ can be represented by the corresponding sequence of images $p = (p(1) \cdots p(k))$. Another very common and equivalent representation, which is the one that we shall be using, describes a permutation by the (unique) sequence of cycles of which the permutation is composed. For instance, the permutation $p = (532461)$, with $p(1) = 5$, $p(2) = 3$, $p(3) = 2$, $p(4) = 4$, $p(5) = 6$, and $p(6) = 1$, can also be represented by the following sequence: $p = (156)(23)(4)$. Here a cycle $c = (x_1 \cdots x_k)$ denotes a permutation of $\{x_1, \ldots, x_k\}$ defined by: $c(x_i) = x_{i+1}$, for all $1 \leq i \leq k-1$, and $c(x_k) = x_1$. We start with a trivial question, for any $k \geq 0$: what is the number s_k of permutations of the set $\{1, \ldots, k\}$ (defining $s_0 = 1$)? The following automaton enumerates all permutations by listing all possible sequences of cycles:

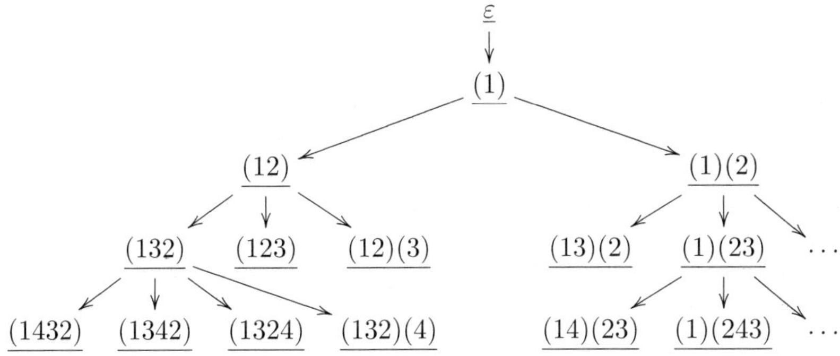

Any state at level k represents a permutation of the set $\{1, \dots k\}$ and is therefore an output state. It can make a transition to a state at the next level, either by adding the number $k+1$ to one of the existing cycles or by adding the new cycle $(k+1)$. There are (for all states at level k) precisely k transitions of the first, and one transition of the second type, $k+1$ transitions in total. This explains the structure of the automaton above, and at the same time indicates that all states of every single level can be identified, yielding the following automaton:

$$\underline{q_0} \xrightarrow{\;1\;} \underline{q_1} \xrightarrow{\;2\;} \underline{q_2} \xrightarrow{\;3\;} \cdots .$$

Applying Theorem 4.6.2, with $r_k = 1$, $l_k = 0$ and $u_k = k+1$ for all $k \geq 0$, we obtain

$$S(q_0) = 1 + 1!X + 2!X^2 + 3!X^3 + \cdots$$

for the stream $(s_0, s_1, s_2, \dots)$ of counts we are after. In other words, there are $k!$ different permutations of the set $\{1, \dots, k\}$, which comes as no surprise.

We have chosen to represent permutations by sequences of cycles in the automaton above, because it can be easily adapted to deal with various related counting problems. A first and straightforward variation is to keep track of the total number of cycles in each permutation. This we can do by fixing any real number (variable) u, which we use as a label for all transitions that represent the addition of a new cycle (recall that all other transitions have label 1, which is as usual omitted):

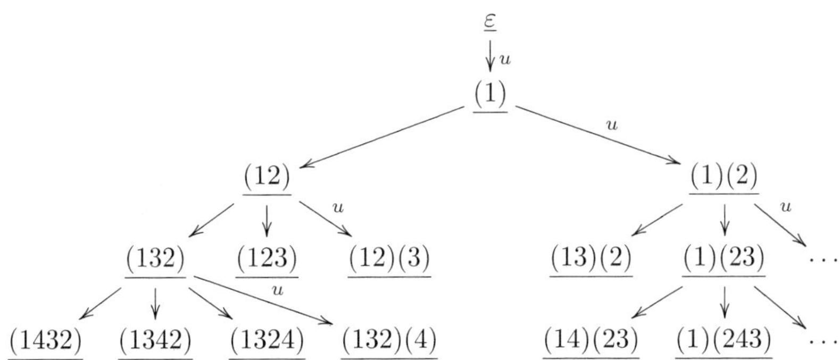

Identifying again all states of the same level gives the following equivalent automaton:

$$\underline{q_0} \xrightarrow{\;u\;} \underline{q_1} \xrightarrow{\;u+1\;} \underline{q_2} \xrightarrow{\;u+2\;} \cdots$$

to which, as before, Theorem 4.6.2 an be applied, now yielding

$$S(q_0) = 1 + u \times X + u(u+1) \times X^2 + u(u+1)(u+2) \times X^3 + \cdots$$
$$= (1 - X)^{-\underline{u}}$$

where the latter equality can be proved using some elementary stream calculus. (Note that this formula generalises the solution of our previous counting, which is obtained by taking $u = 1$.) The above stream can be considered as having u as a parameter. It encodes all numbers $s_{k,n}$, for $k, n \geq 0$, counting all permutations of $\{1, \ldots, k\}$ consisting of n cycles. These numbers are known as the Stirling numbers of the first kind, and can be computed from the stream above as the (Taylor) coefficients of the powers $u^n \times X^k$. (One can also treat both X and u as formal variables; this would bring us to multivariate stream calculus, which is omitted here.)

Next we present yet another way of counting permutations while keeping track of the number of cycles, yielding another representation of the Stirling numbers of the first kind. More specifically, permutations can, alternatively, be enumerated as follows:

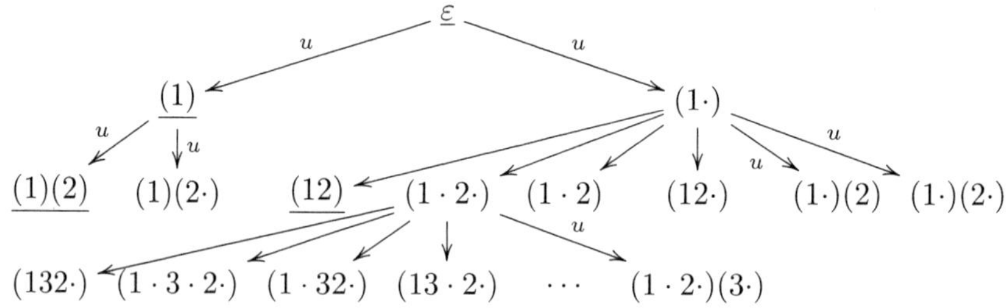

The general pattern in which this tree-like weighted automaton is grown, is as follows. In any state (node) at level $n - 1$, all of the numbers $\{1, \ldots, n-1\}$ occur, possibly together with a number of dots. All states without dots are output states, representing completed permutations. States with dots represent permutations that have not yet been completed, and the dots indicate the places where further growth is possible. More specifically, for each dot, there are four successor states:

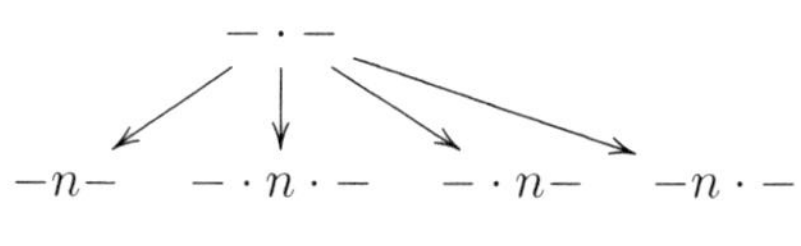

in which the dot has been replaced with n, $\cdot n \cdot$, $\cdot n$, and $n \cdot$, respectively. In addition, any state at level $n - 1$ has two more successors, corresponding to the opening of a new cycle:

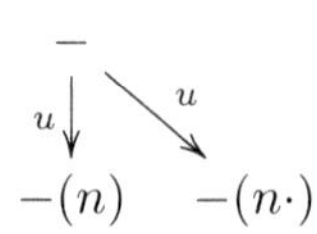

As before, the transitions in which a new cycle is opened, are labelled by u (instead of 1).

States can be identified according to the number of dots they contain, as is indicated by the superscripts below:

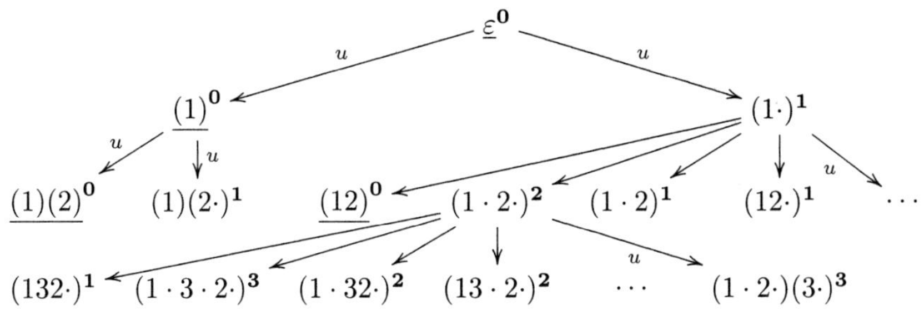

Identification of all equivalent states yields the following well-structured automaton:

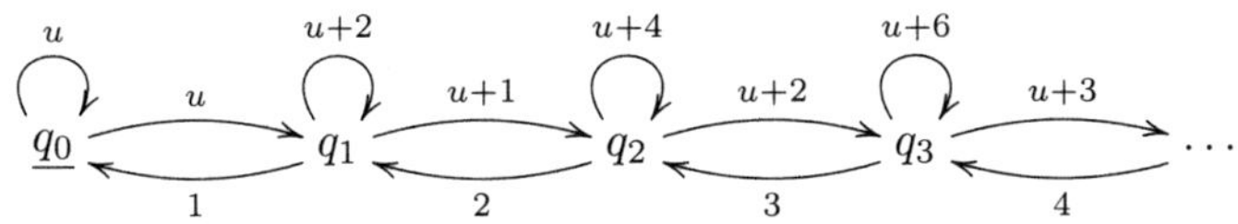

to which Theorem 4.6.1 can be applied, resulting in the following continued fraction:

$$S(q_0) = \cfrac{1}{1 - uX - \cfrac{1uX^2}{1-(u+2)X - \cfrac{2(u+1)X^2}{1-(u+4)X - \cfrac{3(u+2)X^2}{\ddots}}}}$$

$$= (1 - X)^{-\underline{u}}.$$

The second equality is obtained by combining our first and our second way of counting permutations and cycles. Thus we have obtained two different expressions for the Stirling numbers of the first kind. Forgetting about the number of cycles again, that is, taking $u = 1$, we obtain

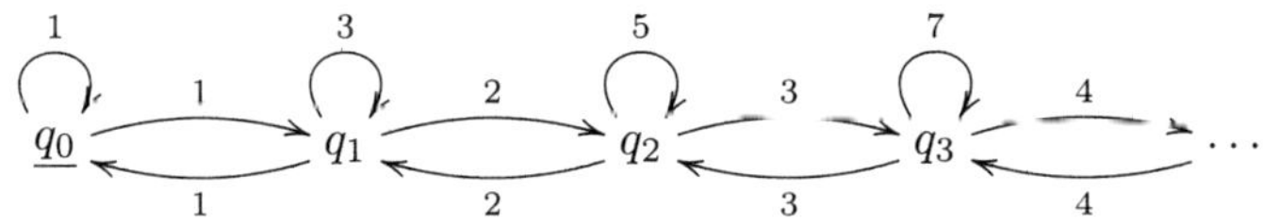

which leads to the following equalities:

$$S(q_0) = \cfrac{1}{1 - X - \cfrac{1^2X^2}{1-3X - \cfrac{2^2X^2}{1-5X - \cfrac{3^2X^2}{\ddots}}}}$$

$$= (1 - X)^{-\underline{1}}$$

$$= (0!, 1!, 2!, 3!, \dots).$$

What is, moving to a next question, for any $k \geq 0$ the number s_k of permutations p of $\{1, \dots, k\}$ such that $p \circ p = 1$ (the so-called involutions)? For this we return to the first automaton of this section, from which we now remove all states but the ones consisting of 1- and 2-cycles only:

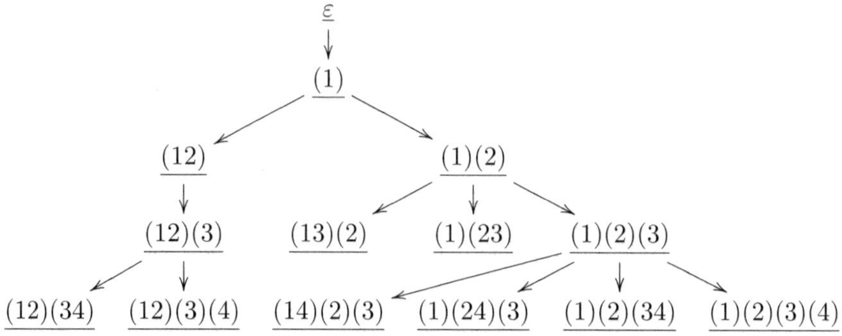

States can be identified according to the number of 1-cycles, which can still become 2-cycles, they contain, as indicated by the superscripts below:

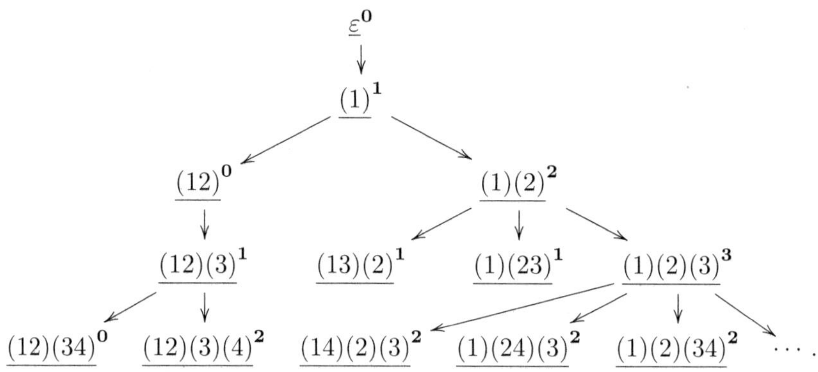

yielding the following automaton

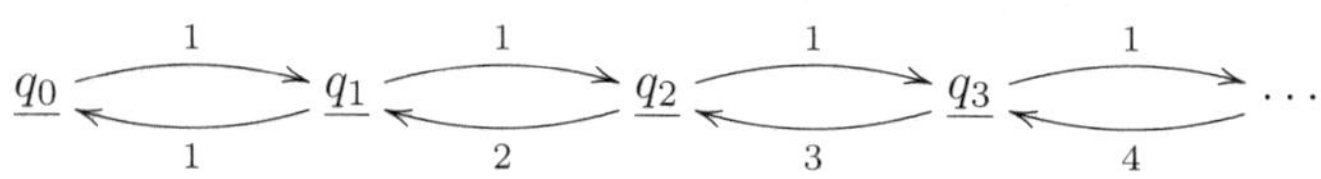

Applying Theorem 4.6.3 yields an expression for $S(q_0)$, but not a very pretty one (which is therefore omitted).

For the special case of permutations consisting of 2-cycles only, the same enumeration as for the involutions can be used, the only difference being that states that contain 1-cycles are no longer output states:

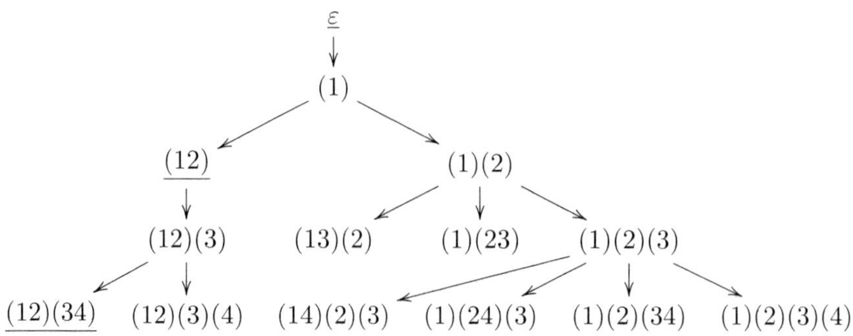

The corresponding reduced automaton now looks like

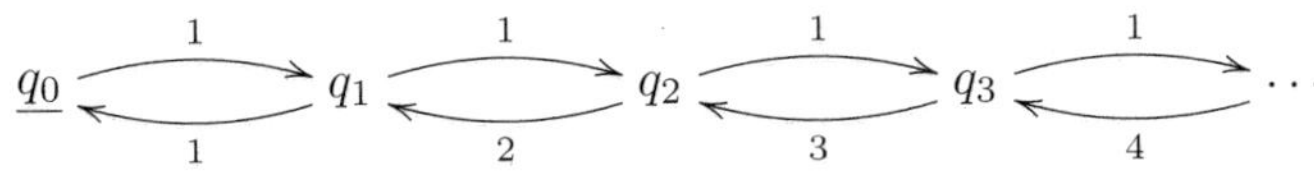

with, according to Theorem 4.6.1, a pleasant expression for the stream of answers:

$$S(q_0) = \cfrac{1}{1 - \cfrac{1 \cdot X^2}{1 - \cfrac{2 \cdot X^2}{1 - \cfrac{3 \cdot X^2}{\ddots}}}}$$

Yet another question: what is, for any $k \geq 0$, the number s_k of permutations $p\colon \{1, \ldots, k\} \to \{1, \ldots, k\}$ such that $|p(i) - i| \leq 1$, for all $1 \leq i \leq k$? These are precisely all permutations consisting of 1-cycles, and of 2-cycles (xy) with $y = x+1$. The following automaton, which happens to be a further pruning of the last tree-like automaton above, lists them all:

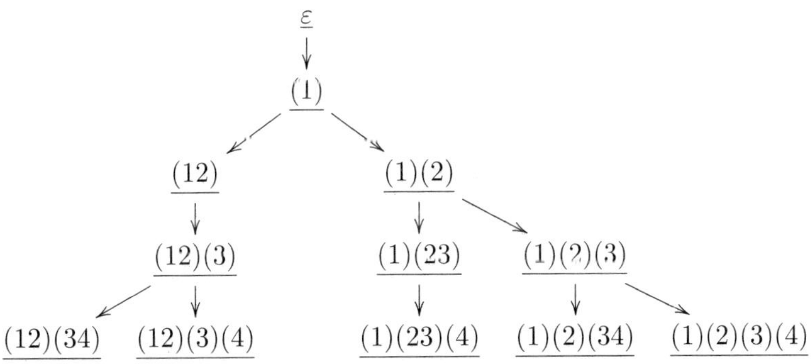

(Note that now all states are output states again.) All states ending with a 1-cycle and all states ending with a 2-cycle can be identified, respectively, leading to the following automaton and corresponding expression for the stream of answers:

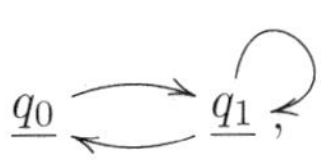

$$; \quad S(q_0) = \frac{1}{1 - X - X^2}.$$

For the last example of this section, we return to the representation of permutations mentioned at the beginning of this section. That is, a permutation (bijection) $p\colon \{1, \ldots, k\} \to \{1, \ldots, k\}$ will be represented below by the corresponding sequence of images $p = \big(p(1) \cdots p(k)\big)$. Referring to this representation, a permutation is called *alternating* whenever either $p(0) > p(1) < p(2) > \cdots p(n)$ or $p(0) < p(1) > p(2) < \cdots p(n)$. What is, we ask ourselves (because we already know that the question will have a pretty answer), for any *odd* natural number $k \geq 0$, the total number s_k of alternating permutations of the set $\{1, \ldots, k\}$ (putting $s_k = 0$

for all even k)? Here is an enumeration of all such odd alternating permutations:

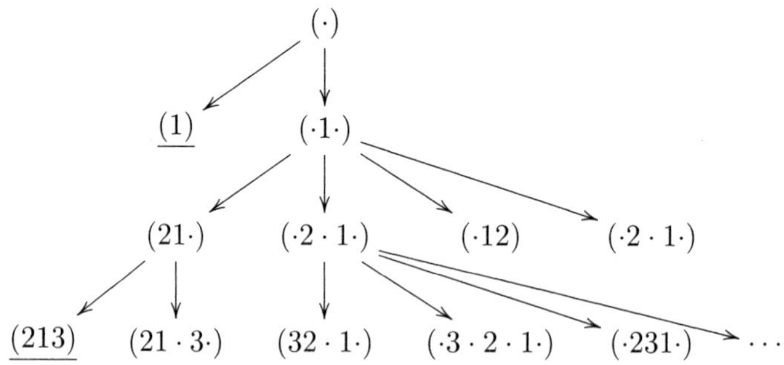

The general pattern according to which this tree is grown is as follows. As before, dots indicate places where further growth is possible. And any dot at level n has two children:

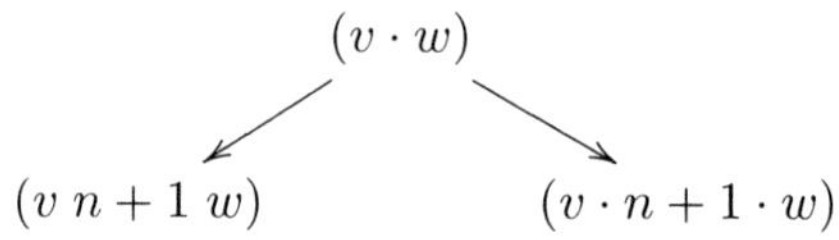

in which the dot has been replaced with $n+1$ or with $n+1$ with dots on both sides. Also as before, states can be identified according to the number of dots they contain:

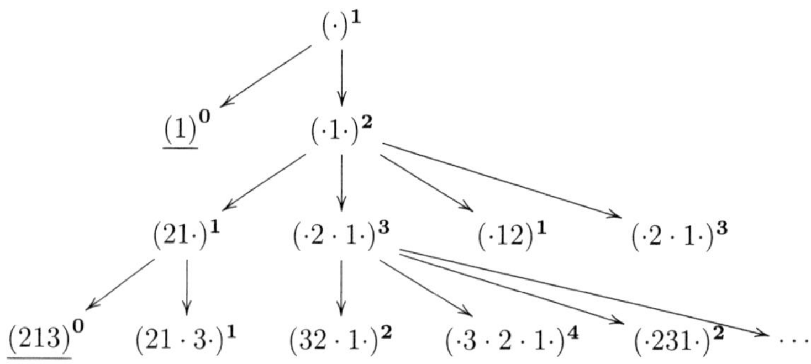

After identification of all equivalent states, the following automaton is obtained:

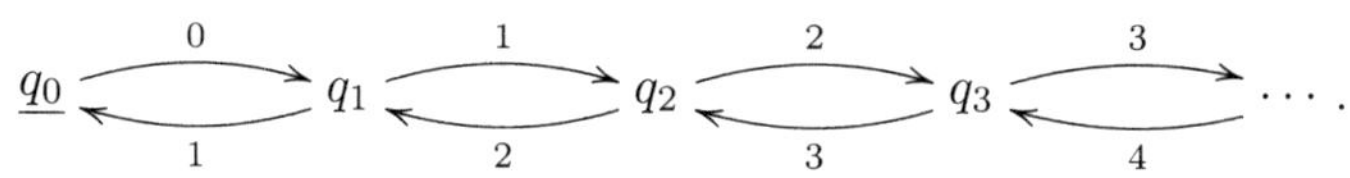

Note that in this automaton, the stream of answers is represented by q_1 and not by q_0. Using (a minor variation of) Theorem 4.6.1, we get

$$S(q_1) = \cfrac{X}{1 - \cfrac{1 \cdot 2 \cdot X^2}{1 - \cfrac{2 \cdot 3 \cdot X^2}{1 - \cfrac{3 \cdot 4 \cdot X^2}{\ddots}}}}$$

One can show that $S(q_1)$ is equal to $\text{Taylor}\big(\tan(x)\big)$, the stream of Taylor coefficients of the function $\tan(x)$, because the automaton above can also be obtained by applying the technique of "splitting derivatives" to Taylor$(\tan(x))$ (see [**39**] for details).

4.9. Set Partitions

Our last series of counting problems deals with set partitions, counted in various ways.

What is, for any $k \geq 0$, the number s_k of ways in which the set $\{1, \ldots, k\}$ can be partitioned into subsets (the so-called Bell numbers)? The following automaton enumerates at level k all such partitions:

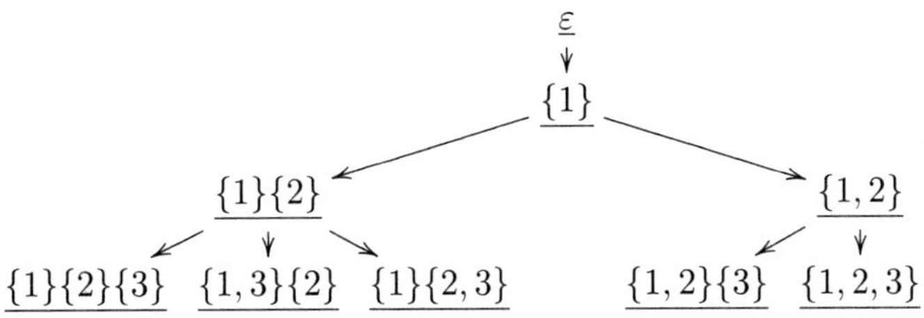

States can be identified according to the number of subsets used in the corresponding partition, as indicated by the superscripts below:

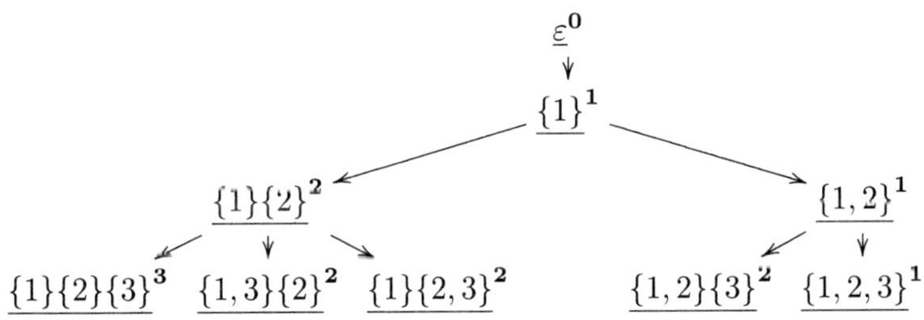

Identifying as usual all equivalent states gives the following reduced automaton:

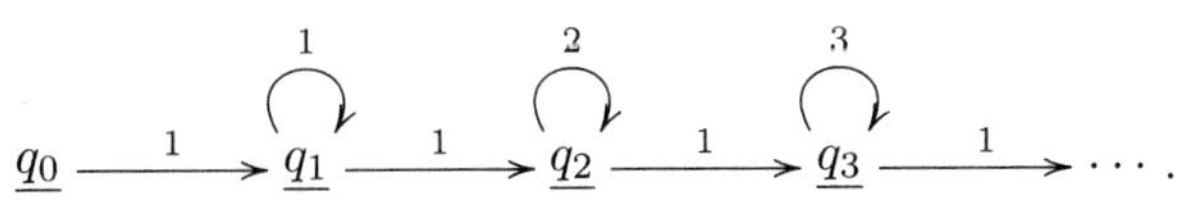

And by formula Theorem 4.6.2, the following expression is obtained for our stream $S(q_0)$ of answers:

$$1 + \frac{X}{1 - X} + \frac{X^2}{(1 - X)(1 - 2X)} + \frac{X^3}{(1 - X)(1 - 2X)(1 - 3X)} + \cdots .$$

It is easy to keep track explicitly of the number of subsets used in any partition, by labelling all transitions in which a new subset is added to the partition by a fixed real number (variable) u:

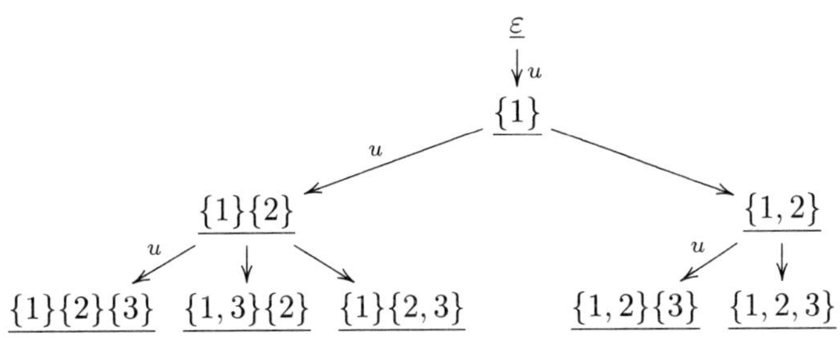

Identification can be done as before, yielding the following reduced automaton

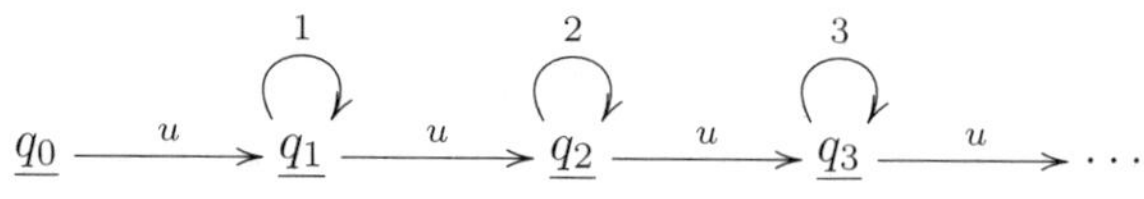

and corresponding expression for the stream $S(q_0)$ of answers:

$$1 + \frac{uX}{1-X} + \frac{u^2X^2}{(1-X)(1-2X)} + \frac{u^3X^3}{(1-X)(1-2X)(1-3X)} + \cdots .$$

This stream encodes, for all $k, n \geq 0$, all numbers $s_{k,n}$ of partitioning the set $\{1, \ldots, k\}$ into n subsets. These numbers are called the Stirling numbers of the second kind. Note that by taking $u = 1$, the previous case is found back.

We have seen various examples where a different way of enumeration, using a different representation of the structures to be counted, leads to a different but equivalent expression for the solution of the counting problem. This phenomenon is here further illustrated by the following, alternative solution for the problem of counting set partitions. For any $k \geq 0$, the following automaton contains at level k all partitions of the set $\{1, \ldots, k\}$ as output states; in addition, each level contains various intermediate states containing dots that indicate possibilities for further growth:

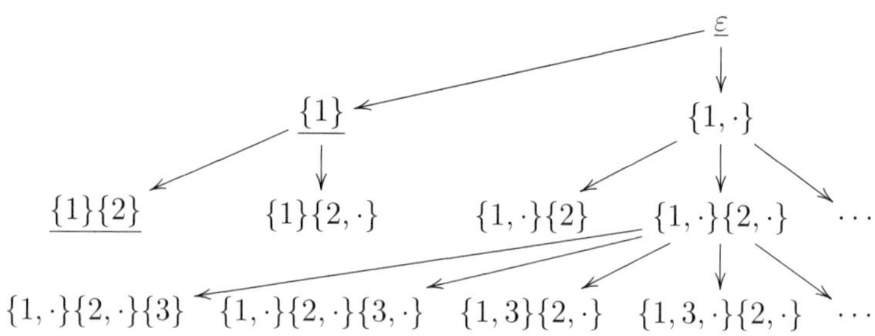

The general pattern is as follows: For any dot occurring in a node at level $n-1$, there are two children, obtained by replacing the dot with either n or with n and a dot (allowing still further growth):

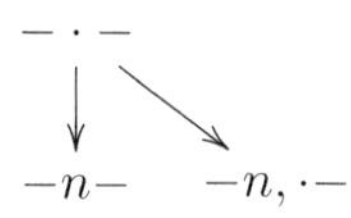

In addition, any node at level $n-1$ has two more children, obtained by opening a new subset containing n, again with or without the presence of a dot:

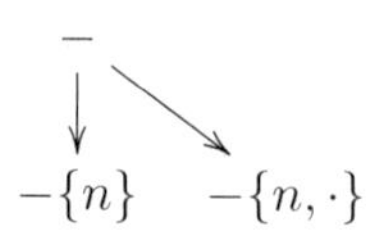

As in all the previous examples of representations involving dots, states can be identified according to the number of dots they contain, yielding the following

reduced automaton

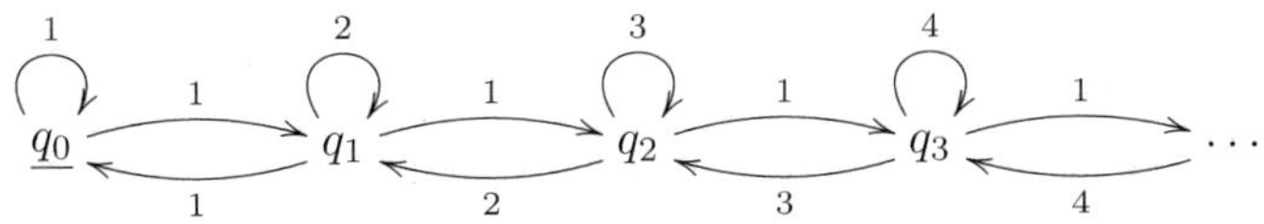

and, by an application of Theorem 4.6.1, the following corresponding expression for the stream $S(q_0)$ of answers:

$$\cfrac{1}{1 - X - \cfrac{X^2}{1 - 2X - \cfrac{2X^2}{1 - 3X - \cfrac{3X^2}{\ddots}}}}$$

$$= 1 + \frac{X}{1-X} + \frac{X^2}{(1-X)(1-2X)} + \frac{X^3}{(1-X)(1-2X)(1-3X)} + \cdots$$

where the latter equality is obtained by identifying the first and the second solution of this problem. Again, it is possible to keep track of the number of subsets used in each partition. Omitting the enumerating automaton, we only present the resulting reduced automaton:

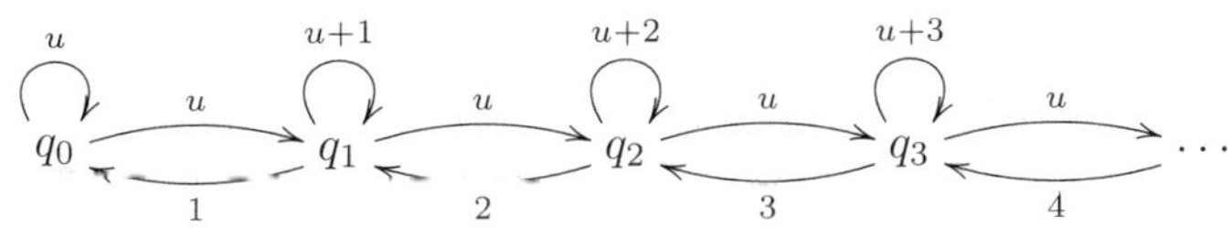

and the corresponding expression for the stream $S(q_0)$ of answers:

$$\cfrac{1}{1 - (u \times X) - \cfrac{uX^2}{1 - ((u+1) \times X)) - \cfrac{2uX^2}{1 - ((u+2) \times X)) - \cfrac{3uX^2}{\ddots}}}}$$

4.10. Special Numbers

The theorem below summarizes the representations by means of a reduced weighted automaton, of the various sequences of so-called special numbers that we have encountered in the preceding sections. It clearly shows that these representations have a great structural similarity.

THEOREM 4.10.1. *In the Table 4.1, the stream of numbers mentioned on the right is represented by the state q_0 of the automaton on the left, except for the stream of the tangent numbers, which is represented by the state q_1.*

4.11. Discussion

For each and every single counting problem that we have dealt with in this chapter, there exist in the literature far more detailed and complete treatments. See for instance the standard works [**43, 44**]. The contribution of the method

TABLE 1. Representations of special numbers.

automaton	represents
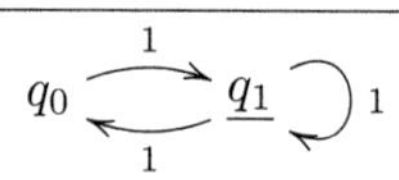	Fibonacci numbers
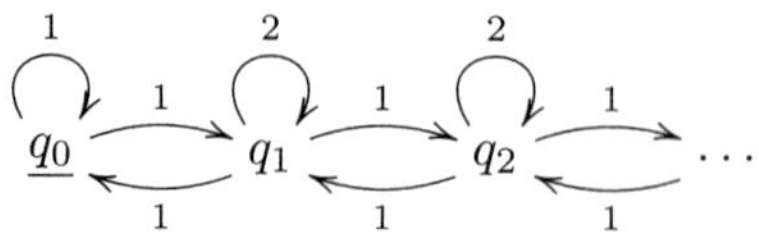	Catalan numbers
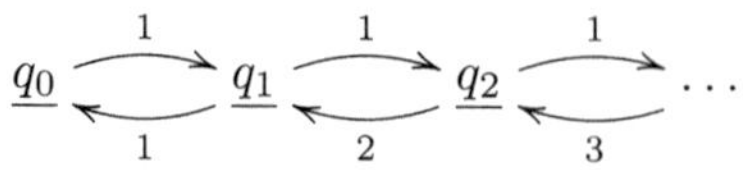	involutions
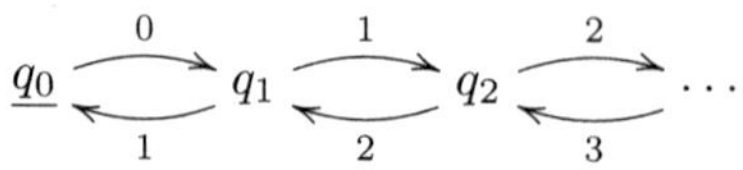	tangent numbers
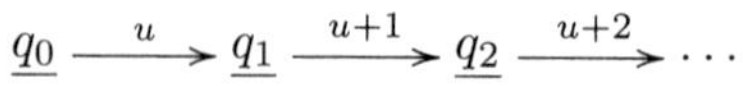	Stirling numbers (1st)
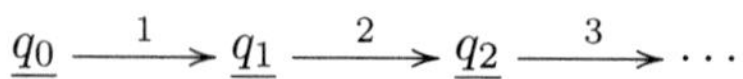	factorial numbers ($u = 1$)
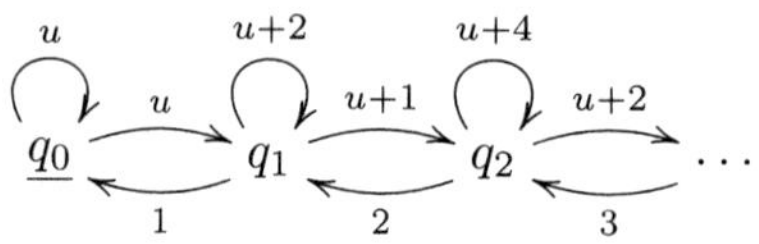	Stirling numbers (1st)
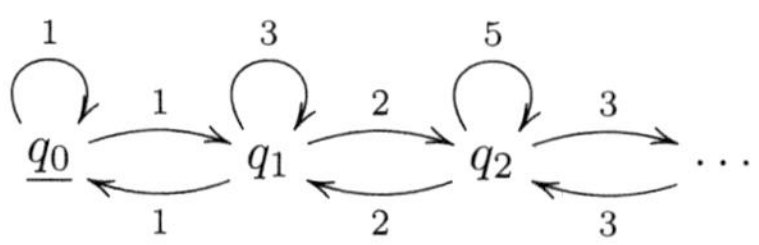	factorial numbers ($u = 1$)
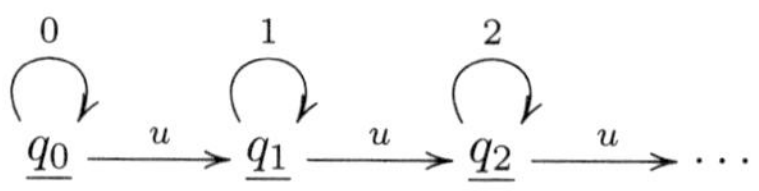	Stirling numbers (2nd) Bell numbers ($u = 1$)
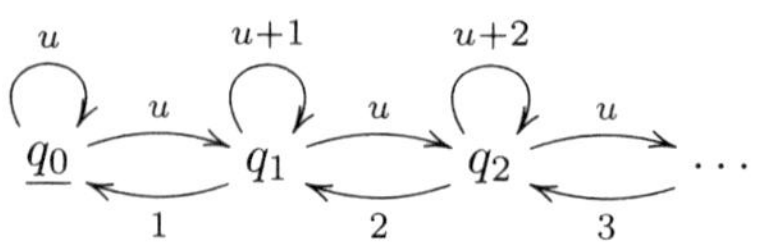	Stirling numbers (2nd)
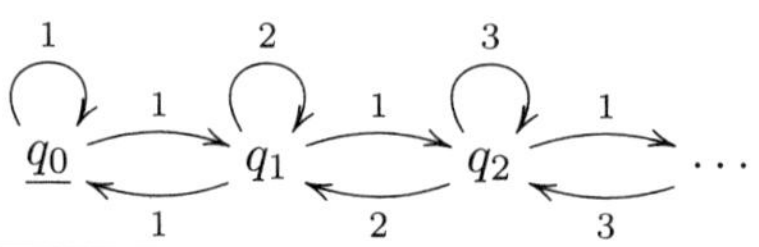	Bell numbers ($u = 1$)

of coinductive counting lies in the fact that it solves all these different counting problems, in the very same way:

(1) The structures to be counted are always enumerated by infinite tree-shaped weighted automata.
(2) These automata are reduced by means of bisimulations.
(3) From the resulting reduced automata, an expression for the stream of answers is derived by means of elementary stream calculus. Traditionally, the situation with respect to these points is as follows:

(a) Many different mathematical structures are used as representations of combinatorial objects, including weighted graphs, sets of walks, various kinds of trees, formal languages (regular and context-free), automata, (transfer-) matrices, and so on.
(b) We know of no *systematic* way, in the sense of being based on one specific notion, of reducing such representations to smaller or better structured ones.
(c) Stream calculus can be viewed as a formalisation and generalisation of the use of both generating functions and formal power series (cf. [**39**] for a detailed comparison).

Technically, the novelty of the coinductive counting method is based on

(i) the systematic use of *infinite* weighted automata, which have received in the literature on automata theory so far no attention;
(ii) the use of the *quantitative* notion of *stream bisimulation*; and
(iii) the consequent use of a *coinductive* calculus of streams in the search for a closed expression for the outcome.

Inspiring references and closely related approaches include [**20–22, 43, 44**]. In fact, the presently proposed method of coinductive counting is far more restricted in scope than these references. Notably, no (classical) analytical methods are being used here. Such methods might be used, though, at the point where our methods stops: many of the obtained stream expressions are suited for further analytical treatment. The use of continued fractions has been inspired by [**19**], and most of the continued fractions that we have computed occur already there. The formal treatment of such continued fractions seems somewhat easier in the present setting of coinductive stream calculus. Also, some of the combinatorial interpretations of the various fractions discussed here, seem to be simpler and more uniform. Another approach to counting is the categorical theory of species [**10**]. It offers a framework that is far more general than the present calculus of streams, but there are many connections. It would be worthwhile, more generally speaking, to investigate the possible role of coinduction in the world of species in some detail.

CHAPTER 5

Component Connectors

Reo (from the Greek word $\rho\varepsilon\omega$ which means "[I] *flow*") is a recently introduced [**3, 4**] channel-based coordination model, wherein complex coordinators, called connectors, are compositionally built out of simpler ones. Reo is intended as a "glue language" for construction of connectors that orchestrate component instances in a component-based system. The emphasis in Reo is on connectors and their composition only, not on the components that are being connected. Using streams and coinduction, we present in this chapter a simple and transparent semantical model of connectors and connector composition, which can be used as a compositional *calculus*, in which properties such as connector equivalence, optimization, and realization can be expressed and proved. The material in this chapter builds only on Chapter 1.

5.1. Introduction

In Reo, the basic connectors are channels, each of which is a point-to-point communication medium with two distinct ends. Channels can be used as the only communication constructs in communication models for concurrent systems, because the primitives of other communication models (e.g., message passing or remote procedure calls) can be easily defined using channels. In contrast to other channel-based models, Reo uses a generalised concept of channel. In addition to the common channel types of synchronous and asynchronous, with bounded or unbounded buffers, and with fifo and other ordering schemes, Reo allows an open ended set of channels, each with its own, sometimes exotic, behaviour. For instance, a channel in Reo need not have both an input and an output end; it can have two input ends or two output ends instead. In addition to channels, Reo has one more basic connector, called the merge operator. More complex connectors can then be constructed from the basic connectors (channels and merge) through an operation of connector composition.

Because Reo is not concerned with the internal activity of the components that it connects, we represent components by their interfaces only. Therefore, we model the input ends and output ends of connectors as streams of abstract (uninterpreted) data items. Moreover, we associate with every such data stream an infinite sequence of (natural or non-negative real) numbers. These numbers stand for the respective moments in time at which their corresponding data items are being input or output. This allows us to describe and reason about the precise timing constraints of connectors (such as synchronous versus asynchronous, and bounded versus unbounded delay). Thus, we model the potential behaviour of connector ends as *timed data streams*, which are pairs consisting of a data stream and a time stream. Note that we use pairs of streams rather than streams of pairs (of timed data elements), since this enables us to reason about time explicitly, which turns out to be particularly useful.

Having modelled connector ends as timed data streams, we then model connectors as *relations* on timed data streams, expressing which combinations of timed data streams are mutually consistent. This relational model is, in spite of its simplicity, already sufficiently expressive to study a number of notions and questions about component connectors, such as equivalence (when do two connectors have the same behavior?), expressiveness (which connectors can I build out of a given set of basic connectors?), optimization (given a connector, can I build an equivalent connector out of a smaller number of basic connectors?), verification (given the specification of a certain connector behavior and given a connector, does the connector meet the specification?), realization (given the specification of a certain connector behavior, can I actually build a connector with precisely that behavior out of a given set of basic connectors?), and the like. In this chapter, we shall mainly focus on connector equivalence and, to a lesser extent, the expressiveness of our connector calculus.

5.2. Timed Data Streams

We model connectors as relations on timed data streams, which we introduce in this section.

Let D be an (arbitrary) set, the elements of which will be called *data* elements. The set DS of *data streams* is defined as

$$\mathsf{DS} = \mathsf{D}^\omega$$

that is, the set of all streams $\alpha = \big(\alpha(0), \alpha(1), \alpha(2), \ldots\big)$ over D. Let $\mathbb{R}_+$ be the set of non-negative real numbers, which play in the present context the role of *time* moments. Let $\mathbb{R}_+^\omega$ be the set of all streams $a = \big(a(0), a(1), a(2), \ldots\big)$ over $\mathbb{R}_+$. Let $<$ and $\leq$ be the relations on $\mathbb{R}_+^\omega$ that are obtained as the pointwise extensions of the corresponding ("strictly smaller" and "smaller than") relations on $\mathbb{R}_+$. That is, for all $a = \big(a(0), a(1), a(2), \ldots\big)$ and $b = \big(b(0), b(1), b(2), \ldots\big)$ in $\mathbb{R}_+^\omega$,

$$a < b \equiv \forall n \geq 0, \quad a(n) < b(n), \qquad a \leq b \equiv \forall n \geq 0, \quad a(n) \leq b(n).$$

The set TS of *time streams* is defined by the following subset of $\mathbb{R}_+^\omega$:

$$\mathsf{TS} = \{a \in \mathbb{R}_+^\omega \mid a < a'\}.$$

Note that time streams $a \in \mathsf{TS}$ satisfy, for all $n \geq 0$,

$$a(n) < a'(n) = a(n+1)$$

and thus consist of increasing time moments $a(0) < a(1) < a(2) < \cdots$.

Finally, the set TDS of *timed data streams* is defined by

$$\mathsf{TDS} = \mathsf{DS} \times \mathsf{TS}$$

and contains pairs $\langle \alpha, a \rangle$ consisting of a data stream

$$\alpha = \big(\alpha(0), \alpha(1), \alpha(2), \ldots\big) \in \mathsf{DS}$$

and a time stream

$$a = \big(a(0), a(1), a(2), \ldots\big) \in \mathsf{TS}.$$

As we shall see shortly, connectors will be modelled as relations on timed data streams. Each of the arguments of such a relation will be viewed as an input or as an output end of the connector that is modelled by the relation. Thus the following operational interpretation of a timed data stream $\langle \alpha, a\rangle$ can be given: the time stream a specifies for each $n \geq 0$ the time moment $a(n)$ at which the nth data element $\alpha(n)$ is being input or output:

α:	$\alpha(0)$	$\alpha(1)$	$\alpha(2)$	$\cdots$	$\alpha(n)$	$\cdots$
a:	$a(0)$	$a(1)$	$a(2)$	$\cdots$	$a(n)$	$\cdots$

Connectors being relations, there are typically many timings a possible at a specific connector's end, which together with a given data stream α form admissible timed data streams $\langle \alpha, a\rangle$, that is, satisfying the connector's relation. A timed data stream $\langle \alpha, a\rangle$ could therefore also be viewed as a *scenario*, one out of many, for the behaviour of a connector end. Connectors, then, relate various such scenarios that together are mutually consistent.

As we already observed in the introduction, one could have, alternatively and equivalently, defined timed data streams as (a subset of) $(\mathsf{D} \times \mathbb{R}_+)^\omega$, because of the existence of an isomorphism

$$\mathsf{D}^\omega \times \mathbb{R}_+^\omega \cong (\mathsf{D} \times \mathbb{R}_+)^\omega,$$
$$\langle \alpha, a\rangle \mapsto \big(\langle \alpha(0), a(0)\rangle, \langle \alpha(1), a(1)\rangle, \langle \alpha(2), a(2)\rangle, \ldots\big).$$

We prefer to work with pairs of streams rather than streams of pairs, because this will allow us to reason about the data streams and time streams separately, which turns out to be of crucial importance for much of what follows.

We could also have used streams of *natural* numbers $0, 1, 2, \ldots$ for our timings, rather than (positive) real numbers. This difference would leave most of our model unaffected. Our model with "continuous time", however, is more abstract than the model with "discrete time" would be, in the sense that more connector equivalences can be proved. (In the world of temporal logic, this observation goes back to at least [**8**].) An example is the equivalence of a fifo_2 buffer (with capacity 2) with the composition of two fifo_1 buffers in Section 5.5.

Finally, it is often useful to require time streams a to be not only increasing: $a < a'$, but also *progressive*: for every $N \geq 0$ there exists $n \geq 0$ with $a(n) > N$. This assumption prevents "Zeno" paradoxes, where infinitely many actions take place in a bounded time interval. In most of what follows, the progressive time assumption is not used, but whenever it is, we shall mention it explicitly.

5.3. Basic Connectors: Channels

The most basic connectors are *channels*, which are formally defined as binary relations

$$R \subseteq \mathsf{TDS} \times \mathsf{TDS}$$

on timed data streams. For such relations, we distinguish between input and output argument positions, called *input ends* and *output ends*, respectively. This information will be relevant for the definition of connector composition in Section 5.5. In the pictures that we draw of channels and connectors, input and output ends are denoted by the following arrow shaft and head:

$$\text{input: } \vdash\!\!\!\longrightarrow \cdots \qquad \text{output: } \cdots \longrightarrow .$$

Here are the (for our purposes) most important examples of channels:

• The *synchronous channel* $\longmapsto$ is defined, for all timed data streams $\langle\alpha, a\rangle$ and $\langle\beta, b\rangle$, by

$$\langle\alpha, a\rangle \longmapsto \langle\beta, b\rangle \;\equiv \alpha = \beta \wedge a = b.$$

This channel inputs the data (elements in the) stream α at times a, and outputs the data stream β at times b. All data elements that come in, come out again (in the same order): $\alpha = \beta$. Moreover, each element enters and exits the channel at the very same time moment: $a = b$.

• The *synchronous drain* $\overset{\text{syndr}}{\longmapsto\!\dashv}$ is defined, for all timed data streams $\langle\alpha, a\rangle$ and $\langle\beta, b\rangle$, by

$$\langle\alpha, a\rangle \overset{\text{syndr}}{\longmapsto\!\dashv} \langle\beta, b\rangle \;\equiv a = b.$$

The corresponding data elements in the streams α and β enter the two input ends of this channel simultaneously: $a = b$. No relation on the data streams is specified (the data elements enter and "disappear").

• The *fifo buffer* $\overset{\text{fifo}}{\longmapsto}$ is defined, for all timed data streams $\langle\alpha, a\rangle$ and $\langle\beta, b\rangle$, by

$$\langle\alpha, a\rangle \overset{\text{fifo}}{\longmapsto} \langle\beta, b\rangle \;\equiv \alpha = \beta \wedge a < b.$$

This is an unbounded fifo (first-in-first-out) buffer. What comes in, comes out (in the same order): $\alpha = \beta$, but later: $a < b$ (which is equivalent to $a(n) < b(n)$, for all $n \geq 0$).

• The *fifo*$_1$ *buffer* $\overset{\text{fifo}_1}{\longmapsto}$ is defined, for all timed data streams $\langle\alpha, a\rangle$ and $\langle\beta, b\rangle$, by

$$\langle\alpha, a\rangle \overset{\text{fifo}_1}{\longmapsto} \langle\beta, b\rangle \;\equiv \alpha = \beta \wedge a < b < a'.$$

This models a 1-bounded fifo buffer. What comes in, comes out: $\alpha = \beta$, but later: $a < b$. Moreover, at any moment the next data item can be input only after the present data item has been output: $b < a'$, which is equivalent to $b(n) < a(n+1)$, for all $n \geq 0$.

• The *fifo*$_k$ *buffer* $\overset{\text{fifo}_k}{\longmapsto}$ for any $k \geq 1$, is defined, for all timed data streams $\langle\alpha, a\rangle$ and $\langle\beta, b\rangle$, by

$$\langle\alpha, a\rangle \overset{\text{fifo}_k}{\longmapsto} \langle\beta, b\rangle \;\equiv \alpha = \beta \wedge a < b < a^{(k)}.$$

(Recall from Section 1 that $a^{(k)}$ denotes the kth derivative of the stream a.) This models a k-bounded fifo buffer, generalizing the fifo$_1$ buffer above. What comes in, comes out: $\alpha = \beta$, but later: $a < b$. Moreover, at any moment the kth-next data item can be input only after the present data item has been output: $b < a^{(k)}$ (which is equivalent to $b(n) < a(n+k)$, for all $n \geq 0$).

• Let $x \in D$ be any fixed data element. The fifo(x) buffer $\overset{\text{fifo}(x)}{\longmapsto}$ is defined, for all timed data streams $\langle\alpha, a\rangle$ and $\langle\beta, b\rangle$, by

$$\langle\alpha, a\rangle \overset{\text{fifo}(x)}{\longmapsto} \langle\beta, b\rangle \;\equiv \beta(0) = x \wedge \alpha = \beta' \wedge a < b'.$$

This channel behaves precisely as the unbounded fifo buffer above, but for the fact that, initially, it contains the data element x, which is the first element to come out: $\beta(0) = x$ (at time $b(0)$).

5.4. More Channels and the Merge Connector

The definitions of the basic (channel) connectors so far have been, mathematically speaking, fairly straightforward. Next, we introduce some further basic connectors, including the merge operator, using greatest fixed point definitions. As we saw in Section 1.5, each of these definitions will come together with its own proof principle, similar to the coinduction principle in Section 1.

• The *asynchronous drain* $\overset{\text{asyndr}}{\longmapsto\!\dashv}$ inputs any two streams of data items at its two input ends, but never at the same time (in contrast to the synchronous drain of Section 5.3). It is defined, for all timed data streams $\langle \alpha, a\rangle$ and $\langle \beta, b\rangle$, by

$$\langle \alpha, a\rangle \overset{\text{asyndr}}{\longmapsto\!\dashv} \langle \beta, b\rangle \;\equiv a \bowtie b$$

where $\bowtie \subseteq \mathsf{TS} \times \mathsf{TS}$ is a relation on time streams, given by

$$a \bowtie b \equiv \big(a(0) < b(0) \wedge a' \bowtie b\big) \vee \big(b(0) < a(0) \wedge a \bowtie b'\big).$$

More precisely, $\bowtie$ is defined as the greatest fixed point of the following monotone operator, $\Phi_{\bowtie}\colon \mathcal{P}(\mathsf{TS} \times \mathsf{TS}) \to \mathcal{P}(\mathsf{TS} \times \mathsf{TS})$, defined for $R \subseteq \mathsf{TS} \times \mathsf{TS}$, by

$$\Phi_{\bowtie}(R) = \big\{\langle a, b\rangle \mid \big(a(0) < b(0) \wedge \langle a', b\rangle \in R\big) \vee \big(b(0) < a(0) \wedge \langle a, b'\rangle \in R\big).$$

Thus $\bowtie = \mathrm{gfp}(\Phi_{\bowtie})$. A $\bowtie$-*bisimulation* is a relation $R \subseteq \mathsf{TS} \times \mathsf{TS}$ with $R \subseteq \Phi_{\bowtie}(R)$. There is, as an immediate consequence of (5) in Section 1.5, the following $\bowtie$-*coinduction* proof principle. For all time streams a and b:

(70) if $a\, R\, b$, for some $\bowtie$-bisimulation R, then $\alpha \bowtie \beta$.

For an example of a proof by $\bowtie$-coinduction, see the end of the present section.

• The connector *merge* is a ternary relation M with two input ends and one output end, and is defined, for timed data streams $\langle \alpha, a\rangle$, $\langle \beta, b\rangle$, and $\langle \gamma, c\rangle$, by

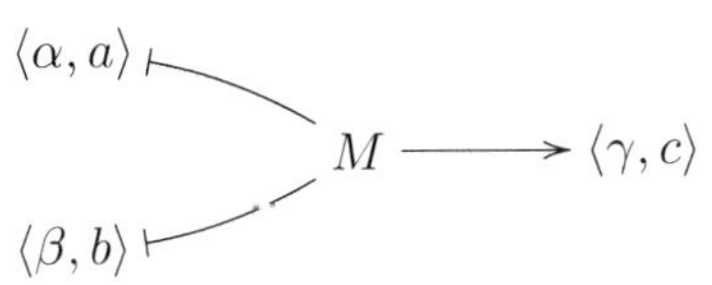

$$\equiv M(\langle \alpha, a\rangle, \langle \beta, b\rangle, \langle \gamma, c\rangle)$$

$$\equiv a(0) \neq b(0) \wedge \begin{cases} \alpha(0) = \gamma(0) \wedge a(0) = c(0) \wedge M(\langle \alpha', a'\rangle, \langle \beta, b\rangle, \langle \gamma', c'\rangle) \\ \hfill \text{if } a(0) < b(0), \\ \beta(0) = \gamma(0) \wedge b(0) = c(0) \wedge M(\langle \alpha, a\rangle, \langle \beta', b'\rangle, \langle \gamma', c'\rangle) \\ \hfill \text{if } b(0) < a(0). \end{cases}$$

This connector merges the two data streams α and β on its input ends into a stream γ on its output end, on a "first come first served" basis. It inputs one data element at a time: $a(0) \neq b(0)$. The data element that is handled first, say $\alpha(0)$ at time $a(0) < b(0)$, is the first element to come out: $\alpha(0) = \gamma(0)$, at exactly the same moment: $a(0) = c(0)$. After that, the connector handles the remainder of the streams in the same manner again: $M(\langle \alpha', a'\rangle, \langle \beta, b\rangle, \langle \gamma', c'\rangle)$. (Similarly for the case that $\beta(0)$ is handled first, at time $b(0) < a(0)$.) The relation M can be formally defined as the greatest fixed point of a monotone operator Φ_M defined, for any $R \subseteq \mathsf{TDS} \times \mathsf{TDS} \times \mathsf{TDS}$, by

$$\Phi_M(R)(\langle\alpha, a\rangle, \langle\beta, b\rangle, \langle\gamma, c\rangle)$$

$$\equiv a(0) \neq b(0) \wedge \begin{cases} \alpha(0) = \gamma(0) \wedge a(0) = c(0) \wedge R(\langle\alpha', a'\rangle, \langle\beta, b\rangle, \langle\gamma', c'\rangle) \\ \qquad\qquad\qquad\qquad \text{if } a(0) < b(0), \\ \beta(0) = \gamma(0) \wedge b(0) = c(0) \wedge R(\langle\alpha, a\rangle, \langle\beta', b'\rangle, \langle\gamma', c'\rangle) \\ \qquad\qquad\qquad\qquad \text{if } b(0) < a(0). \end{cases}$$

An M-*bisimulation* is a relation R with $R \subseteq \Phi_M(R)$ and we have an M-*coinduction* proof principle: if $R(\langle\alpha, a\rangle, \langle\beta, b\rangle, \langle\gamma, c\rangle)$, for some M-bisimulation R, then $M(\langle\alpha, a\rangle, \langle\beta, b\rangle, \langle\gamma, c\rangle)$.

Under the assumption that our time streams are progressive (defined at the end of Section 5.2), the merge operator is *fair*: from both input ends, infinitely many data elements will be input.

Next, we illustrate the use of the coinduction proof principles that were introduced above. A look at the definition of M and one moment's thought suffice to see that, for all timed data streams $\langle\alpha, a\rangle, \langle\beta, b\rangle, \langle\gamma, c\rangle$,

$$\text{if } M(\langle\alpha, a\rangle, \langle\beta, b\rangle, \langle\gamma, c\rangle) \text{ then } a \bowtie b \tag{71}$$

since the merge connector never inputs two data items at its two input ends at the same time. But how to prove it formally? The answer is provided by what we have called $\bowtie$-coinduction above. Consider the following relation:

$$R = \{\langle k, l\rangle \mid \exists \kappa, , \lambda, \mu, m : M(\langle\kappa, k\rangle, \langle\lambda, l\rangle, \langle\mu, m\rangle)\}.$$

Using the definition of M, it is straightforward to prove that this is a $\bowtie$-bisimulation. As a consequence of (70), $R \subseteq \bowtie$, which implies (71). For a second example, consider the following equivalence, for all timed data streams $\langle\alpha, a\rangle, \langle\beta, b\rangle, \langle\gamma, c\rangle$:

$$M(\langle\alpha, a\rangle, \langle\beta, b\rangle, \langle\gamma, c\rangle) \equiv M(\langle\beta, b\rangle, \langle\alpha, a\rangle, \langle\gamma, c\rangle). \tag{72}$$

The implication from left to right (and thus the equivalence) follows by M-coinduction from the trivial observation that

$$S = \{(\langle\alpha, a\rangle, \langle\beta, b\rangle, \langle\gamma, c\rangle) \mid M(\langle\beta, b\rangle, \langle\alpha, a\rangle, \langle\gamma, c\rangle)\}$$

is an M-bisimulation, which implies $S \subseteq M$.

5.5. Composing Connectors

Connectors are relations and their composition can therefore be naturally modelled by relational composition. For instance, the composition of two copies of the synchronous channel yields the following binary relation, defined for all timed data streams $\langle\alpha, a\rangle$ and $\langle\beta, b\rangle$, by

$$\begin{aligned} \langle\alpha, a\rangle \longmapsto \circ \longmapsto \langle\beta, b\rangle &\equiv \exists\langle\gamma, c\rangle\colon\ \langle\alpha, a\rangle \longmapsto \langle\gamma, c\rangle \ \wedge\ \langle\gamma, c\rangle \longmapsto \langle\beta, b\rangle \\ &\equiv \exists\langle\gamma, c\rangle\colon (\alpha = \gamma \wedge a = c) \wedge (\gamma = \beta \wedge c = b) \end{aligned}$$

(which happens to be equivalent to $\langle\alpha, a\rangle \longmapsto \langle\beta, b\rangle \equiv \alpha = \beta \wedge a = b$). Composition essentially does two things at the same time: the output end (argument) of the first connector is identified with the input end of the second, and the resulting "mixed" end is moreover hidden (encapsulated) by the existential quantification. We shall also use the following picture to denote connector composition:

$$\langle \alpha, a\rangle \longmapsto \exists\langle \gamma, c\rangle \longmapsto \langle \beta, b\rangle\ .$$

Here $\exists\langle \gamma, c\rangle$ is used to indicate that this is an internal end of the connector, which is no longer accessible (for further compositions) from outside. The two aspects of connector composition: identification and hiding, could, and for the full version of Reo actually should, be separated. But for the basic examples we shall be dealing with, this type of composition is sufficient.

Note that the identification of connector ends, which are timed data streams, includes the identification of the respective time streams, thus synchronising the timings of the two connectors.

The general definition of the composition of an arbitrary n-ary connector R and an m-ary connector T, is essentially the same. One has to select a number of (distinct) output ends and input ends from R, and equal numbers of input ends and output ends from T, which then are connected in pairs in precisely the same manner as in the example above. To describe this in full generality, one would have to be slightly more formal and explicit about the (input and output) types of argument positions. Although not very difficult, such a formalisation would not be very interesting. Moreover, it will not be necessary for the instances of connector composition that will be presented here. In all cases, the relevant typing information will be contained in the pictorial representations with which connector compositions will be introduced.

We shall also allow an output end to be connected to *several* input ends at the same time, to each of which the output is copied. Here is an example, in which the output end of a synchronous channel is connected to the input ends of two other synchronous channels:

$$\begin{array}{ccccc} \langle \alpha, a\rangle & \longmapsto & \exists\langle \delta, d\rangle & \longmapsto & \langle \beta, b\rangle \\ & & \Big\downarrow & & \\ & & \langle \gamma, c\rangle & & \end{array}$$

$$\begin{aligned} &\equiv \exists\langle \delta, d\rangle : (\alpha = \delta \wedge a = d) \wedge (\delta = \beta \wedge d = b) \wedge (\delta = \gamma \wedge d = c) \\ &\equiv \alpha = \beta = \gamma \wedge a = b = c. \end{aligned}$$

The connection of several *output ends* to one and the same input end, can be modelled by means of the merge operator introduced in Section 5.4.

Finally, it is relevant to note that nothing in the above prevents us from connecting an output end to an input end of one and the same connector, simply by connecting them to the input end and the output end of a synchronous channel. In other words, we have in passing included the possibility of *feedback* loops into our calculus.

Here are various examples of composite connectors (including examples of feedback) built out of a number of basic connectors. As usual, $\langle \alpha, a\rangle, \langle \beta, b\rangle, \langle \gamma, c\rangle, \ldots$ are arbitrary timed data streams.

• The composition of two unbounded fifo buffers yields again an unbounded fifo buffer:

$$\xmapsto{\text{fifo}} \circ \xmapsto{\text{fifo}} = \xmapsto{\text{fifo}}$$

because of the following equivalences:

$$\begin{aligned}
\langle\alpha,a\rangle \overset{\text{fifo}}{\longmapsto} \circ \overset{\text{fifo}}{\longmapsto} \langle\beta,b\rangle \;&\equiv\; \langle\alpha,a\rangle \overset{\text{fifo}}{\longmapsto} \exists\langle\gamma,c\rangle \overset{\text{fifo}}{\longmapsto} \langle\beta,b\rangle \\
&\equiv \exists\langle\gamma,c\rangle : \alpha=\gamma \wedge a<c \wedge \gamma=\beta \wedge c<b \\
&\equiv \alpha=\beta \wedge a<b \\
&\equiv\; \langle\alpha,a\rangle \overset{\text{fifo}}{\longmapsto} \langle\beta,b\rangle \ .
\end{aligned}$$

- The composition of two fifo_1 buffers yields a fifo_2 buffer:

$$\overset{\text{fifo}_1}{\longmapsto} \circ \overset{\text{fifo}_1}{\longmapsto} = \overset{\text{fifo}_2}{\longmapsto}$$

because

$$\begin{aligned}
&\langle\alpha,a\rangle \overset{\text{fifo}_1}{\longmapsto} \circ \overset{\text{fifo}_1}{\longmapsto} \langle\beta,b\rangle \\
&\qquad\equiv\; \langle\alpha,a\rangle \overset{\text{fifo}_1}{\longmapsto} \exists\langle\gamma,c\rangle \overset{\text{fifo}_1}{\longmapsto} \langle\beta,b\rangle \\
&\qquad\equiv \exists\langle\gamma,c\rangle : \alpha=\gamma \wedge a<ca' \wedge \gamma=\beta \wedge c<b<c' \\
&\qquad\Rightarrow \alpha=\beta \wedge a<b<a''=a^{(2)}\text{since } c<a' \text{ implies } c'<a'' \\
&\qquad\equiv\; \langle\alpha,a\rangle \overset{\text{fifo}_2}{\longmapsto} \langle\beta,b\rangle \ .
\end{aligned}$$

Given $\alpha=\beta$ and $a<b<a''$, the converse of the above implication can be proved by defining $\gamma=\alpha$ and

$$c(n)=\tfrac{1}{2}\times\left(\max\{a(n),b(n-1)\}+\min\{a(n+1),b(n)\}\right)$$

for all $n\geq 0$ (where $b(n-1)=0$ for $n=0$).

- Consider the following composition of three synchronous channels:

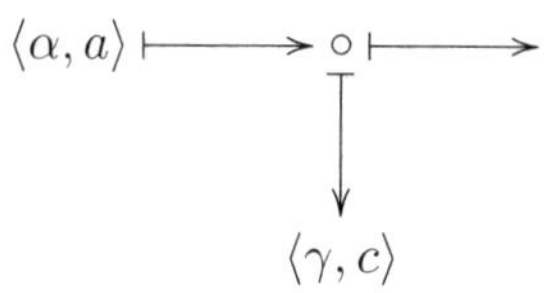

$$\equiv \alpha=\beta=\gamma \wedge a=b=c$$

This connector can be viewed as a "take-cue" regulator: any time a data item is taken from γ (by some future context), that same data item flows from left to right. This constitutes one of the most basic examples of what could be called *exogenous coordination*, that is, coordination from outside.

- The following connector is a variation on the previous one, in that the lower channel now is a synchronous drain:

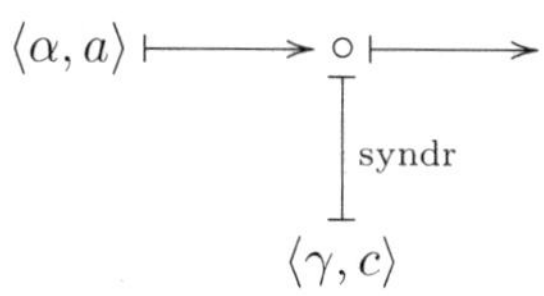

$$\equiv \alpha=\beta \wedge a=b=c$$

It is a "write-cue" regulator that regulates the flow of data items from left to right by inputs or writes on the lower channel end. Note that what is being input there is irrelevant. What matters is that such inputs are synchronised, through the synchronous drain, with both the channels above: $a=b=c$.

• With four synchronous channels and one synchronous drain, the following barrier synchroniser can be constructed:

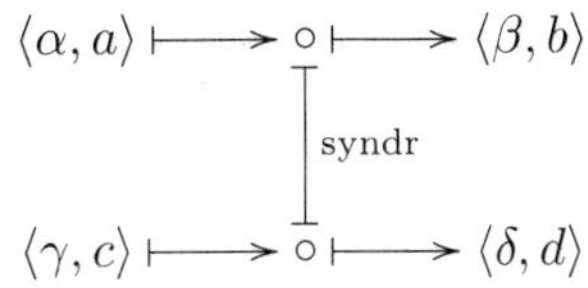

$$\equiv \alpha = \beta \wedge \gamma = \delta \wedge a = b = c = d$$

The synchronous drain in the middle ensures that data items pass through the upper and lower channels simultaneously.

• Here is a simple example of a feedback loop, consisting of one (unbounded) fifo buffer containing an initial data element $x \in D$, and two synchronous channels:

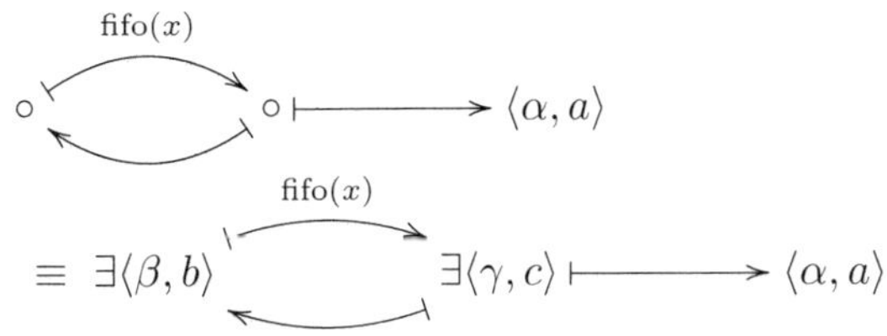

$$\equiv \exists\langle \beta, b\rangle \exists\langle \gamma, c\rangle : \big(\gamma(0) = x \wedge \beta = \gamma' \wedge b < c'\big) \wedge (\gamma = \beta \wedge c = b) \wedge (\gamma = \alpha \wedge c = a)$$
$$\equiv \exists\langle \beta, b\rangle \exists\langle \gamma, c\rangle : \alpha = \beta = \gamma \wedge a = b = c \wedge \gamma(0) = x \wedge \gamma = \gamma' \wedge c < c'$$
$$\equiv \exists\langle \beta, b\rangle \exists\langle \gamma, c\rangle : \alpha = \beta = \gamma \wedge a = b = c \wedge \gamma = (x, x, x, \dots)$$
$$\equiv \alpha = (x, x, x, \dots).$$

For the last but one equivalence, note that $\big(\gamma(0) = x \wedge \gamma = \gamma'\big)$ is equivalent to $\gamma = (x, x, x, \dots)$; moreover, the inequality $c < c'$ is redundant, because c is by assumption a time sequence and hence satisfies this inequality by definition. The behaviour of this connector is thus pretty much what it should be. It outputs perpetually the data element x. Note that there are no constraints on the time stream a $(= b = c)$. The only requirement is that it is indeed a time stream, that is, satisfies $a < a'$.

• Let $x \in D$ be again some fixed data element. The following connector acts as a *sequencer* on its two connector ends:

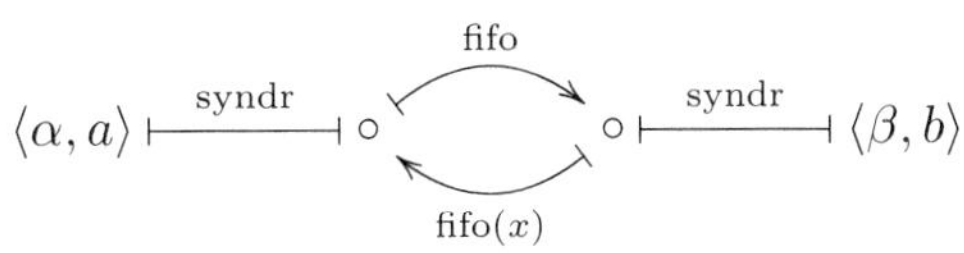

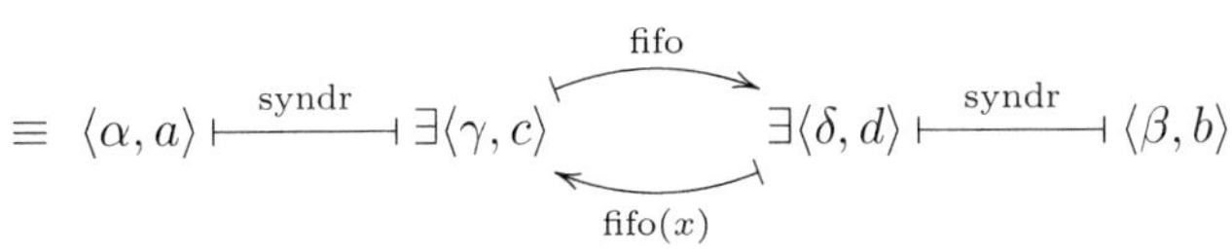

$$\equiv \exists\langle \gamma, c\rangle \exists\langle \delta, d\rangle : a = c \wedge d = b \wedge (\gamma = \delta \wedge c < d) \wedge \big(\gamma(0) = x \wedge \gamma' = \delta \wedge d < c'\big)$$
$$\equiv \exists\langle \gamma, c\rangle \exists\langle \delta, d\rangle \gamma = \delta = (x, x, x, \dots) \wedge c = a \wedge d = b \wedge (c < d < c')$$
$$= a < b < a'.$$

Thus, arbitrary data elements can be input alternately from the left and the right channel ends, at times

$$a(0) < b(0) < a(1) < b(1) < \cdots .$$

For future reference, we introduce the following notation for the sequencer connector:

$$\langle \alpha, a\rangle \overset{\text{seq}}{\vdash\!\!\!-\!\!\!-\!\!\!\dashv} \langle \beta, b\rangle \equiv a < b < a'. \tag{73}$$

• One can construct a connector that serialises any number k of channel ends by combining $k+1$ sequencers. For instance,

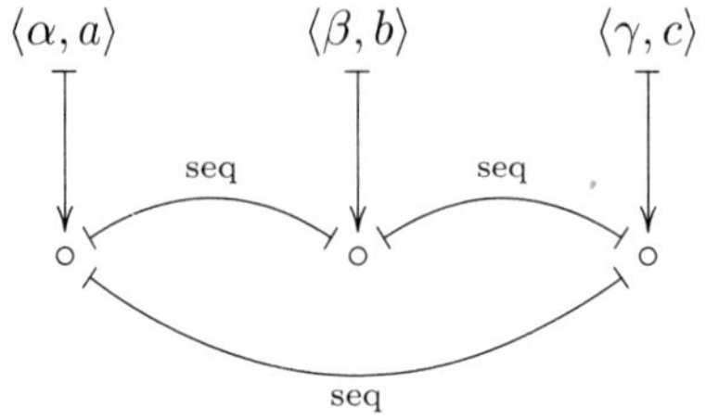

$$\equiv (a < b < a') \wedge (a < c < a') \wedge (b < c < b')$$
$$\equiv a < b < c < a'.$$

5.6. Connector Equivalence

In Section 5.5, we already saw some elementary examples of connector equivalence, such as:

$$\vdash\!\!\!\longrightarrow \;\circ\; \vdash\!\!\!\longrightarrow \;=\; \vdash\!\!\!\longrightarrow$$

$$\overset{\text{fifo}}{\vdash\!\!\!\longrightarrow} \;\circ\; \overset{\text{fifo}}{\vdash\!\!\!\longrightarrow} \;=\; \overset{\text{fifo}}{\vdash\!\!\!\longrightarrow}$$

$$\overset{\text{fifo}_1}{\vdash\!\!\!\longrightarrow} \;\circ\; \overset{\text{fifo}_1}{\vdash\!\!\!\longrightarrow} \;=\; \overset{\text{fifo}_2}{\vdash\!\!\!\longrightarrow}$$

Below we present some further, slightly less elementary examples.

(1) Recall the definition of the sequencer in Section 5.5:

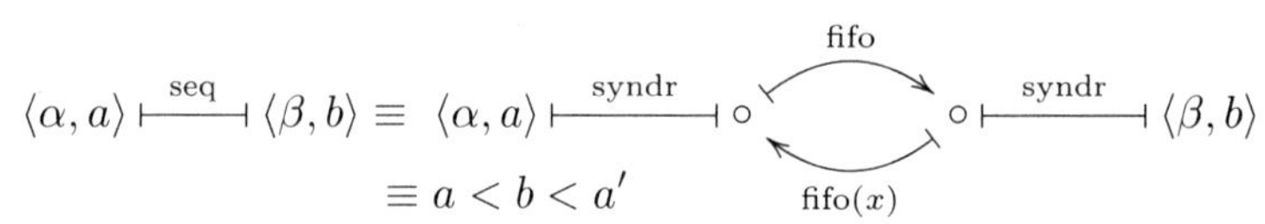

(where $x \in D$ is some fixed data item). Here is an alternative way of constructing the sequencer, now with a 1-bounded fifo buffer and a synchronous drain:

$$\begin{aligned}\langle \alpha, a\rangle \overset{\text{fifo}_1}{\vdash\!\!\!\longrightarrow} \circ \overset{\text{syndr}}{\vdash\!\!\!-\!\!\!-\!\!\!\dashv} \langle \beta, b\rangle \;&\equiv\; \langle \alpha, a\rangle \overset{\text{fifo}_1}{\vdash\!\!\!\longrightarrow} \exists\langle \gamma, c\rangle \overset{\text{syndr}}{\vdash\!\!\!-\!\!\!-\!\!\!\dashv} \langle \beta, b\rangle\\ &\equiv \exists\langle \gamma, c\rangle : (\alpha = \gamma \wedge a < c < a') \wedge c = b\\ &\equiv a < b < a'\\ &\equiv\; \langle \alpha, a\rangle \overset{\text{seq}}{\vdash\!\!\!-\!\!\!-\!\!\!\dashv} \langle \beta, b\rangle\;.\end{aligned}$$

(2) Conversely, a 1-bounded fifo buffer can be constructed using two synchronous channels, a sequencer, and an unbounded fifo buffer:

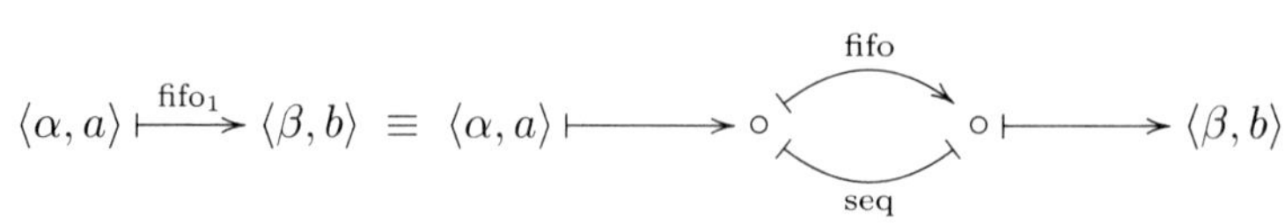

because we have the following equivalences:

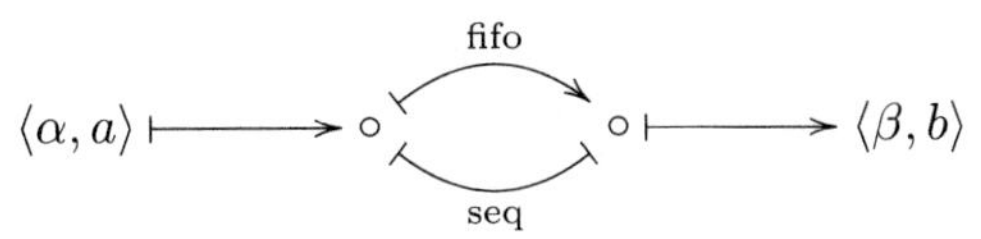

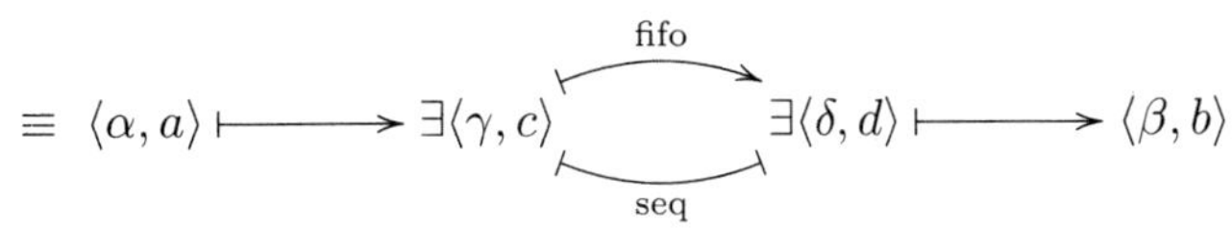

$$
\begin{aligned}
&\equiv \exists\langle\gamma,c\rangle, \exists\langle\delta,d\rangle : \langle\alpha,a\rangle = \langle\gamma,c\rangle \wedge \langle\delta,d\rangle = \langle\beta,b\rangle \wedge (\gamma = \delta \wedge c < d) \wedge (c < d < c')\\
&\equiv \alpha = \beta \wedge a < b < a'\\
&\equiv \langle\alpha,a\rangle \overset{\text{fifo}_1}{\longmapsto} \langle\beta,b\rangle\ .
\end{aligned}
$$

(3) Recall the definition of asynchronous drain in Section 5.4:

$$\langle\alpha,a\rangle \overset{\text{asyndr}}{\longmapsto} \langle\beta,b\rangle \ \equiv a \bowtie b.$$

Here is another way of constructing the asynchronous drain, using the merge connector and a synchronous drain:

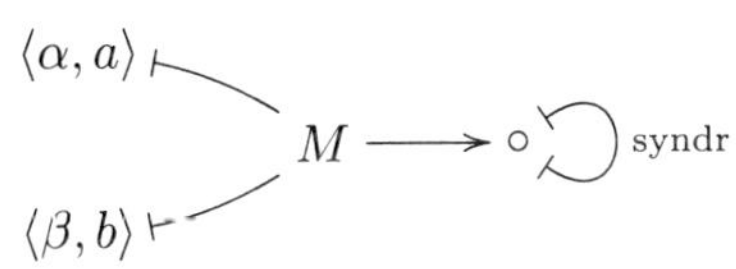

$$
\begin{aligned}
&\equiv \exists\langle\gamma,c\rangle : M(\langle\alpha,a\rangle, \langle\beta,b\rangle, \langle\gamma,c\rangle) \wedge c = c\\
&\equiv a \bowtie b\\
&\equiv \ \langle\alpha,a\rangle \overset{\text{asyndr}}{\longmapsto} \langle\beta,b\rangle\ .
\end{aligned}
$$

For the middle equivalence, the implication from left to right follows from (71) in Section 5.4, which was proved by $\bowtie$-coinduction. The converse implication can be proved in a similar fashion, using M-coinduction.

(4) Next we look at a connector that is built from two unbounded fifo buffers, the sequencer, and the merge operator:

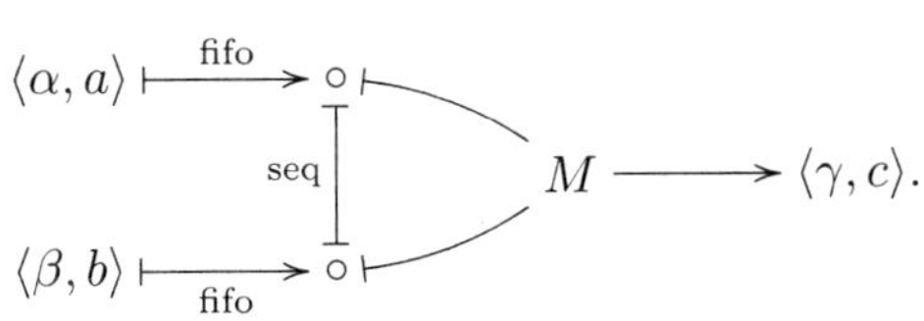

Using the coinduction proof principle, we shall prove that this connector has the following behaviour (with the operations zip, even and odd as in Section 1):

$$\text{zip}(\alpha,\beta) = \gamma \wedge a < \text{even}(c) \wedge b < \text{odd}(c).$$

The proof consists of the following sequence of equivalences:

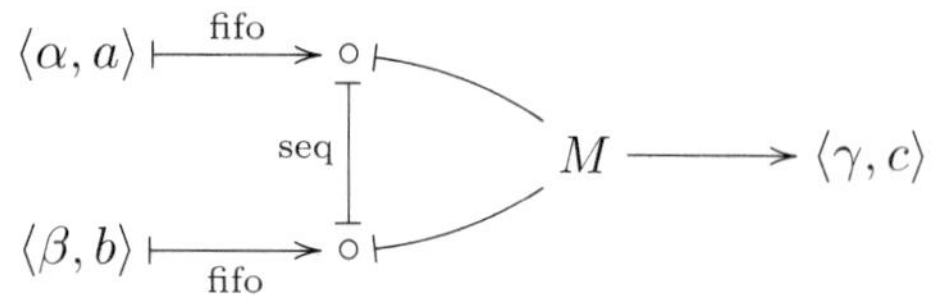

≡

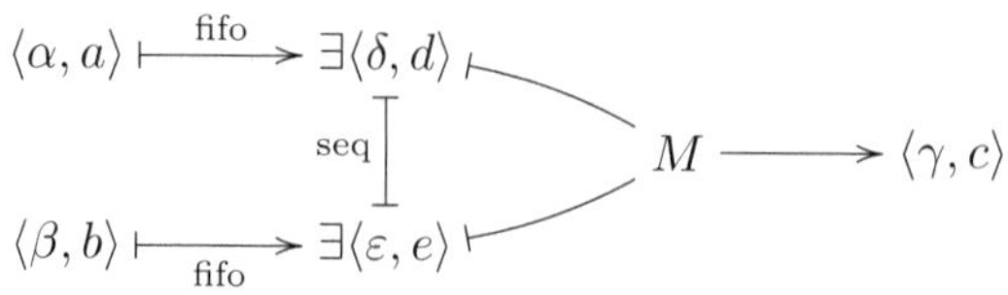

$$\begin{aligned}
&\equiv \exists\langle\delta, d\rangle\exists\langle\varepsilon, e\rangle : (\alpha = \delta \wedge a < d) \wedge (\beta = \varepsilon \wedge b < e) \wedge (d < e < d') \\
&\qquad\qquad\qquad \wedge M(\langle\delta, d\rangle, \langle\varepsilon, e\rangle, \langle\gamma, c\rangle) \\
&\equiv \exists d, e : a < d \wedge b < e \wedge (d < e < d') \wedge M(\langle\alpha, d\rangle, \langle\beta, e\rangle, \langle\gamma, c\rangle) \\
&\equiv \exists d, e : a < d \wedge b < e \wedge \mathrm{zip}(\alpha, \beta) = \gamma \wedge \mathrm{zip}(d, e) = c [\text{using (74) below}] \\
&\equiv \exists d, e : a < d \wedge b < e \wedge \mathrm{zip}(\alpha, \beta) = \gamma \wedge d = \mathrm{even}(c) \wedge e = \mathrm{odd}(c) \\
&\equiv \mathrm{zip}(\alpha, \beta) = \gamma \wedge a < \mathrm{even}(c) \wedge b < \mathrm{odd}(c).
\end{aligned}$$

We have used the following equivalence, which will be proved by coinduction:

$$(74) \qquad (d < e < d') \wedge M(\langle\alpha, d\rangle, \langle\beta, e\rangle, \langle\gamma, c\rangle) \equiv \big(\mathrm{zip}(\alpha, \beta) = \gamma \wedge \mathrm{zip}(d, e) = c\big).$$

For the implication from left to right, define the following relation on streams:

$$R = \langle\mathrm{zip}(\kappa, \lambda), \mu\rangle \mid \exists k, l, m : (k < l < k') \wedge M(\langle\kappa, k\rangle, \langle\lambda, l\rangle, \langle\mu, m\rangle).$$

We show that R is a bisimulation. Consider a pair $\langle\mathrm{zip}(\kappa, \lambda), \mu\rangle$ in R, with corresponding time streams k, l, m. Because $k < l$ it follows from $M(\langle\kappa, k\rangle, \langle\lambda, l\rangle, \langle\mu, m\rangle)$ that $\mu(0) = \kappa(0)$, and since $\kappa(0) = \mathrm{zip}(\kappa, \lambda)(0)$, this proves the first of the two bisimulation conditions. Next consider the pair of derivatives $\langle\mathrm{zip}(\kappa, \lambda)', \mu'\rangle = \langle\mathrm{zip}(\lambda, \kappa'), \mu'\rangle$. It follows from the definition of M that the latter pair is again in R, since

$$\begin{aligned}
&(k < l < k') \wedge M(\langle\kappa, k\rangle, \langle\lambda, l\rangle, \langle\mu, m\rangle) \\
&\Rightarrow (l < k' < l') \wedge M(\langle\kappa', k'\rangle, \langle\lambda, l\rangle, \langle\mu', m'\rangle) \equiv (l < k' < l') \wedge M(\langle\lambda, l\rangle, \langle\kappa', k'\rangle, \langle\mu', m'\rangle) \\
&\hfill [\text{by equivalence (72)}].
\end{aligned}$$

This proves that R is a bisimulation. Assuming now $(d < e < d')$ and $M(\langle\alpha, d\rangle, \langle\beta, e\rangle, \langle\gamma, c\rangle)$, $\mathrm{zip}(\alpha, \beta) = \gamma$ follows by coinduction. In the same manner, one shows $\mathrm{zip}(d, e) = c$. This proves the implication from left to right of equivalence (74). The implication from right to left can be proved along similar lines, using M-coinduction.

5.7. Protocol Verification

Our calculus of component connectors also allows the formulation and formal verification of communication protocols. We present a simple example, taken from [**13**, pp. 29–36]. It consists of an (unbounded fifo) lossy buffer composed with a driver that corrects the lossiness of the buffer. Below we specify both connectors, and prove that their composition is equivalent to an ordinary (correct) unbounded fifo buffer.

Let $\langle \alpha, a \rangle$, $\langle \beta, b \rangle$, and $\langle \delta, d \rangle$ be timed data streams over an arbitrary data set D and let $\langle \gamma, c \rangle$ be a timed data stream with $\gamma \in \{0,1\}^\omega$. We define the lossy buffer as a ternary relation L on timed data streams with one input end and two output ends as follows:

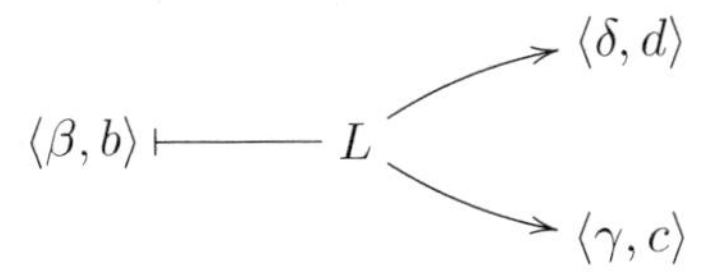

$$\equiv L(\langle \beta, b \rangle, \langle \gamma, c \rangle, \langle \delta, d \rangle)$$

$$\equiv b < c \wedge b < d \wedge \begin{cases} \gamma(0) = 1 \wedge \delta(0) = \beta(0) \wedge L(\langle \beta', b' \rangle, \langle \gamma', c' \rangle, \langle \delta', d' \rangle) \\ \vee \gamma(0) = 0 \wedge L(\langle \beta', b' \rangle, \langle \gamma', c' \rangle, \langle \delta, d \rangle) \end{cases}$$

This connector inputs data items at the input end β. For every data item that is input, there are two possible scenarios:

(1) the data item is stored successfully and is output (at some later moment) at the upper output end δ together (not necessarily simultaneously) with a success signal 1 along γ, after which the connector proceeds as before with the remainder of all streams involved;

(2) storage of the data item fails, no data item is output along the end β, a 0 signalling the failure is output along γ, and the connector proceeds as before, now with $\langle \beta', b' \rangle$ and $\langle \gamma', c' \rangle$, but with $\langle \delta, d \rangle$ unchanged.

It follows from the definition of the lossy buffer that eventually some data item gets successfully stored and output. In other words, there exists $n \geq 0$ with $\gamma(n) = 1$. In order to prove this, assume $L(\langle \beta, b \rangle, \langle \gamma, c \rangle, \langle \delta, d \rangle)$ and suppose that $\gamma(n) = 0$, for all $n \geq 0$. Then

$$L(\langle \beta^{(n)}, b^{(n)} \rangle, \langle \gamma^{(n)}, c^{(n)} \rangle, \langle \delta, d \rangle)$$

for all $n \geq 0$ (recall that the superscript (n) stands for the nth derivative). As a consequence, $b^{(n)}(0) = b(n) < d(0)$, for all n. Under the assumption that our time streams are progressive (cf. the end of Section 5.2): for any $N \geq 0$ there exists $n \geq 0$ with $b(n) > N$, this is a contradiction. Therefore, there exists $n \geq 0$ with $\gamma(n) = 1$. We see that, somewhat surprisingly, the fact that all our streams are infinite and the assumption that time streams are progressive, together imply here the right type of fairness (or liveness) behaviour.

Next we turn to the driver, which has two input ends and one output end, and is defined as the following ternary relation D:

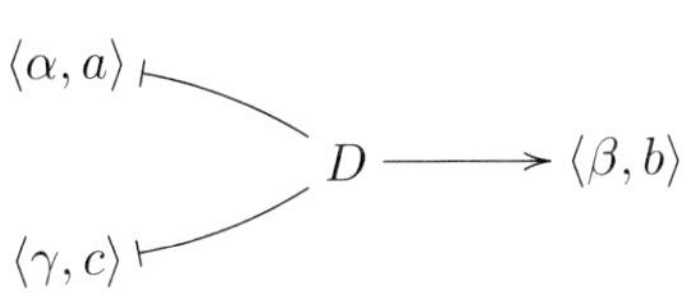

$$\equiv D(\langle \alpha, a \rangle, \langle \beta, b \rangle, \langle \gamma, c \rangle)$$

$$\equiv a = b \wedge a < c < a' \wedge \beta(0) = \alpha(0) \wedge \begin{cases} \gamma(0) = 1 \wedge D(\langle \alpha', a' \rangle, \langle \beta', b' \rangle, \langle \gamma', c' \rangle) \\ \vee \gamma(0) = 0 \wedge D(\langle \alpha, a' \rangle, \langle \beta', b' \rangle, \langle \gamma', c' \rangle) \end{cases}$$

The driver inputs data items at α and outputs them at β. Before proceeding with the next data item, it checks its input at γ. If $\gamma(0) = 1$ then the last data item that has been output is considered to have been handled correctly (by the lossy buffer in the composition below), and D proceeds as before with the remainder of all streams involved. If $\gamma(0) = 0$, however, something has gone wrong (the buffer has lost the data item), and D sends the data item again. This is modelled here by $D(\langle\alpha, a'\rangle, \langle\beta', b'\rangle, \langle\gamma', c'\rangle)$, in which all streams have progressed to their derivatives but for α, which remains unchanged. As a consequence, $\alpha(0)$ is (again) the next data item that D will output (but note that the time stream a *has* changed into a').

Composing the driver and the lossy buffer as below yields a connector that is equivalent with the (non-lossy) unbounded fifo buffer: for all timed data streams $\langle\alpha, a\rangle$ and $\langle\delta, d\rangle$,

$$\langle\alpha, a\rangle \longmapsto D \;\; \underset{\exists\langle\gamma, c\rangle}{\overset{\exists\langle\beta, b\rangle}{\rightleftarrows}} \;\; L \longrightarrow \langle\delta, d\rangle \;\equiv\; \langle\alpha, a\rangle \overset{\text{fifo}}{\longmapsto} \langle\delta, d\rangle.$$

For the implication from left to right, we have to show that $\alpha = \delta$ (and $a < d$). To this end, define the following relation on data streams:

$$R = \{\langle\alpha, \delta\rangle \mid \exists a, d, \langle\beta, b\rangle, \langle\gamma, c\rangle : D(\langle\alpha, a\rangle, \langle\beta, b\rangle, \langle\gamma, c\rangle) \wedge L(\langle\beta, b\rangle, \langle\gamma, c\rangle, \langle\delta, d\rangle)\}.$$

In order to prove that R is a bisimulation relation, consider a pair $\langle\alpha, \delta\rangle$ in R with "witnesses" $a, d, \langle\beta, b\rangle, \langle\gamma, c\rangle$ such that $D(\langle\alpha, a\rangle, \langle\beta, b\rangle, \langle\gamma, c\rangle)$ and $L(\langle\beta, b\rangle, \langle\gamma, c\rangle, \langle\delta, d\rangle)$. Let n be the smallest natural number such that $\gamma(n) = 1$ (which exists by the remark above). It follows that

$$D(\langle\alpha, a^{(n)}\rangle, \langle\beta^{(n)}, b^{(n)}\rangle, \langle\gamma^{(n)}, c^{(n)}\rangle) \wedge L(\langle\beta^{(n)}, b^{(n)}\rangle, \langle\gamma^{(n)}, c^{(n)}\rangle, \langle\delta, d\rangle).$$

Together with $\gamma^{(n)}(0) = \gamma(n) = 1$, this implies $\alpha(0) = \beta^{(n)}(0) = \delta(0)$ and, moreover,

$$D(\langle\alpha', a^{(n+1)}\rangle, \langle\beta^{(n+1)}, b^{(n+1)}\rangle, \langle\gamma^{(n+1)}, c^{(n+1)}\rangle)$$
$$\wedge\, L(\langle\beta^{(n+1)}, b^{(n+1)}\rangle, \langle\gamma^{(n+1)}, c^{(n+1)}\rangle, \langle\delta', d'\rangle).$$

Thus, $\langle\alpha', \delta'\rangle \in R$, which concludes the proof that R is a bisimulation. It now follows by coinduction that $\alpha = \delta$. (A minor variation on this argument proves that $a < d$.)

For the implication from right to left, choose $\langle\beta, b\rangle = \langle\alpha, a\rangle$, $\gamma = (1, 1, 1, \ldots)$, and $c = \frac{1}{2} \times (a + a')$ (in a hopefully self explanatory notation). It is now a straightforward proof by (D- and L-)coinduction to show that $D(\langle\alpha, a\rangle, \langle\beta, b\rangle, \langle\gamma, c\rangle)$ and $L(\langle\beta, b\rangle, \langle\gamma, c\rangle, \langle\delta, d\rangle)$.

5.8. Discussion

Reo is more general than data-flow models, Kahn-networks, and Petri-nets, which can be viewed as specialised channel-based models that incorporate certain basic constructs for primitive coordination. While Reo is designed to deal with the flow of data, it, more specifically, differs fundamentally from classical data-flow models in four important aspects:

(1) Although not treated here, the topology of connections in "full" Reo is inherently dynamic and accommodates mobility.
(2) Reo supports a much more general notion of channels.
(3) The model of Reo is based on a clear separation of data and time.
(4) Coinduction is the main reasoning principle.

More details about Reo can be found in [**3, 4**]. The model presented in this chapter is described in [**5**]. Of related work, Broy and Stølen's book [**13**] deserves special mention, since it is also based on (timed) data streams. However, the points mentioned above distinguish our model also from theirs. In particular, the separation of data and time, in our model, in combination with the use of coinduction, leads to simpler specifications (definitions) and proofs. See Section 5.7 for a concrete example. Finally, coalgebra and coinduction have been used in models of component-based systems in [**7, 17**]. Also these models are distinguished from ours by the (first three) points above. Moreover, our model is far more concrete, and therefore allows actual equivalence proofs.

Our work on Reo and this model is on-going. One of the first questions to address is to decide what set(s) of basic channels and connectors to choose as the basis for a connector calculus (or calculi). Another plan is to look at more instances of connector protocol verification. On the basis of the example of Section 5.7 (and other examples not included in the present chapter, such as the alternating bit protocol), we expect that the present model will be competitive with both traditional data-flow networks and with process algebra, by combining the best of those two worlds. Like data flow and unlike process algebra, Reo is channel-based and models the (communication) topology of connectors explicitly. Like process algebra and unlike data flow, (our model of) Reo is a calculus in which complex connectors are compositionally built out of simpler ones. Moreover, unlike data flow, the model for Reo that we presented is both simple and formal enough to allow actual verification. And unlike process algebra, there is no need to use nondeterministic transition systems and computationally complicated notions such as weak or branching bisimulation. Instead, streams and coinduction are all that is needed.

APPENDIX A

Key Differential Equations

The following table contains the most important behavioural differential equations used in this book:

derivative	initial value	side conditions
$[r]' = [0]$	$[r](0) = r$	
$X' = [1]$	$X(0) = 0$	
$(\sigma + \tau)' = \sigma' + \tau'$	$(\sigma + \tau)(0) = \sigma(0) + \tau(0)$	
$(\sigma \times \tau)' = (\sigma' \times \tau) + (\sigma(0) \times \tau')$	$(\sigma \times \tau)(0) = \sigma(0) \times \tau(0)$	
$(\sigma^{-1})' = -\sigma(0)^{-1} \times (\sigma' \times \sigma^{-1})$	$(\sigma^{-1})(0) = \sigma(0)^{-1}$	$\sigma(0) \neq 0$
$(\sigma \circ \tau)' = \tau' \times (\sigma' \circ \tau)$	$(\sigma \circ \tau)(0) = \sigma(0)$	$\tau(0) = 0$
$\left(\sum_{n=0}^{\infty} \sigma_n\right)' = \sum_{n=0}^{\infty} (\sigma_n')$	$\left(\sum_{n=0}^{\infty} \sigma_n\right)(0) = \sum_{n=0}^{\infty} \sigma_n(0)$	$\{\sigma_n\}_{n=0}^{\infty}$ is summable
$(\sigma \otimes \tau)' = (\sigma' \otimes \tau) + (\sigma \otimes \tau')$	$(\sigma \otimes \tau)(0) = \sigma(0) \times \tau(0)$	
$(\sigma^{-\underline{1}})' = -\sigma' \otimes (\sigma^{-\underline{1}} \otimes \sigma^{-\underline{1}})$	$\sigma^{-\underline{1}}(0) = \sigma(0)^{-1}$	$\sigma(0) \neq 0$
$\exp(\sigma)' = \sigma' \otimes \exp(\sigma)$	$\exp(\sigma)(0) = e^{\sigma(0)}$	
$\Lambda_C(\sigma)' = \Lambda_C\left(\frac{d\sigma}{dX}\right)$	$\Lambda_C(\sigma)(0) = \sigma(0)$	
$\mathcal{T}(f)' = \mathcal{T}(f')$	$\mathcal{T}(f)(0) = f(0)$	f is analytical
$(\sqrt{\sigma})' = \frac{\sigma'}{\sqrt{\sigma(0)} + \sqrt{\sigma}}$	$\sqrt{\sigma}(0) = \sqrt{\sigma(0)}$	$\sigma(0) > 0$

Bibliography

1. P. Aczel, *Non-well-founded sets*, CSLI Lecture Notes, vol. 14, Center for the Study of Language and Information, Stanford University, Stanford, CA, 1988.
2. P. Aczel and N. Mendler, *A final coalgebra theorem*, Category Theory and Computer Science (Manchester, 1989), (D. H. Pitt, D. E. Ryeheard, P. Dybjer, A. M. Pitts, and A. Poigne, eds.), Lecture Notes in Comput. Sci., vol. 389, 1989, Springer, Berlin, pp. 357–365.
3. F. Arbab, *A channel-based coordination model for component composition*, Report SEN-R0203, CWI, 2002; `http://www.cwi.nl`.
4. F. Arbab and F. Mavaddat, *Coordination through channel composition*, 5th Internat. Conf. on Coordination Models and Languages (F. Arbab and C. Talcott, eds.), Lecture Notes in Comput. Sci., vol. 2315, Springer-Verlag, London, UK, 2002, pp. 22–39.
5. F. Arbab and J. J. M. M. Rutten, *A coinductive calculus of component connectors*, Proceedings of WADT 2002 (to appear); Report SEN-R0216, CWI, 2002; `http://www.cwi.nl`.
6. M. A. Arbib and E. G. Manes, *Machines in a category*, J. Pure Appl. Algebra **19** (1980), 9–20.
7. L. Barbosa, *Components as coalgebras*, Ph.D. thesis, Universidade do Minho, Braga, Portugal, 2001.
8. H. Barringer, R. Kuiper, and A. Pnueli, *A really abstract concurrent model and its temporal logic*, 13th Annual ACM Symposium on Principles of Programming Languages, ACM Press, New York, USA, 1986, pp. 173–183.
9. F. Bartels, *Generalised coinduction*, Report SEN-R0043, CWI, 2000; `http://www.cwi.nl`.
10. F. Bergeron, G. Labelle, and P. Leroux, *Combinatorial species and tree-like structures*, Encyclopedia Math. Appl., vol. 67, Cambridge University Press, Cambridge, 1998.
11. J. Berstel and C. Reutenauer, *Rational series and their languages*, Monogr. Theoret. Comput. Sci. EATCS Ser., vol. 12, Springer-Verlag, Berlin, 1988.
12. G. Birkhoff and G.-C. Rota, *Ordinary differential equations*, 3rd ed., John Wiley and Sons, New York–Chichester–Brisbane, 1978.
13. M. Broy and K. Stølen, *Specification and development of interactive systems*, Monogr. Comput. Sci., vol. 62, Springer-Verlag, New York, 2001.
14. J. A. Brzozowski, *Derivatives of regular expressions*, J. ACM **11** (1964), no. 4, 481–494.
15. L. Comtet, *Advanced combinatorics. The art of finite and infinite expansions*, D. Reidel Publishing Co., Dordrecht, 1974.
16. J. H. Conway, *Regular algebra and finite machines*, Chapman and Hall, 1971.
17. E.-E. Doberkat, *Pipes and filters: Modelling a software architecture through relations*, Report 123, Chair for software technology, University of Dortmund, 2002.
18. M. Forti and F. Honsell, *Set theory with free construction principles*, Ann. Scuola Norm. Sup. Pisa Cl. Sci. (4) **10** (1983), no. 3, 493–522.
19. P. Flajolet, *Combinatorial aspects of continued fractions*, Discrete Math. **32** (1980), 125–161.
20. P. Flajolet and R. Sedgewick, *The average case analysis of algorithms: Counting and generating functions*, Research Report 1888, INRIA Rocquencourt, 1993.
21. ———, *Analytical combinatorics: Functional equations, rational and algebraic functions*, Research Report 4103, INRIA Rocquencourt, 2001.
22. R. L. Graham, D. E. Knuth, and O. Patashnik. *Concrete mathematics. A foundation for computer science*, 2nd ed., Addison–Wesley Publishing Company, Reading, MA, 1994.

23. Bart Jacobs and J. J. M. M. Rutten, *A tutorial on (co)algebras and (co)induction*, Bull. Eur. Assoc. Theor. Comput. Sci. EATCS **62** (1997), 222–259; `http://www.cwi.nl/~janr`.
24. J. Karczmarczuk, *Generating power of lazy semantics*, Theoret. Comput. Sci. **187** (1997), 203–219.
25. ______, *Lazy processing and optimization of discrete sequences*, JFLA'2000, 2000 (French) (to appear); `http://www.info.unicaen.fr/ karczma`.
26. D. C. Kozen, *Automata and computability*, Undergrad. Texts in Comput. Sci., Springer-Verlag, New York, 1997.
27. E. G. Manes and M. A. Arbib, *Algebraic approaches to program semantics*, Texts Monogr. Comput. Sci., AKM Series in Theoretical Computer Science, Springer-Verlag, New York, 1986.
28. M. D. McIlroy, *Power series, power serious*, J. Funct. Programming **9** (1999), 323–335.
29. ______, *The music of streams*, Inform. Process. Lett. **77** (2001), 189–195.
30. J. Mikusinski, *Operational calculus*. I, 2nd ed., Int. Ser. Monogr. Pure Appl. Math., vol. 109, Pergamon Press, Oxford; PWN—Polish Scientific Publishers, Warsaw, 1983.
31. R. Milner, *A calculus of communicating cystems*, Lecture Notes Comput. Sci., vol. 92, Springer-Verlag, Berlin-New York, 1980.
32. ______, *Communication and concurrency*, Prentice Hall, 1989.
33. D. M. R. Park, *Concurrency and automata on infinite sequences*, 5th GI Conference (Karlsruhe, Germany) (P. Deussen, ed.), Lecture Notes in Comput. Sci., vol. 104, Springer-Verlag, 1981, pp. 167–183.
34. D. Pavlović and M. Escardó, *Calculus in coinductive form*, 13th Annual IEEE Symposium on Logic in Computer Science (Indianapolis, 1998), IEEE Computer Soc., Los Alamitos, 1998, pp. 408–417.
35. J. J. M. M. Rutten, *Automata and coinduction (an exercise in coalgebra)*, CONCUR'98: Concurrency Theory (Nice) (D. Sangiorgi and R. de Simone, eds.), Lecture Notes in Comput. Sci., vol. 1466, Springer, Berlin, 1998, pp. 194–218.
36. ______, *Automata, power series, and coinduction: Taking input derivatives seriously (extended abstract)*, Automata, Languages and Programming (Prague, 1999) (J. Wiedermann, P. van Emde Boas, and M. Nielsen, eds.), Lecture Notes in Comput. Sci., vol. 1644, Springer, Berlin, 1999, pp. 645–654.
37. ______, *Behavioural differential equations: A coinductive calculus of streams, automata, and power series*, Theoret. Comput. Sci., vol. 308 (1–3), Elsevier Science Publishers, 2003, pp. 1–53.
38. ______, *Universal coalgebra: A theory of systems*, Theoret. Comput. Sci. **249** (2000), no. 1, 3–80.
39. ______, *Elements of stream calculus (an extensive exercise in coinduction)*, Technical Report SEN-R0120, CWI, Amsterdam, 2001, pp. 1–54; Proceedings MFPS '01, ENTCS, vol. 45, Elsevier Science B.V., 2001; Math. Structures in Comput. Sci. (to appear).
40. ______, *Coinductive counting with weighted automata*, J. Automat. Lang. Comb. **8** (2003), no. 2, 319–352.
41. D. Sangiorgi, *On the bisimulation proof method*, Math. Structures Comput. Scie. **8** (1998), no. 5, 447–479.
42. I. N. Sneddon, *The use of integral transforms*, McGraw-Hill, 1972.
43. R. P. Stanley, *Enumerative combinatorics*. I, Cambridge Stud. Adv. Math., vol. 49, Cambridge University Press, Cambridge, 1997.
44. ______, *Enumerative Combinatorics*. II, Cambridge Stud. Adv. Math., vol. 62, Cambridge University Press, Cambridge, 1999.
45. A. Tarski, *A lattice-theoretical fixpoint theorem and its applications*, Pacific J. Math. **5** (1955), 285–309.
46. J. van Benthem, *Modal correspondence theory*, Ph.D. thesis, University of Amsterdam, Amsterdam, 1976.
47. H. S. Wilf, *Generatingfunctionology*, Academic Press, 1994.

Modelling and Verification of Probabilistic Systems

Marta Kwiatkowska, Gethin Norman, and David Parker

Preface

These lecture notes contain an expanded version of the lecture course on Probabilistic Model Checking taught as part of the Workshop on Mathematical Models and Techniques for Analysing Systems, September 30–October 4, 2002, held at the Centre de Recherches Mathématiques, University of Montréal. The main emphasis of the course was to feature both theory (models, specification languages and model checking algorithms) as well as practice (implementation, symbolic model checking, case studies) of probabilistic model checking. The lecture material is inevitably influenced by the experience of implementing and applying the probabilistic model checking tool PRISM (`www.cs.bham.ac.uk/~dxp/prism`) developed at the University of Birmingham.

The authors are indebted to Prakash Panangaden for the invitation to lecture in Montreal and his patience with the course material itself, and hope that this book will be of use to theoreticians and practitioners alike interested in the modelling and analysis of probabilistic systems.

Acknowledgements

These achievements would not have been possible without the many collaborators who contributed in various ways, including encouragement and finding errors. The following deserve specific mention: Rashid Mehmood, Stephen Gilmore and Andrew Hinton (PRISM tool development), Jeremy Sproston, Roberto Segala and Conrado Daws (probabilistic timed automata), Christel Baier, Luca de Alfaro and Sylvain Peyronnet (PCTL algorithms), Markus Siegle and Holger Hermanns (MTBDDs), Joost-Pieter Katoen, António Pacheco and Håkan Younes (CSL algorithms), and Vitaly Shmatikov and Sandeep Shukla (contribution to case studies). This work was supported in part by EPSRC grants GR/M04617, GR/N22960 and GR/S11107 and QinetiQ.

Marta Kwiatkowska
Gethin Norman
and David Parker

CHAPTER 1

Introduction

Probability plays an increasingly important role in the design and analysis of software and hardware systems. In distributed algorithms, randomness, through the mechanism of electronic coin tossing, is used as a symmetry breaker and a means to derive efficient algorithms, for example, randomised leader election [**76**], randomised consensus [**7**] and root contention in IEEE 1394 FireWire [**74**]. Probability is also used to model unreliable or unpredictable behaviour, as in e.g. fault-tolerant systems and multi-media protocols where properties such as frame loss of 1 in every 100 can be described probabilistically. Traditionally, it has also been used as a tool to analyse system performance, where typically queueing theory is applied to obtain steady-state probabilities in order to arrive at estimates of measures such as throughput and mean waiting time.

In this book, we consider techniques for the modelling and verification of probabilistic systems. In particular, we focus on *probabilistic model checking*, a method for calculating the likelihood of the occurrence of certain events during the execution of a system. Conventional model checkers input a description of a model, represented as a state transition system, and a specification, typically a formula in some temporal logic, and return 'yes' or 'no', indicating whether or not the model satisfies the specification. In the case of probabilistic model checking, the models are probabilistic (typically variants of Markov chains), in the sense that they encode the *probability* of making a transition between states instead of simply the existence of such a transition. Probabilistic specifications include:

- for a randomised leader election algorithm:
  - "leader election is eventually resolved with probability 1".
  - "the expected time until a leader is elected is at most 50ms".
- for a system that can suffer failures:
  - "the chance of shutdown occurring is at most 1%".
  - "the probability that the system fails within 100s is at most 0.001".
- for a multi-media protocol:
  - "the chance of a video frame being delivered within 5ms is at least 89%".
  - "the probability that N video frames are lost within 1s is at most 0.1".
- for a wireless cell:
  - "the chance that a hand-off call is dropped within 100ms is less than 6%".
  - "in the long run, the chance of a call being blocked is less than 0.03".
- for a lift system:
  - "the probability of the lift failing within the first 6 months is below 0.001".

– "in the long run, mean waiting time in the lift queue is at most 30s".

In common with conventional model checking, probabilistic model checking involves reachability analysis of the underlying transition system, but, in addition, it must entail the calculation of the actual likelihoods through appropriate numerical or analytical methods.

Since model checking was first proposed, it has become well established as a means to analyse and debug protocols and designs. Several model checkers are now available, for example SMV, NuSMV, SPIN, FDR2 and, for real-time systems, UPPAAL and KRONOS.

Probabilistic model checking is still in the early stages of development: although the fundamental algorithms were developed in the 1980s, work on implementation and tools did not begin until recently. The first tools reported were TPWB (Time and Probability WorkBench) and ProbVerus. A number of performance analysis tools exist, for process algebras, Petri nets and other graphical model description languages; these support steady-state solutions (and in some cases transient analysis) but no probabilistic temporal logic specifications. More recently, three tools which admit probabilistic temporal logic specifications have been proposed: $\mathsf{E} \vdash \mathsf{MC}^2$, which supports analysis of discrete- and continuous-time Markov chains; RAPTURE, which implements model checking of Markov decision processes; and PRISM, which supports all three models.

In these lecture notes, we summarise the theory and practice of automatic verification of probabilistic systems. We begin by motivating the uses of probability in modelling and designing systems (1.1), then subsequently discuss four kinds of probabilistic models: discrete-time Markov chains (which feature probabilistic choice only), Markov decision processes (which feature both nondeterministic and probabilistic choice), continuous time Markov chains (which model continuous time, but not nondeterminism) and finally probabilistic timed automata (which admit continuous time, probabilistic choice and nondeterminism). For each type of model we also introduce a specification language, a temporal logic extended with a probabilistic quantifier and expectation operators, PCTL (Probabilistic Computation Tree Logic) for discrete-time Markov chains and Markov decision processes, CSL (Continuous Stochastic Logic) for continuous-time Markov chains and PTCTL (Probabilistic Timed Computation Tree Logic) for probabilistic timed automata, together with the corresponding model checking methods. We illustrate the usefulness of these techniques through a number of case studies analysing real-world probabilistic protocols using the tool PRISM: the IPv4 ZeroConf self-configuration protocol, self-stabilisation algorithms, stochastic policies for dynamic power management and the IEEE 1394 FireWire root contention protocol.

1.1. Motivation

There are many uses of probability in the design and analysis of systems. They can be roughly grouped into three broad classes:

As a symmetry breaker in distributed algorithms. It is a well known result [**56**] that there are no symmetric solutions to certain distributed systems problems in the presence of failures, for example distributed consensus among processes if Byzantine or even stopping failures are possible. M. Rabin pioneered the central observation that if one weakens termination from "the algorithm eventually terminates" to "the algorithm terminates with probability 1", then using random choices, such as the outcome of electronic coin tosses, ensures that such solutions exist, see e.g. randomised mutual exclusion [**115**] and randomised consensus of [**7**]. Using random choices often yields simple, elegant and efficient symmetric algorithmic solutions to distributed problems, for example, leader election [**76**] and IPv4 ZeroConf protocol [**37**]. In many such algorithms, modelling nondeterminism as well as probabilistic choices is essential. The correctness arguments for such randomised distributed algorithms include both non-probabilistic statements, such as "no two processes are simultaneously in the critical section", as well as probabilistic statements such as "termination occurs with probability 1".

To model failures. Probability is useful to quantify the chance of an action failing to terminate correctly; for example, in a multimedia protocol, frame loss rate of 1 in every 100 represents 99% chance of successful transmission. More generally, a discrete probability distribution on the target states is needed to model such behaviour. In a continuous time setting, mean time to failure characteristics of hardware components, such as the processor, router or communication link, can also be described probabilistically, this time with a continuous time (exponential) distribution, yielding the probability of the component breaking down within time t. If the probability of some actions failing is known, it is important to estimate the impact of this on the overall system behaviour.

To model system performance. Traditionally, queueing theory and continuous time Markov chains have been used to analyse performance of systems such as computer networks, multi-processor systems and wireless communication networks. First, a model of a system is built, and then the steady-state probabilities vector calculated and used as input into formulas which obtain various measures, including throughput, average response time, and mean waiting time. The results of such an analysis help predict the behaviour of the system under certain conditions, and can assist in the re-design of the system.

1.2. The models

There are a number of probabilistic models, of which we will consider four types, all variants of Markov chains with *discrete* states; for an approach which deals with *continuous* state spaces see [**52**]. The simplest class of models we consider are discrete-time Markov chains (DTMCs), which admit *probabilistic choice* only, in the sense that one can specify the probability of making a transition from state to another. Markov decision processes (MDPs) allow for probabilistic and *nondeterministic choice*; they arise, for example, through the asynchronous parallel composition of DTMCs and are particularly appropriate in the context of a distributed computation. Both DTMCs and MDPs do not model continuous (dense) time, only discrete time. Continuous-time Markov chains (CTMCs), well known in performance analysis, model *continuous real time* and probabilistic choice: one can specify the rate of making a transition from one state to another. CTMCs do not exhibit nondeterminism but one can simulate concurrency though the race conditions arising between transitions with different rates. The combination of dense real time, nondeterminism and probabilistic choice is possible in probabilistic timed automata (PTAs), an extension of timed automata [**5**] with *probabilistic transitions* and resets of clocks to values drawn from a certain class of continuous-time probability distributions.

1.3. Properties of Interest

When analysing non-probabilistic system models, we are interested in establishing the truth of statements such as "eventually event A occurs" and "nowhere event B occurs", for some good event A and bad event B, with additional quantification "in every (some) future path" in the branching time models. When analysing probabilistic models, we can calculate *the probability* of events, and thus a typical property is "the probability that event A occurs is at least p (at most p)", where p is a value in the interval $[0, 1]$. Probability bounds 0 and 1 are allowed. In the case of MDP models, the presence of nondeterminism means that we cannot calculate the exact probability, only the minimum/maximum over all the possible ways of resolving nondeterminism. The properties in this case are of the form "*under any scheduling of processes*, the probability that event A occurs is at least p (at most p)". It is often useful to establish the probability of event A occurring within a given time t: for DTMCs and MDPs this time t must be discrete, whereas for CTMCs it can be real-valued. For PTAs the time bounds can be rational (via scaling).

1.4. Probabilistic Model Checking and Other Analysis Methods

Once a probabilistic model has been built, either as an annotated graph, a stochastic Petri net or a program written in some textual system description language, it can be subjected to various types of analysis: *simulation*, *analytical solution*, *reachability*, and *model checking* against temporal logic specification. Simulation and analytical approaches result in performance and quality of service estimates, typically represented as functions of system parameters. Reachability analysis and model checking rely on performing traversal of the underlying transition system graph of the model in order to establish e.g. if it is possible for an unsafe state to be reached from the initial state. *Probabilistic* model checking combines probabilistic analysis and conventional reachability in a single tool. On the one hand, it offers an alternative to simulation and analytical approaches, with the former often being time consuming and the latter not representing the system at the correct level of detail, and, on the other, conventional reachability and model checking. The specification language is a *probabilistic* temporal logic, capable of expressing temporal relationships between events and likelihood of events. A probabilistic model checker inputs a description of the model and a specification and checks if the specification is satisfied by the model; additionally, it can output the probability or expected cost (for each state) of it being satisfied.

CHAPTER 2

Discrete-Time Markov Chains

2.1. Introduction

In this chapter we investigate the probabilistic model checking of *discrete-time Markov chains* (DTMCs), the simplest of the four probabilistic models we consider. A DTMC is a family of random variables $\{X(k) \mid k = 0, 1, 2, \ldots\}$, where $X(k)$ are observations made at discrete time-steps. The values taken by $X(k)$ are called *states*. The set of all possible states, known as the *state space*, is discrete. We will restrict our attention to the case where the state space is finite. A DTMC satisfies the *Markov property* (sometimes known as *memoryless property*), meaning that its state at time-step k depends only on its state at the previous time-step $k-1$, that is, for any integer $k \geq 0$ and states $s_0, \ldots, s_k$:

$$P[X(k)=s_k \mid X(k-1)=s_{k-1}, \ldots, X(0)=s_0] = P[X(k)=s_k \mid X(k-1)=s_{k-1}].$$

Furthermore, we consider only *homogeneous* DTMCs, for which the transition probabilities are independent of the time at which they occur. This means that, to describe a DTMC, it is sufficient to give the probability that it will make a transition from one state to another. We define a DTMC as follows.

DEFINITION 2.1.1. A DTMC is a tuple $\mathcal{D} = (S, \bar{s}, \mathbf{P})$ where:

- S is a finite set of *states*;
- $\bar{s} \in S$ is the *initial state*;
- $\mathbf{P}\colon S \times S \to [0,1]$ is a *transition probability matrix*, such that:

$$\sum_{s' \in S} \mathbf{P}(s, s') = 1$$

  for all states $s \in S$.

Each element $\mathbf{P}(s, s')$ of the transition probability matrix gives the probability of making a transition from state s to state s', that is, for any $k \geq 0$:

$$\mathbf{P}(s, s') = P[X(k+1) = s' \mid X(k) = s].$$

Note that the probabilities on transitions emanating from a single state must sum to one. Terminating states, i.e. those from which the system cannot move to another state, can be modelled by adding a self-loop (a single transition going back to the same state with probability 1).

In order to be able to analyse DTMCs, we have to annotate them with additional information. The labelling function $L\colon S \to 2^{\mathrm{AP}}$ maps states to sets of atomic propositions from a set AP. We use these atomic propositions to label states with properties of interest. We also annotate a DTMC with a cost or reward structure, namely a function $\mathbf{C}\colon S \times S \to \mathbb{R}^{\geq 0}$ which maps pairs of states to non-negative real values. Intuitively, $\mathbf{C}(s, s')$ denotes the cost incurred when a transition from s to

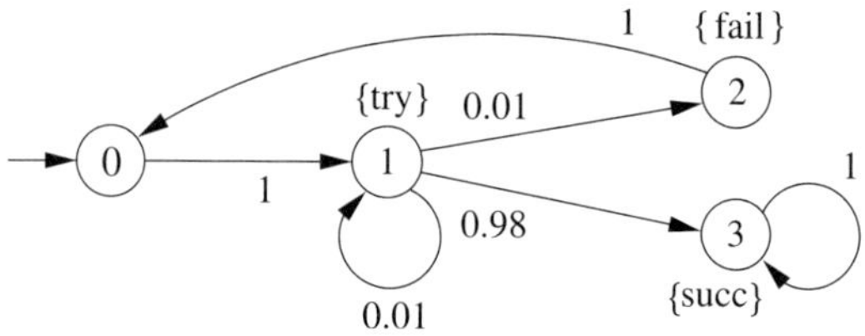

FIGURE 1. A four state DTMC.

s' is taken. A cost structure can be used to represent additional information about the system the DTMC represents, for example the power consumption, number of packets sent or the number of lost customers.

DEFINITION 2.1.2. A *labelled* DTMC is a tuple $(\mathcal{D}, L, \mathbf{C})$ where:

- $\mathcal{D} = (S, \bar{s}, \mathbf{P})$ is a DTMC;
- $L\colon S \to 2^{\mathrm{AP}}$ is a *labelling function*;
- $\mathbf{C}\colon S \times S \to \mathbb{R}^{\geq 0}$ is a *cost structure*.

EXAMPLE 2.1.3. Figure 1 shows a simple example of a DTMC. It has four states: $S = \{0, 1, 2, 3\}$. In our graphical notation, states are drawn as circles and transitions as arrows, labelled with their associated probabilities. The initial state is indicated by an additional incoming arrow. The transition probability matrix $\mathbf{P}$ is:

$$\mathbf{P} = \begin{pmatrix} 0 & 1 & 0 & 0 \\ 0 & 0.01 & 0.01 & 0.98 \\ 1 & 0 & 0 & 0 \\ 0 & 0 & 0 & 1 \end{pmatrix}.$$

The atomic propositions used to label states are taken from the set $\{\mathrm{try}, \mathrm{fail}, \mathrm{succ}\}$. Here, the DTMC models a simple process which tries to send a message. After one time-step, it enters a state from which, with probability 0.01 it waits another time-step, with probability 0.98 it successfully sends the message, and with probability 0.01 it tries but fails to send the message. In the latter case, the process restarts. The labelling function allows us to assign meaningful names to states of the DTMC:

$$L(1) = \{\mathrm{try}\}, \quad L(2) = \{\mathrm{fail}\}, \quad L(3) = \{\mathrm{succ}\}.$$

We also associate a cost structure (represented as a matrix $\mathbf{C}$) with the DTMC:

$$\mathbf{C} = \begin{pmatrix} 0 & 1 & 0 & 0 \\ 0 & 1 & 0 & 0 \\ 0 & 1 & 0 & 0 \\ 0 & 1 & 0 & 0 \end{pmatrix}.$$

This example assigns value one to all transitions that enter state 1 and zero to all others. This particular cost structure would be useful when we are interested in the number of times that state 1 is entered. For clarity, this information is omitted from the graphical representation in Fig. 1.

2.2. Paths and Probability Measures

An execution of the system which is being modelled by a DTMC is represented by a *path*. Formally, a path ω is a non-empty sequence of states $s_0 s_1 s_2 \ldots$ where $s_i \in S$ and $\mathbf{P}(s_i, s_{i+1}) > 0$ for all $i \geq 0$. A path can be either finite or infinite. We denote by $\omega(i)$ the ith state of a path ω, $|\omega|$ the length of ω (number of transitions) and for a finite path ω_{fin}, the last state by $\text{last}(\omega_{\text{fin}})$. We say that a finite path ω_{fin} of length n is a *prefix* of the infinite path ω if $\omega_{\text{fin}}(i) = \omega(i)$ for $0 \leq i \leq n$. The sets of all finite and infinite paths starting in state s are denoted $\text{Path}_s^{\text{fin}}$ and Path_s, respectively. Unless stated explicitly, we always deal with infinite paths.

In order to reason about the probabilistic behaviour of the DTMC, we need to be able to determine the probability that certain paths are taken. This is achieved by defining, for each state $s \in S$, a probability measure Prob_s over Path_s. Below, we give an outline of this construction. For further details, see [**79**]. The probability measure is induced by the transition probability matrix $\mathbf{P}$ as follows. First, for any finite path $\omega_{\text{fin}} \in \text{Path}_s^{\text{fin}}$, we define the probability $\mathbf{P}_s(\omega_{\text{fin}})$:

$$\mathbf{P}_s(\omega_{\text{fin}}) \stackrel{\text{def}}{=} \begin{cases} 1 & \text{if } n = 0 \\ \mathbf{P}(\omega(0), \omega(1)) \cdots \mathbf{P}(\omega(n-1), \omega(n)) & \text{otherwise} \end{cases}$$

where $n = |\omega_{\text{fin}}|$. Next, we define the *cylinder set* $C(\omega_{\text{fin}})$ as:

$$C(\omega_{\text{fin}}) \stackrel{\text{def}}{=} \{\omega \in \text{Path}_s \mid \omega_{\text{fin}} \text{ is a prefix of } \omega\}$$

that is, the set of all infinite paths with prefix ω_{fin}. Then, let Σ_s be the smallest σ-algebra (see Appendix A) on Path_s which contains all the sets $C(\omega_{\text{fin}})$ where ω_{fin} ranges over paths in $\text{Path}_s^{\text{fin}}$. We define the probability measure Prob_s on Σ_s as the unique measure such that:

$$\text{Prob}_s(C(\omega_{\text{fin}})) = \mathbf{P}_s(\omega_{\text{fin}}) \quad \text{for all } \omega_{\text{fin}} \in \text{Path}_s^{\text{fin}}.$$

We can now quantify the probability that a DTMC behaves in a specified fashion by identifying the set of paths which satisfy this specification and, assuming that it is measurable, using the associated measure Prob_s.

EXAMPLE 2.2.1. Consider again the DTMC in (2.1.3) (see Fig. 1). There are five distinct finite paths of length 3 starting in state 0. The probability measure of the cylinder sets associated with each of these is:

$$\begin{aligned}
\text{Prob}_0(C(0111)) &= 1 \cdot 0.01 \cdot 0.01 = 0.0001, \\
\text{Prob}_0(C(0112)) &= 1 \cdot 0.01 \cdot 0.01 = 0.0001, \\
\text{Prob}_0(C(0113)) &= 1 \cdot 0.01 \cdot 0.98 = 0.0098, \\
\text{Prob}_0(C(0120)) &= 1 \cdot 0.01 \cdot 1 = 0.01, \\
\text{Prob}_0(C(0133)) &= 1 \cdot 0.98 \cdot 1 = 0.98.
\end{aligned}$$

Using the probability measure Prob_s we can now define expectations over paths. In particular, for a given target set of states F, we can calculate the expected cost of reaching a state in F. More precisely, letting E_s be the usual expectation with respect to the measure Prob_s and for any path $\omega \in \text{Path}_s$:

$$\text{cost}(F)(\omega) \stackrel{\text{def}}{=} \begin{cases} \sum_{i=1}^{\min\{j \mid \omega(j) \in F\}} \mathbf{C}(\omega(i-1), \omega(i)) & \text{if } \exists j \in \mathbb{N} \,.\, \omega(j) \in F \\ \infty & \text{otherwise} \end{cases}$$

then $E_s(\text{cost}(F))$ equals the expected cost of reaching F. Note that we define the cost of a path which does not reach F to be ∞, even though the total cost of the path may not be infinite. Hence, the expected cost of reaching F from s is finite if and only if a state in F is reached from s with probability 1.

2.3. Probabilistic Computation Tree Logic (PCTL)

To write specifications of DTMCs, we use PCTL (Probabilistic Computation Tree Logic) [**62**], a probabilistic extension of the temporal logic CTL. PCTL is essentially the same as the logic PCTL of [**9**]. For this presentation, we have added an operator to reason about expected cost in DTMCs.

DEFINITION 2.3.1. The syntax of PCTL is as follows:

$$\phi ::= \text{true} \mid a \mid \neg\phi \mid \phi \wedge \phi \mid \mathcal{P}_{\bowtie p}[\psi] \mid \mathcal{E}_{\bowtie c}[\phi], \quad \psi ::= \phi \mid \phi \mid \mathcal{U}^{\leq k}\phi \mid \phi\mathcal{U}\phi$$

where a is an atomic proposition, $\bowtie \in \{\leq, <, \geq, >\}$, $p \in [0, 1]$, $c \in \mathbb{R}^{\geq 0}$ and $k \in \mathbb{N}$.

PCTL formulas are interpreted over the states of a DTMC. For the presentation of the syntax, above, we distinguish between state formulas ϕ and path formulas ψ, which are evaluated over states and paths, respectively. To specify a property of a DTMC, we always use a state formula: path formulas only occur as the parameter of the $\mathcal{P}_{\bowtie p}[\psi]$ operator. Intuitively, a state s satisfies $\mathcal{P}_{\bowtie p}[\psi]$ if the probability of taking a path from s satisfying ψ is in the interval specified by $\bowtie p$. For this, we use the probability measure over sets of infinite paths introduced in the previous section.

As path formulas we allow the X ("next"), $\mathcal{U}^{\leq k}$ ("bounded until") and $\mathcal{U}$ ("until") operators which are standard in temporal logic. Intuitively, ϕ is true if ϕ is satisfied in the next state; $\phi_1\mathcal{U}^{\leq k}\phi_2$ is true if ϕ_2 is satisfied within k time-steps and ϕ_1 is true up until that point; and $\phi_1\mathcal{U}\phi_2$ is true if ϕ_2 is satisfied at some point in the future and ϕ_1 is true up until then.

The $\mathcal{E}_{\bowtie c}[\phi]$ operator takes as a parameter a state formula ϕ. A state satisfies $\mathcal{E}_{\bowtie c}[\phi]$ if the expected cost, from s, to reach a state satisfying ϕ is in the interval specified by $\bowtie c$. We use the definitions of probability measure and expectation given in the previous section.

For a state s and PCTL formula ϕ, we write $s \models \phi$ to indicate that s satisfies ϕ. Alternatively, we say that "ϕ is satisfied in s", "ϕ holds in s", "ϕ is true in s" or that "s is a ϕ-state". Similarly, for a path ω satisfying path formula ψ, we write $\omega \models \psi$. The semantics of PCTL over DTMCs are as follows.

DEFINITION 2.3.2. Let $\mathcal{D} = (S, \bar{s}, \mathbf{P}, L, \mathbf{C})$ be a labelled DTMC. For any state $s \in S$, the satisfaction relation $\models$ is defined inductively by:

$$\begin{aligned}
s &\models \text{true} && \text{for all } s \in S,\\
s &\models a && \Leftrightarrow a \in L(s),\\
s &\models \neg\phi && \Leftrightarrow s \not\models \phi,\\
s &\models \phi_1 \wedge \phi_2 && \Leftrightarrow s \models \phi_1 \wedge s \models \phi_2,\\
s &\models \mathcal{P}_{\bowtie p}[\psi] && \Leftrightarrow p_s(\psi) \bowtie p,\\
s &\models \mathcal{E}_{\bowtie c}[\phi] && \Leftrightarrow e_s(\phi) \bowtie c
\end{aligned}$$

where:

$$p_s(\psi) \stackrel{\text{def}}{=} \text{Prob}_s(\{\omega \in \text{Path}_s \mid \omega \models \psi\}), \quad e_s(\phi) \stackrel{\text{def}}{=} E_s\big(\text{cost}(\{s' \in S \mid s' \models \phi\})\big)$$

and for any path $\omega \in \text{Path}_s$:

$$\begin{aligned} \omega \models \phi &\Leftrightarrow \omega(1) \models \phi. \\ \omega \models \phi_1 \mathcal{U}^{\leq k} \phi_2 &\Leftrightarrow \exists i \leq k \,.\, \big(\omega(i) \models \phi_2 \wedge \omega(j) \models \phi_1 \; \forall j < i\big), \\ \omega \models \phi_1 \mathcal{U} \phi_2 &\Leftrightarrow \exists k \geq 0 \,.\, \omega \models \phi_1 \mathcal{U}^{\leq k} \phi_2. \end{aligned}$$

Note that, for any PCTL path formula ψ, the set $\{\omega \in \text{Path}_s \mid \omega \models \psi\}$ is a *measurable* set of Path_s (see for example [**127**]), and hence Prob_s is well defined over this set.

From the basic syntax of PCTL, given above, we can derive a number of additional useful operators. Among these are the well known logical equivalences:

$$\begin{aligned} \text{false} &\equiv \neg\, \text{true}, \\ \phi_1 \vee \phi_2 &\equiv \neg(\neg\phi_1 \wedge \neg\phi_2), \\ \phi_1 \to \phi_2 &\equiv \neg\phi_1 \vee \phi_2. \end{aligned}$$

We also allow path formulas to contain the $\Diamond$ ("diamond" or "eventually") operator, which is common in temporal logic. Intuitively, $\Diamond\phi$ means that ϕ is eventually satisfied and its bounded variant $\Diamond^{\leq k}\phi$ means that ϕ is satisfied within k time units. These can be expressed in terms of the PCTL "until" and "bounded until" operators as follows:

$$\mathcal{P}_{\bowtie p}[\Diamond\phi] \equiv \mathcal{P}_{\bowtie p}[\text{true}\,\mathcal{U}\phi], \quad \mathcal{P}_{\bowtie p}[\Diamond^{\leq k}\phi] \equiv \mathcal{P}_{\bowtie p}[\text{true}\,\mathcal{U}^{\leq k}\phi].$$

Another common temporal logic operator is $\Box$ ("box" or "always"). Intuitively, if a path satisfies $\Box\phi$, then ϕ is true along every state of the path. Similarly, the bounded variant $\Box^{\leq k}\phi$ means that ϕ is true in the first k steps of the path. In theory, one can express $\Box$ in terms of $\Diamond$ as follows:

$$\Box\phi \equiv \neg\Diamond\neg\phi, \quad \Box^{\leq k}\phi \equiv \neg\Diamond^{\leq k}\neg\phi.$$

However, the syntax of PCTL does not allow negation of path formulas. Observing, though, that for a state s and path formula ψ, $p_s(\neg\psi) = 1 - p_s(\psi)$, we can show for example that:

$$\begin{aligned} \mathcal{P}_{\geq p}[\Box\phi] &\Leftrightarrow p_s(\Box\phi) \geq p \\ &\Leftrightarrow 1 - p_s(\Diamond\neg\phi) \geq p \\ &\Leftrightarrow p_s(\Diamond\neg\phi) \leq 1 - p \\ &\Leftrightarrow \mathcal{P}_{\leq 1-p}[\Diamond\neg\phi]. \end{aligned}$$

In fact, we have the following equivalences:

$$\mathcal{P}_{\bowtie p}[\Box\phi] \equiv \mathcal{P}_{\overline{\bowtie} 1-p}[\Diamond\neg\phi], \quad \mathcal{P}_{\bowtie p}[\Box^{\leq k}\phi] \equiv \mathcal{P}_{\overline{\bowtie} 1-p}[\Diamond^{\leq k}\neg\phi]$$

where $\overline{\leq} \equiv \geq$, $\overline{<} \equiv >$, $\overline{\geq} \equiv \leq$ and $\overline{>} \equiv <$.

The $\mathcal{P}_{\bowtie p}[\psi]$ operator of PCTL can be seen as the probabilistic analogue of the temporal operators of CTL. For example, the PCTL formula $\mathcal{P}_{\bowtie p}[\Diamond\phi]$, which states that the probability of reaching a ϕ-state is $\bowtie p$, is closely related to the CTL formulas $\forall\Diamond\phi$ and $\exists\Diamond\phi$ (sometimes written AFϕ and EFϕ), which assert that

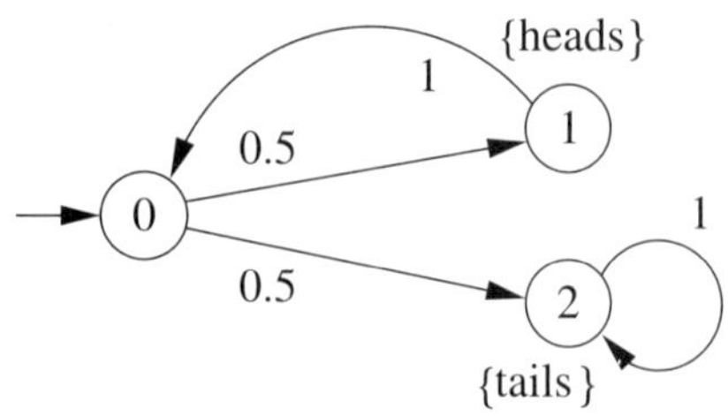

FIGURE 2. Example demonstrating difference between $\mathcal{P}_{\geq 1}[\Diamond\phi]$ and $\forall\Diamond\phi$.

all paths or *at least one* path reach a ϕ-state, respectively. In fact, we have the following equivalence:

$$\exists\Diamond\phi \equiv \mathcal{P}_{>0}[\Diamond\phi]$$

since the probability can only be greater than zero if there exists a finite path leading to a ϕ-state. Conversely, the following are not equivalent:

$$\forall\Diamond\phi \not\equiv \mathcal{P}_{\geq 1}[\Diamond\phi].$$

To see an example of this, consider the DTMC in Fig. 2. This models a process which repeatedly tosses a fair coin until the result is "tails". State 0 satisfies $\mathcal{P}_{\geq 1}[\Diamond\text{tails}]$ since the probability of reaching state 2 is one. There is, though, an (infinite) path 0101... which never reaches state 2. Hence, $\forall\Diamond$tails is not satisfied in state 0.

EXAMPLE 2.3.3. Below are some typical examples of PCTL formulas:

- $\mathcal{P}_{<0.1}[\Diamond^{\leq 10}\text{error}]$—the probability that the system reaches an error state within 10 time-steps is strictly less than 0.1;
- $\mathcal{P}_{\geq 0.9}[\neg\text{down}\,\mathcal{U}\,\text{served}]$—the probability the system does not go down until after the request has been served is greater than or equal to 0.9;
- $\mathcal{E}_{<10}[\text{finished}]$—the expected number of lost customers before completion is less than 10;
- $\mathcal{E}_{\geq 14}[\text{done}]$—the expected number of correctly delivered messages is at least 14.

A perceived weakness of PCTL is that it is not possible to determine the actual probability with which a certain path formula is satisfied, only whether or not the probability meets a particular bound. In fact, this restriction is in place purely to ensure that each PCTL formula evaluates to a Boolean. In practice, this constraint can be relaxed: if the outermost operator of a PCTL formula is either the $\mathcal{P}_{\bowtie p}$ or $\mathcal{E}_{\bowtie c}$ operator, we can omit the bound $\bowtie p$ or $\bowtie c$ and simply compute the probability or expected cost instead. Since the algorithm for PCTL model checking proceeds by computing the actual probability and then comparing it to the bound, no additional work is needed. It is also often useful to study a range of such values by varying one or more parameters, either of the model or of the property.

2.4. PCTL Model Checking

We now summarise the model checking algorithm for PCTL over DTMCs, which was first presented in [**41, 43, 62**]. The inputs to the algorithm are a labelled DTMC $(S, \bar{s}, \mathbf{P}, L, \mathbf{C})$ and a PCTL formula ϕ. The output is the set $\text{Sat}(\phi) = \{s \in$

$S \mid s \models \phi\}$, i.e. the set containing all the states of the model which satisfy ϕ. In a typical scenario, we may only be interested in whether the initial state $\bar{s}$ of the DTMC satisfies ϕ. However, the algorithm works by checking whether each state in S satisfies it.

The overall structure of the algorithm is identical to the model checking algorithm for CTL [**38**], the non-probabilistic temporal logic of which PCTL is an extension. We first construct the parse tree of the formula ϕ. Each node of this tree is labelled with a subformula of ϕ, the root node is labelled with ϕ itself and leaves of the tree will be labelled with either true or an atomic proposition a. Working upwards towards the root of the tree, we recursively compute the set of states satisfying each subformula. By the end, we have determined whether each state in the model satisfies ϕ. The algorithm for PCTL operators (including the derived ones) can be summarised as follows:

$$\begin{aligned}
\mathrm{Sat}(\mathrm{true}) &= S,\\
\mathrm{Sat}(\mathrm{false}) &= \emptyset,\\
\mathrm{Sat}(a) &= \{s \mid a \in L(s)\},\\
\mathrm{Sat}(\neg\phi) &= S \backslash \mathrm{Sat}(\phi),\\
\mathrm{Sat}(\phi_1 \wedge \phi_2) &= \mathrm{Sat}(\phi_1) \cap \mathrm{Sat}(\phi_2),\\
\mathrm{Sat}(\phi_1 \vee \phi_2) &= \mathrm{Sat}(\phi_1) \cup \mathrm{Sat}(\phi_2),\\
\mathrm{Sat}(\phi_1 \Rightarrow \phi_2) &= \big(S \backslash \mathrm{Sat}(\phi_1)\big) \cup \mathrm{Sat}(\phi_2),\\
\mathrm{Sat}(\mathcal{P}_{\bowtie p}[\psi]) &= \{s \in S \mid p_s(\psi) \bowtie p\},\\
\mathrm{Sat}(\mathcal{E}_{\bowtie c}[\phi]) &= \{s \in S \mid e_s(\phi) \bowtie c\}.
\end{aligned}$$

Model checking for the majority of these operators is trivial to implement and is, in fact, the same as for the non-probabilistic logic CTL. The exceptions are the $\mathcal{P}_{\bowtie p}[\psi]$ and $\mathcal{E}_{\bowtie c}[\phi]$ operators. For these, we must calculate, for all states s of the DTMC, either the probability $p_s(\psi)$ or expected cost $e_s(\phi)$ and then compare these values to the bound in the formula. In the following sections, we describe how to compute these values for the four cases: $\mathcal{P}_{\bowtie p}[\phi]$, $\mathcal{P}_{\bowtie p}[\phi_1 \mathcal{U}^{\leq k} \phi_2]$, $\mathcal{P}_{\bowtie p}[\phi_1 \mathcal{U} \phi_2]$ and $\mathcal{E}_{\bowtie c}[\phi]$. Because of the recursive nature of the PCTL model checking algorithm, we can assume that the relevant sets, $\mathrm{Sat}(\phi)$, $\mathrm{Sat}(\phi_1)$ or $\mathrm{Sat}(\phi_2)$, are already known.

2.4.1. The $\mathcal{P}_{\bowtie p}[\phi]$ operator. In this case, we need to compute the probability $p_s(\phi)$ for each state $s \in S$. This only requires the probabilities of the immediate transitions from s:

$$p_s(\phi) = \sum_{s' \in \mathrm{Sat}(\phi)} \mathbf{P}(s, s').$$

We determine the vector $\underline{p(\phi)}$ of probabilities for all states as follows. Assuming that we have a state-indexed column vector $\underline{\phi}$ with $\underline{\phi}(s) = 1$ if $s \models \phi$ and 0 otherwise, $\underline{p(\phi)}$ is computed using a single matrix-vector multiplication:

$$\underline{p(\phi)} = \mathbf{P} \cdot \underline{\phi}.$$

EXAMPLE 2.4.1. Consider the DTMC from Fig. 1 and the PCTL formula $\mathcal{P}_{\geq 0.9}[X(\neg \mathrm{try} \vee \mathrm{succ})]$. Proceeding recursively from the innermost subformulas, we compute:

$$\begin{aligned} \text{Sat}(\text{try}) &= \{1\}, \\ \text{Sat}(\neg\text{try}) &= S \backslash \text{Sat}(\text{try}) = \{0,2,3\}, \\ \text{Sat}(\text{succ}) &= \{3\}, \\ \text{Sat}(\neg\text{try} \vee \text{succ}) &= \text{Sat}(\neg\text{try}) \cup \text{Sat}(\text{succ}) = \{0,2,3\}. \end{aligned}$$

This leaves the outermost X operator. If we take $\mathbf{P}$ to be as in Fig. 1 and define $\underline{\phi} = [1,0,1,1]$, then:

$$\underline{p(\phi)} = \mathbf{P} \cdot \underline{\phi} = [0, 0.99, 1, 1].$$

Hence, $\text{Sat}\big(\mathcal{P}_{\geq 0.9}[X(\neg\text{try} \vee \text{succ})]\big) = \{1,2,3\}$.

2.4.2. The $\mathcal{P}_{\bowtie p}[\phi_1 \mathcal{U}^{\leq k} \phi_2]$ operator. In this case, we compute the probabilities $p_s(\phi_1 \mathcal{U}^{\leq k} \phi_2)$ for all $s \in S$. We first split the set of states S into three disjoint subsets, S^{no}, S^{yes} and $S^{?}$ where:

$$S^{\text{no}} = S \backslash \big(\text{Sat}(\phi_1) \cup \text{Sat}(\phi_2)\big), \quad S^{\text{yes}} = \text{Sat}(\phi_2), \quad S^{?} = S \backslash (S^{\text{no}} \cup S^{\text{yes}}).$$

The set S^{no} contains the states s for which $p_s(\phi_1 \mathcal{U}^{\leq k} \phi_2)$ is trivially 0. This is because neither ϕ_1 nor ϕ_2 is satisfied in s, and hence no path starting in s can possibly satisfy the path formula $\phi_1 \mathcal{U}^{\leq k} \phi_2$. In similar fashion, the set S^{yes} contains the states s for which $p_s(\phi_1 \mathcal{U}^{\leq k} \phi_2)$ is trivially 1 since all paths leaving s clearly satisfy $\phi_1 \mathcal{U}^{\leq k} \phi_2$. The set $S^{?}$ contains all remaining states. For these states, we can define the probability $p_s(\phi_1 \mathcal{U}^{\leq k} \phi_2)$ recursively as follows:

$$p_s(\phi_1 \mathcal{U}^{\leq k} \phi_2) = \begin{cases} 0 & \text{if } k = 0 \\ \sum_{s' \in S} \mathbf{P}(s,s') \cdot p_{s'}(\phi_1 \; \mathcal{U}^{\leq k-1} \; \phi_2) & \text{if } k \geq 1. \end{cases}$$

If we define the matrix $\mathbf{P}'$ as:

$$\mathbf{P}'(s,s') = \begin{cases} \mathbf{P}(s,s') & \text{if } s \in S^{?} \\ 1 & \text{if } s \in S^{\text{yes}} \quad \text{and} \quad s = s' \\ 0 & \text{otherwise} \end{cases}$$

the computation required for all states can be expressed as follows. For $k = 0$, $\underline{p(\phi_1 \mathcal{U}^{\leq 0} \phi_2)}(s) = 1$ if $s \in S^{\text{yes}}$ and 0 otherwise. For $k > 0$:

$$\underline{p(\phi_1 \mathcal{U}^{\leq k} \phi_2)} = \mathbf{P}' \cdot \underline{p(\phi_1 \mathcal{U}^{\leq k-1} \phi_2)}.$$

In total, this requires k matrix-vector multiplications. An alternative is to precompute the matrix power $(\mathbf{P}')^k$ and then perform a single matrix vector multiplication. In theory, this could be more efficient since the matrix power could be computed by repeatedly squaring $\mathbf{P}$, requiring approximately $\log_2 k$, as opposed to k, multiplications. In practice, however, the matrix $\mathbf{P}'$ is large and sparse. Such an approach can dramatically increase the number of non-zero matrix elements which has serious implications for both time and memory usage.

EXAMPLE 2.4.2. Let us consider the DTMC in Fig. 1 and the PCTL formula $\mathcal{P}_{>0.98}[\Diamond^{\leq 2}\text{succ}]$. This is equivalent to the formula $\mathcal{P}_{>0.98}[\text{true } U^{\leq 2}\text{succ}]$. We have:

$$\begin{aligned} \text{Sat}(\text{true}) &= \{0,1,2,3\}, \\ \text{Sat}(\text{succ}) &= \{3\}, \\ S^{\text{no}} &= S \backslash \big(\text{Sat}(\text{true}) \cup \text{Sat}(\text{succ})\big) = \emptyset, \\ S^{\text{yes}} &= \text{Sat}(\text{succ}) = \{3\}. \end{aligned}$$

```
PROB0(Sat(φ1), Sat(φ2))
R := Sat(φ2)
done := false
while (done = false)
        R' := R ∪ {s ∈ Sat(φ1) | ∃s' ∈ R . P(s, s') > 0}
        if (R' = R) then done := true
        R := R'
endwhile
return S\R
```

ALGORITHM 1. The PROB0 algorithm.

The matrix $\mathbf{P}'$, as defined above, is in this case identical to $\mathbf{P}$. We compute:

$$\begin{aligned}
\underline{p}(\phi_1\mathcal{U}^{\leq 0}\phi_2) &= [0, 0, 0, 1], \\
\underline{p}(\phi_1\mathcal{U}^{\leq 1}\phi_2) &= \mathbf{P}' \cdot [0, 0, 0, 1] = [0, 0.98, 0, 1], \\
\underline{p}(\phi_1\mathcal{U}^{\leq 2}\phi_2) &= \mathbf{P}' \cdot [0, 0.98, 0, 1] = [0.98, 0.9898, 0, 1].
\end{aligned}$$

Hence, $\text{Sat}(\mathcal{P}_{>0.98}[\Diamond^{<2}\text{succ}]) = \{1, 3\}$.

2.4.3. The $\mathcal{P}_{\bowtie p}[\phi_1\mathcal{U}\phi_2]$ operator. Here, we compute $p_s(\phi_1\mathcal{U}\phi_2)$ for all $s \in S$. As with the "bounded until" operator, we can identify the sets for which $p_s(\phi_1\mathcal{U}\phi_2)$ is trivially 0 or 1. In this case, we actually extend these sets to contain *all* states for which $p_s(\phi_1\mathcal{U}\phi_2)$ is exactly 0 or 1. As above, we denote these sets S^{no} and S^{yes}. They can be determined with the algorithms PROB0 and PROB1, respectively, which are described in Algorithms 1, 2:

$$\begin{aligned}
S^{\text{no}} &= \text{PROB0}\big(\text{Sat}(\phi_1), \text{Sat}(\phi_2)\big), \\
S^{\text{yes}} &= \text{PROB1}\big(\text{Sat}(\phi_1), \text{Sat}(\phi_2), S^{\text{no}}\big), \\
S^{?} &= S\backslash(S^{\text{no}} \cup S^{\text{yes}}).
\end{aligned}$$

PROB0 computes all the states from which it is possible, with *non-zero* probability, to reach a state satisfying ϕ_2 without leaving states satisfying ϕ_1. It then subtracts these from S to determine the states which have a *zero* probability. PROB1 computes the set S^{yes} in a similar fashion. It first determines the set of states for which the probability is *less than* 1. These are the states from which there is a non-zero probability of reaching a state in S^{no}, passing only through states satisfying ϕ_1 but not ϕ_2. It then subtracts this set from S to produce S^{yes}. Note that both algorithms are based on the computation of a fixpoint operator and will hence require at most $|S|$ iterations.

The probabilities $p_s(\phi_1\mathcal{U}\phi_2)$ for all states can now be computed by solving the linear equation system in variables $\{x_s \mid s \in S\}$:

$$x_s = \begin{cases} 0 & \text{if } s \in S^{\text{no}} \\ 1 & \text{if } s \in S^{\text{yes}} \\ \sum_{s' \in S} \mathbf{P}(s, s') \cdot x_{s'} & \text{if } s \in S^{?} \end{cases}$$

and then letting $p_s(\phi_1\mathcal{U}\phi_2) = x_s$. To rewrite this in the traditional $\mathbf{A} \cdot \underline{x} = \underline{b}$ form, we let $\mathbf{A} = \mathbf{I} - \mathbf{P}'$ where $\mathbf{I}$ is the identity matrix and $\mathbf{P}'$ is given by:

$$\mathbf{P}'(s, s') = \begin{cases} \mathbf{P}(s, s') & \text{if } s \in S^{?} \\ 0 & \text{otherwise} \end{cases}$$

PROB1$\big(\text{Sat}(\phi_1), \text{Sat}(\phi_2), S^{\text{no}}\big)$
$R := S^{\text{no}}$
done := **false**
while (done = **false**)
$R' := R \cup \{s \in (\text{Sat}(\phi_1) \setminus \text{Sat}(\phi_2)) \mid \exists s' \in R \,.\, \mathbf{P}(s, s') > 0\}$
if $(R' = R)$ **then** done := **true**
$R := R'$
endwhile
return $S \setminus R$

ALGORITHM 2. The PROB1 algorithm.

and $\underline{b}$ is a state-indexed column vector with $\underline{b}(s)$ equal to 1 if $s \in S^{\text{yes}}$ and 0 otherwise. Since the probabilities for states in S^{yes} and S^{no} are known, it is possible to instead solve the linear equation system in variables $\{x_s \mid s \in S^?\}$:

$$x_s = \sum_{s' \in S^?} \mathbf{P}(s, s') \cdot x_{s'} + \sum_{s' \in S^{\text{yes}}} \mathbf{P}(s, s')$$

reducing the number of unknowns from $|S|$ to $|S^?|$. In practice, the former system may be preferred in order to avoid modification of the matrix $\mathbf{P}$.

The linear equation system can be solved by any standard approach. These include direct methods, such as Gaussian elimination and L/U decomposition, or iterative methods, such as Jacobi and Gauss–Seidel. Since we are aiming to solve these problems for very large models, we will concentrate on iterative methods which are discussed further in Appendix B.

The algorithms PROB0 and PROB1 form the first part of the calculation of $p_s(\phi_1 \mathcal{U} \phi_2)$. For this reason, we refer to them as *precomputation algorithms*. For *qualitative* PCTL properties (i.e. where the bound p in the $\mathcal{P}_{\bowtie p}[\phi_1 \mathcal{U} \phi_2]$ operator is either 0 or 1) or for cases where $p_s(\phi_1 \mathcal{U} \phi_2)$ happens to be either 0 or 1 for all states (i.e. $S^{\text{no}} \cup S^{\text{yes}} = S$), it suffices to use these precomputation algorithms alone. For *quantitative* properties with an arbitrary bound $\bowtie p$, numerical computation is also usually required. The precomputation algorithms are still valuable in this case. Firstly, they can reduce the number of states for which numerical computation is required. Secondly, they determine the exact probability for the states in S^{no} and S^{yes}, whereas numerical computation typically only computes an approximation and may also be subject to round-off errors.

Finally we note that, if desired, the PROB1 algorithm can be omitted and S^{yes} set to be equal to $\text{Sat}(\phi_2)$. The set S^{no}, however, must be computed as described above to ensure that the linear equation system has a unique solution.

EXAMPLE 2.4.3. Consider again the DTMC in Fig. 1 and the PCTL formula $\mathcal{P}_{>0.99}[\text{try}\,\mathcal{U}\,\text{succ}]$. We have $\text{Sat}(\text{try}) = \{1\}$ and $\text{Sat}(\text{succ}) = \{3\}$. PROB0 determines in two iterations that $S^{\text{no}} = \{0, 2\}$. PROB1 determines that $S^{\text{yes}} = \{3\}$. In this example, these are trivial anyway since neither try nor succ is satisfied in states 0 and 2 and succ is only true in state 3. The resulting system of linear equations is as follows:

$$x_0 = 0, \quad x_1 = 0.01x_1 + 0.01x_2 + 0.98x_3, \quad x_2 = 0, \quad x_3 = 1.$$

This yields the solution $(x_0, x_1, x_2, x_3) = \big(0, \frac{98}{99}, 0, 1\big)$. Letting $p_s(\text{try}\,\mathcal{U}\,\text{succ}) = x_s$, we see that the formula $\mathcal{P}_{>0.99}[\text{try}\,\mathcal{U}\,\text{succ}]$ is satisfied only in state 3.

2.4.4. The $\mathcal{E}_{\bowtie c}[\phi]$ operator. In this case, we compute the expected cost $e_s(\phi)$ for all states $s \in S$. Similarly to the "until" operator, we can identify the sets of states for which $e_s(\phi)$ is either 0 or ∞. We denote these S^0 and S^∞. The set S^0 is simply those states that satisfy ϕ while the set S^∞ is those states from which the probability of reaching a ϕ state is less than 1. We compute the latter using the precomputation algorithms PROB1 and PROB0, described in the previous section:

$$\begin{aligned} S^0 &= \mathrm{Sat}(\phi), \\ S^\infty &= S \backslash \textsc{Prob1}\big(S, \mathrm{Sat}(\phi), \textsc{Prob0}(S, \mathrm{Sat}(\phi))\big), \\ S^? &= S \backslash (S^0 \cup S^\infty). \end{aligned}$$

We still need to compute $e_s(\phi)$ for the remaining states $S^?$. This can be done by solving the linear equation system in variables $\{x_s \mid s \in S^?\}$:

$$x_s = \sum_{s' \in S^?} \mathbf{P}(s, s') \cdot \big(\mathbf{C}(s, s') + x_{s'}\big)$$

and then letting $e_s(\phi) = x_s$ for $s \in S^?$. As for "until" formulas, this can be solved using any standard direct or iterative method.

EXAMPLE 2.4.4. Consider (2.1.3) which described the DTMC in Fig. 1. Recall that we assigned a cost of 1 to each transition that entered state 1 and a cost of 0 to all others. The PCTL formula $\mathcal{E}_{<2}[\mathrm{succ}]$ then asserts that the expected number of times state 1 is entered before reaching a state satisfying succ is less than 2. Following the procedure outlined above, we compute:

$$\begin{aligned} S^0 &= \mathrm{Sat}(\mathrm{succ}) - \{3\}, \\ \textsc{Prob0}\big(S, \mathrm{Sat}(\phi)\big) &= \emptyset, \\ \textsc{Prob1}\big(S, \mathrm{Sat}(\phi), \emptyset\big) &= \{0, 1, 2, 3\}, \\ S^\infty &= S \backslash \{0, 1, 2, 3\} = \emptyset, \\ S^? &= S \backslash \{3\} = \{0, 1, 2\}. \end{aligned}$$

This leads to the linear equation system:

$$x_0 = 1 \cdot (1 + x_1), \quad x_1 = 0.01 \cdot (1 + x_1) + 0.01 \cdot (0 + x_2), \quad x_2 = 1 \cdot (0 + x_0)$$

which yields the solution $(x_0, x_1, x_2) = \big(\frac{198}{98}, \frac{100}{98}, \frac{198}{98}\big)$. Hence, the vector of all expected costs is $\underline{e(\mathrm{succ})} = \big(\frac{198}{98}, \frac{100}{98}, \frac{198}{98}, 0\big)$ and $\mathrm{Sat}(\mathcal{E}_{<2}[\mathrm{succ}]) = \{1, 3\}$.

2.4.5. Complexity of PCTL model checking. The overall time complexity for model checking a PCTL formula ϕ against a DTMC $\mathcal{D} = (S, \bar{s}, \mathbf{P})$ is linear in $|\phi|$ and polynomial in $|S|$. The size $|\phi|$ of a formula ϕ is, as defined in [**62**], equal to the number of logical connectives and temporal operators in the formula plus the sum of the sizes of the temporal operators. Because of the recursive nature of the algorithm, we perform model checking for each of the $|\phi|$ operators and each one is at worst polynomial in $|S|$. The most expensive of these are the "until" and "expected cost" operators, for which a system of linear equations of size $|S|$ must be solved. This can be done with Gaussian elimination, the complexity of which is cubic in the size of the system. Strictly speaking, the complexity of model checking is also linear in $k_{\max}$, the maximum value of k found in formulas of type $\mathcal{P}_{\bowtie p}[\phi_1 \mathcal{U}^{\leq k} \phi_2]$. In practice, k is usually much smaller than $|S|$.

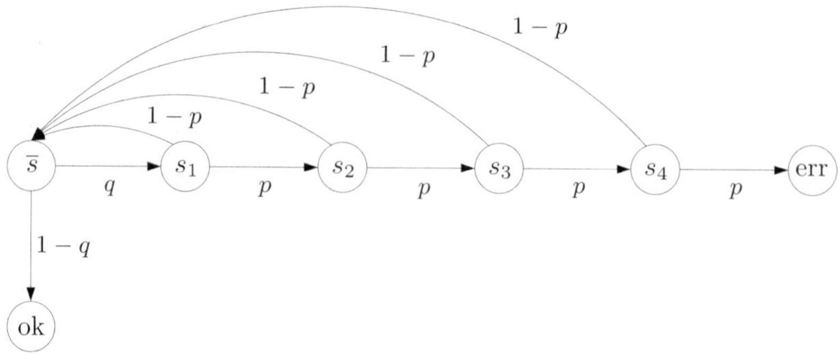

FIGURE 3. DTMC model of the IPv4 protocol ($N = 4$).

2.5. Case Study: The IPv4 ZeroConf Protocol

In this section, we demonstrate the utility of PCTL model checking for DTMCs by describing a real case study: the dynamic configuration protocol for IPv4 link-local addresses of [**37**]. This protocol offers a distributed, "plug-and-play" solution for the problem of IP address configuration in a network, where address allocation is managed by the individual devices which connect to the network. The protocol is designed for small, ad-hoc networks of appliances such as VCRs, DVD-players, laptops and hand-held devices. Below, we outline the protocol, present an abstract DTMC model of the protocol taken from [**26**] and show how PCTL can be used for its analysis.

The basic structure of the protocol is as follows. Upon being connected to the network, a device (henceforth called a *host*) is first required to randomly select an IP address from a pool of 65024 available addresses. The Internet Assigned Number Authority (IANA) has allocated the IP addresses 169.254.1.0–169.254.254.255 for exactly this purpose. The host then sends four ARP (Address Resolution Protocol) packets, called *probes*, to all of the other hosts in the network. These probes, containing the IP address selected by the host, constitute requests to use the address and are sent at two second intervals. If another host has already started using this IP address, it must respond with an ARP reply packet, asserting its claim to the address. In this case, the original host will restart the protocol, selecting another address and sending new probes.

We study the behaviour of a single host trying to configure an IP address in a network where M hosts are already connected. Since there are 65024 addresses to choose from, when the host randomly picks an IP address the probability that the address is already in use is $q = M/65024$. We suppose that the host sends N probes and that, when a host sends a probe containing an IP address which is already in use, with probability p there is no reply. This probability represents the chance that either the probe gets lost, the replying host is busy or the reply gets lost.

We now explain the DTMC model of the protocol due to Bohnenkamp et al. [**26**]. The DTMC has $N+3$ states: $\{\bar{s}, s_1, \ldots, s_N, \mathrm{ok}, \mathrm{err}\}$. In the initial state $\bar{s}$ the host randomly picks an IP address; in states $\{s_i \mid 1 \leq i \leq N\}$ the host is sending its ith probe; in state ok the host has successfully selected a fresh IP address; and in state err it has chosen an address already in use by another host.

In state $\bar{s}$, there are two possibilities: with probability q the randomly chosen IP address is already in use and the host moves to state s_1; and with probability $1-q$ the address is fresh and the host moves to state ok. In state s_i, for $1 \leq i < N$, the host sends a probe regarding an IP address already in use. With probability p the host will not get a reply and will send another probe, moving to state s_{i+1}; with probability $1-p$ the host will get a reply and reconfigure its IP address, returning to state $\bar{s}$. State s_N is similar except that, if the host does not get a reply to its probe, then it will start using an IP address which is already in use by another host, moving to state err. The DTMC for the case $N = 4$ is shown in Fig. 3. The probability transition matrix of the DTMC in the general case has the following form:

$$\mathbf{P} = \begin{pmatrix} 0 & q & 0 & 0 & \cdots & 0 & 1-q & 0 \\ 1-p & 0 & p & 0 & \cdots & 0 & 0 & 0 \\ 1-p & 0 & 0 & p & \cdots & 0 & 0 & 0 \\ \vdots & \vdots & \vdots & \vdots & \ddots & \vdots & \vdots & \vdots \\ 1-p & 0 & 0 & 0 & \cdots & p & 0 & 0 \\ 1-p & 0 & 0 & 0 & \cdots & 0 & 0 & p \\ 0 & 0 & 0 & 0 & \cdots & 0 & 1 & 0 \\ 0 & 0 & 0 & 0 & \cdots & 0 & 0 & 1 \end{pmatrix}.$$

We associate a cost structure with the DTMC which corresponds to the time taken to choose an IP address plus an additional large cost in the case that the host starts using an address which is being used by another host. When a host sends out probes, it sends them at 2 second intervals. Hence, our cost structure is defined as follows. Transitions from states $s_1, \ldots, s_{N-1}$ have cost 2, since a single probe is sent out in each of these states. Similarly, the transition from state s_N to $\bar{s}$ has cost 2. The transition from s_N to err is assigned a large cost R since, once this transition has been performed, the host starts using an IP address which is already in use by another host. The choice of a value for R will depend on how damaging it is for two hosts to use the same IP address which, in turn, depends on the network and the nature of the devices in it. Finally, for the transition between states $\bar{s}$ and ok we assign a cost of $2N$. In this state, the host will send out N probes at 2 second intervals. Since the IP address is in this case fresh, we can safely assume that there will be no reply to the probes. Any other transitions in the DTMC have zero cost. The cost structure of the DTMC, in matrix form, is thus:

$$\mathbf{c} = \begin{pmatrix} 0 & 0 & 0 & 0 & \cdots & 0 & 2N & 0 \\ 2 & 0 & 2 & 0 & \cdots & 0 & 0 & 0 \\ 2 & 0 & 0 & 2 & \cdots & 0 & 0 & 0 \\ \vdots & \vdots & \vdots & \vdots & \ddots & \vdots & \vdots & \vdots \\ 2 & 0 & 0 & 0 & \cdots & 2 & 0 & 0 \\ 2 & 0 & 0 & 0 & \cdots & 0 & 0 & R \\ 0 & 0 & 0 & 0 & \cdots & 0 & 0 & 0 \\ 0 & 0 & 0 & 0 & \cdots & 0 & 0 & 0 \end{pmatrix}.$$

Letting a_s represent the atomic proposition which is true only in state s (i.e. $L(a_s) = \{s\}$), the properties we consider are the following:

- $\mathcal{P}_{\bowtie p}[\Diamond a_{\mathrm{err}}]$—the probability of using an IP address already in use is $\bowtie p$;
- $\mathcal{E}_{\bowtie r}[a_{\mathrm{ok}} \vee a_{\mathrm{err}}]$—the expected cost of choosing an IP address is $\bowtie r$.

TABLE 1. Probability of using an address already in use.

Number of probes sent (N)	Probability of message loss (p)			
	0	0.001	0.01	0.01
1	0	1.56e-5	1.56e-4	0.001560
2	0	1.56e-8	1.56e-6	1.56e-4
3	0	1.56e-11	1.56e-8	1.56e-5
4	0	1.56e-14	1.56e-10	1.56e-6
5	0	1.56e-17	1.56e-12	1.56e-7
6	0	1.56e-20	1.56e-14	1.56e-8

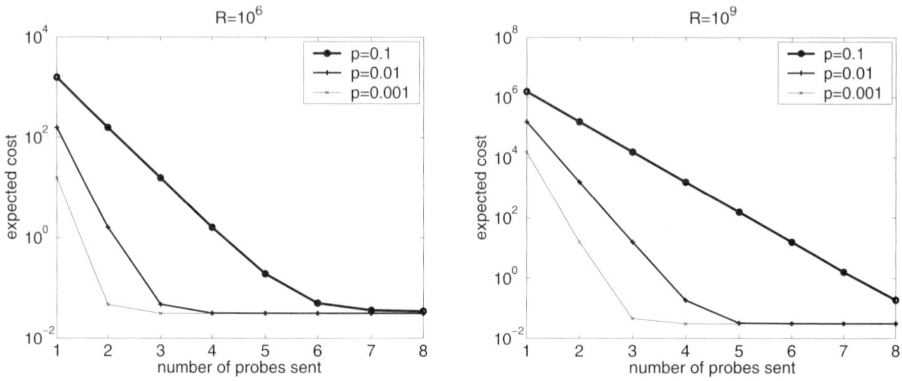

FIGURE 4. Expected cost results for the IPv4 ZeroConf protocol.

We investigate these properties while varying the model parameters N (number of probes sent) and p (probability of message loss).

In Table 1 we report on the probabilities obtained for the initial state when checking the property $\mathcal{P}_{\bowtie p}[\Diamond a_{\mathrm{err}}]$. The results obtained are as we would expect. Firstly, increasing the number of probes sent decreases the probability of the host using an IP address which is in use by another host. Secondly, as the probability of message loss decreases, the probability of using an IP address which is in use by another host also decreases.

In Fig. 4, we present the expected costs obtained when checking the property $\mathcal{E}_{\bowtie r}[a_{\mathrm{ok}} \vee a_{\mathrm{err}}]$ for the cases $R = 10^6$ and $R = 10^9$. These results are similar to those of [**26**]: as the probability of message loss increases, one must increase the number of probes sent out in order to reduce the expected cost; however, if too many probes are sent then the expected cost may start to increase. The rationale is that, although increasing the number of probes sent decreases the probability of the host using an IP address which is already in use (i.e. decreases the chance of incurring the additional cost), it increases the expected time to choose an IP address (because sending more probes takes more time).

We now explain how it is possible to, by extending the model, analyse several properties which are not expressible in PCTL. More precisely, we show how to calculate the probability of a host configuring a fresh IP address within some time bound T. The method essentially relies on an encoding of the elapsed time in the

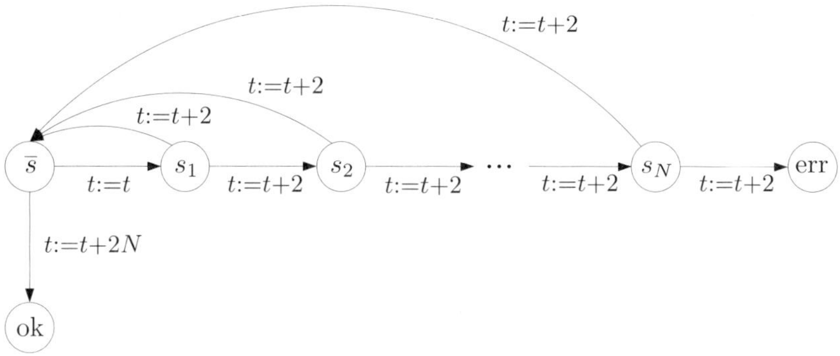

FIGURE 5. Time elapsing in the IPv4 protocol.

states of the model. A state of the extended model is a pair (s, t), where s is a state of the original DTMC and t is the elapsed time. Consider the time elapsed when each transition of the original model is performed, then, since the time between sending probes is two seconds, we enhance the model as in Fig. 5, where t increases as the transitions of the original model are taken. For example, from state $(\bar{s}, t)$ the system moves to state (s_1, t) with probability q and to state $(\text{ok}, t+2N)$ with probability $1-q$. To keep the model finite (and since we are only interested in whether the elapsed time exceeds T), we place an upper bound of $T+1$ on t. Thus, a transition of the form $t := t+k$ can actually be replaced with $t := \min(t+k, T+1)$.

In this extended model, the formula $\mathcal{P}_{\bowtie p}[\Diamond(a_{\text{ok}} \wedge t \leq T)]$ is true if and only if, in the original model, the probability of using a fresh IP address within time T is $\bowtie p$. In Fig. 6, we present the probabilities obtained when checking this property for the cases $T = 10$ and $T = 30$. The results demonstrate that, for small time bounds, the probability is higher when N is smaller. This is to be expected since sending more probes takes more time. However, for larger time bounds the probability is larger when more probes are sent. This is due to the fact that, when more probes are sent, there is less chance of using an IP address already in use, and hence not reaching a state where the host uses a fresh IP address.

Finally, note that this approach can be adapted for modelling other constraints. For example, the model can be extended to calculate the probability of the host configuring a fresh IP address while sending at most K probes, or to calculate probabilities when there are bounds both on the elapsed time and the total number of probes sent. For further details about this case study, see the PRISM website [**111**].

2.6. Discussion

The logic PCTL discussed in this chapter is due to [**62**], and is essentially the same as the logic pCTL of [**9**]. We have extended it with the expectation operator, but its expressive power is still relatively limited. For example, conjunction of probabilistic quantifier formulas cannot be expressed. More expressive logics that allow this have been proposed, including LTL and PCTL* [**9, 25**], a probabilistic extension of CTL*. Model checking algorithms for these logics have been proposed [**9, 17, 25, 41, 127**], based on a transformation of the underlying DTMC to a larger

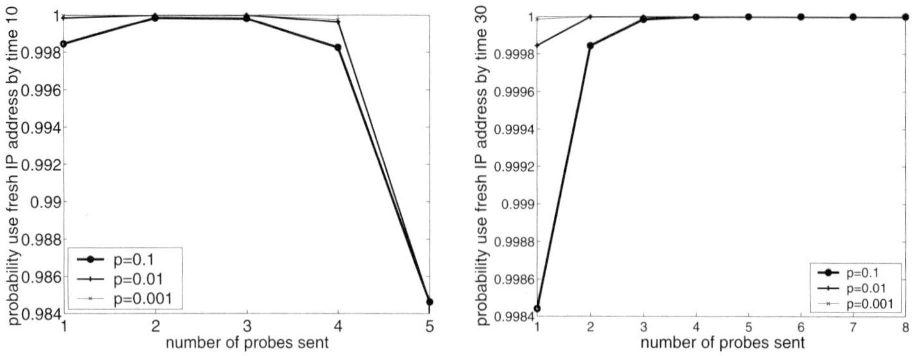

FIGURE 6. Results for the extended model of the IPv4 protocol.

one. Finally, we mention work on probabilistic generalisations of bisimulation and simulation relations [**99, 117**] which were omitted from our discussion.

CHAPTER 3

Markov Decision Processes

3.1. Introduction

In this chapter we consider *Markov decision processes* (MDPs), which can be seen as a generalisation of DTMCs. An MDP can describe both nondeterministic and probabilistic behaviour. It is well known that nondeterminism is a valuable tool for modelling concurrency: an MDP allows us to describe the behaviour of a number of probabilistic systems operating in parallel. Nondeterminism is also useful when the exact probability of a transition is not known, or when it is known but not considered relevant.

DEFINITION 3.1.1. An MDP is a tuple $\mathcal{M} = (S, \bar{s}, \mathrm{Act}, \mathrm{Steps})$ where:

- S is a finite set of *states*;
- $\bar{s} \in S$ is the *initial state*;
- Act is a set of *actions*;
- Steps: $S \to 2^{\mathsf{Act} \times \mathsf{Dist}(S)}$ is the *probabilistic transition function.*

Like a DTMC, the basic component of an MDP is a finite set of states S, one of which, $\bar{s}$, is designated as the initial state. The transition probability matrix $\mathbf{P}$, however, is replaced by Steps, a function mapping each state $s \in S$ to a finite, non-empty subset of Act $\times \mathsf{Dist}(S)$, where Act is a set of actions and $\mathsf{Dist}(S)$ is the set of all probability distributions over the set S, i.e. the set of functions $\mu \colon S \to [0,1]$ such that $\sum_{s \in S} \mu(s) = 1$. While in a state s of a DTMC the next state is determined by the probabilities $\mathbf{P}(s, s')$ for $s' \in S$, i.e. a single probability distribution over S, in an MDP there is a *nondeterministic choice* between $|\mathrm{Steps}(s)|$ action-labelled probability distributions. Once an action-distribution pair $(a, \mu) \in \mathrm{Steps}(s)$ has been selected, the action a is performed and the probability of the next state being s' is $\mu(s')$. Note that this satisfies the Markov property, i.e. the next state of the system depends only on the current state.

As in the DTMC case, to analyse MDPs we annotate them with additional information. We use a labelling function L to associate states with atomic propositions from some set AP and a cost function $\mathbf{c}$ to assign non-negative, real-valued costs to transitions. For an MDP, we actually associate each pair of states and actions with a cost, i.e. the cost is the same for all transitions in a probability distributions labelled with the same action.

DEFINITION 3.1.2. A labelled MDP as a tuple $(\mathcal{M}, L, \mathbf{c})$ where:

- $\mathcal{M}$ is an MDP;
- $L \colon S \to 2^{\mathrm{AP}}$ is a *labelling function*;
- $\mathbf{c} \colon S \times \mathrm{Act} \to \mathbb{R}^{\geq 0}$ is the *cost function.*

EXAMPLE 3.1.3. Figure 1 shows an example of an MDP and how it can be used to model the parallel composition of two DTMCs. The figure includes two three

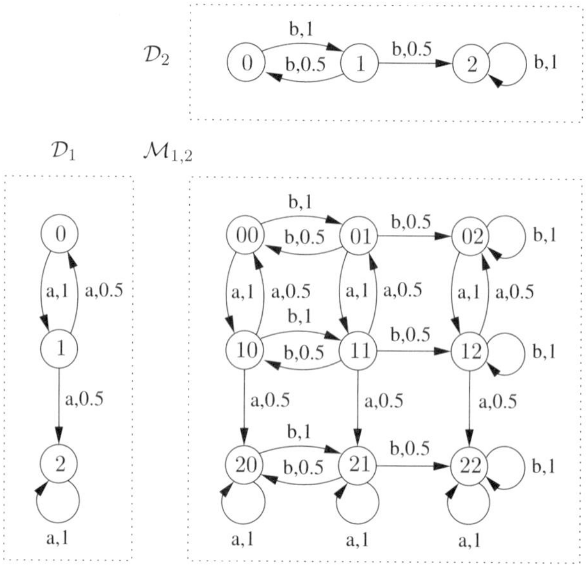

FIGURE 1. An MDP $\mathcal{M}_{1,2}$ representing the parallel composition of DTMCs $\mathcal{D}_1$ and $\mathcal{D}_2$.

state DTMCs, $\mathcal{D}_1$ and $\mathcal{D}_2$, and the corresponding nine state MDP $\mathcal{M}_{1,2}$. The state space of $\mathcal{M}_{1,2}$ is the product of $\mathcal{D}_1$ and $\mathcal{D}_2$: $\{0,1,2\} \times \{0,1,2\}$. The two DTMCs are composed asynchronously, i.e. for each state s of $\mathcal{M}_{1,2}$, the set Steps(s) contains 2 probability distributions, one corresponding to $\mathcal{D}_1$ and one to $\mathcal{D}_2$. For example, when $\mathcal{D}_1$ is in state 1 it moves to either state 0 or 2, each with probability 0.5. When $\mathcal{D}_2$ is in state 2, it remains in that state with probability 1. Hence, in state 12 of the MDP, there are two probability distributions, one selecting states 02 and 22, each with probability 0.5, and one selecting state 12 with probability 1.

Our graphical notation for MDPs is almost identical to that used for DTMCs. The only difference is the addition of action labels. Actions a and b correspond to the behaviour of DTMCs $\mathcal{D}_1$ and $\mathcal{D}_2$, respectively. Due to the way the transitions are arranged in this example, we have labelled every individual transition with its associated action. Typically, it suffices to use a single label for all the transitions belong to one probability distribution. For this example, transitions of the DTMCs $\mathcal{D}_1$ and $\mathcal{D}_2$ have also been labelled with a and b, respectively. For clarity, in this example we omit details of labelling and cost functions.

3.2. Paths and Probability Measures

A *path* in the MDP is a non-empty sequence of the form

$$s_0 \xrightarrow{(a_1,\mu_1)} s_1 \xrightarrow{(a_2,\mu_2)} s_2 \ldots$$

where $s_i \in S$, $(a_{i+1}, \mu_{i+1}) \in \text{Steps}(s_i)$ and $\mu_{i+1}(s_{i+1}) > 0$ for all $i \geq 0$. As for DTMCs, $\omega(i)$ denotes the ith state of a path ω, $|\omega|$ is the length of ω and $\text{last}(\omega_{\text{fin}})$ is the last state of a finite path ω_{fin}. In addition, we let $\text{step}(\omega, i)$ denote the ith action of the path ω. The sets of all finite and infinite paths starting in a state s are $\text{Path}_s^{\text{fin}}$ and Path_s, respectively.

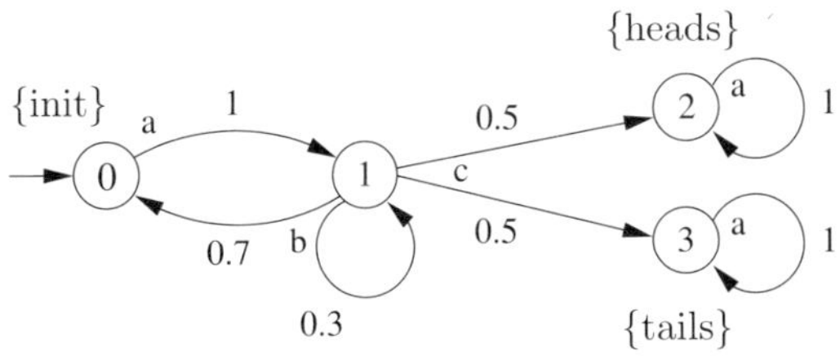

FIGURE 2. A four state MDP.

To trace a path through an MDP, both the nondeterministic and probabilistic choices have to be resolved. We assume that the nondeterministic choices are made by an *adversary* (also known as a "scheduler" or "policy"), which selects a choice based on the history of choices made so far. Formally, we have the following definition.

DEFINITION 3.2.1. An adversary A of an MDP $\mathcal{M}$ is a function mapping every finite path ω_{fin} of $\mathcal{M}$ onto an element $A(\omega_{\text{fin}})$ of the set $\text{Steps}\big(\text{last}(\omega_{\text{fin}})\big)$. Let $\text{Adv}_{\mathcal{M}}$ denote the set of all possible adversaries of the MDP and, for any adversary A, let Path_s^A denote the subset of Path_s which corresponds to A.

EXAMPLE 3.2.2. Figure 2 shows a simple MDP with four states, $\{0, 1, 2, 3\}$. From state 0, the MDP always moves directly to state 1. In state 1, there is a non-deterministic choice between b and c: the MDP will either choose probabilistically between returning to state 0 and remaining in the same state (b), or it will choose probabilistically between moving to states 2 and 3 (c). Since 1 is the only state s for which $|\text{Steps}(s)| > 1$, it is the only state in which an adversary actually has to make a choice. Let us denote the two probability distributions in state 1 as μ_b and μ_c, according to the actions b and c associated with them. We can now describe two example adversaries, A_1 and A_2, of this MDP as follows:

- Adversary A_1 : $A_1(01) = \mu_c$.
- Adversary A_2 : $A_2(01) = \mu_b$, $A_2(011) = \mu_c$, $A_2(0101) = \mu_c$.

Note that, for this simple example, we can unambiguously write paths in an abbreviated form, e.g. we have:

$$011 \equiv 0 \xrightarrow{(a,\mu_a)} 1 \xrightarrow{(b,\mu_b)} 1$$

where μ_a is the single probability distribution available in state 0.

The behaviour of an MDP under a given adversary A is purely probabilistic. For an MDP $\mathcal{M} = (S, \bar{s}, \text{Act}, \text{Steps})$ and state $s \in S$, the behaviour from state s can be described with the infinite-state DTMC $\mathcal{D}^A = (S^A, \bar{s}^A, \mathbf{P}^A)$ where:

- $S^A = \text{Path}_s^{\text{fin}}$,
- $\bar{s}^A = s$

and for two finite paths $\omega, \omega' \in S^A$:

- $\mathbf{P}^A(\omega, \omega') = \begin{cases} \mu(s') & \text{if } \omega' \text{ is of the form } \omega \xrightarrow{A(\omega)} s' \text{ and } A(\omega) = (a, \mu) \\ 0 & \text{otherwise.} \end{cases}$

There is a one-to-one correspondence between the paths of $\mathcal{D}^A$ and the set of paths Path_s^A in the MDP. Hence, using the probability measure over DTMCs given in

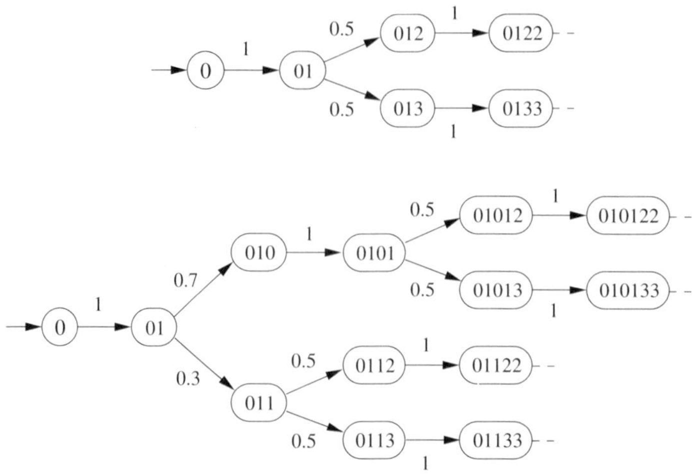

FIGURE 3. Fragments of the DTMCs $\mathcal{D}^{A_1}$ (top) and $\mathcal{D}^{A_2}$ (bottom) for the adversaries A_1 and A_2 of (3.2.2).

the previous section (which extends quite naturally to the case of infinite state DTMCs), we can define a probability measure Prob_s^A over the set of paths Path_s^A.

EXAMPLE 3.2.3. Consider the MDP from (3.2.2) (see Fig. 2) and the two example adversaries A_1 and A_2. For state 0 of the MDP, Fig. 3 shows fragments of the DTMCs $\mathcal{D}^{A_1}$ and $\mathcal{D}^{A_2}$ corresponding to the two adversaries.

Similarly to the DTMC case, for a given adversary A and using the measure Prob_s^A, we can define the expected cost of reaching a set of target states F as $E_s^A(\mathrm{cost}(F))$, where E_s^A is the usual expectation with respect to the measure Prob_s^A and for any $\omega \in \mathrm{Path}_s^A$:

$$\mathrm{cost}(F)(\omega) \stackrel{\text{def}}{=} \begin{cases} \sum_{i=0}^{\min\{i \mid \omega(i) \in F\}} \mathbf{c}(\mathrm{step}(\omega, i-1)) & \text{if } \exists i \in \mathbb{N} \,.\, \omega(i) \in F \\ \infty & \text{otherwise.} \end{cases}$$

Reasoning about the precise probabilistic behaviour of an MDP for a single adversary is of limited use. We can, however, still verify meaningful properties of the MDP by computing the *maximum* or *minimum* probability/expected cost of some specified behaviour being observed *over all possible adversaries*.

3.3. Probabilistic Computation Tree Logic (PCTL)

To write specifications of MDPs, we use exactly the same logic, PCTL, as for DTMCs (see Section 2.3). Recall that the syntax of PCTL is:

$$\phi ::= \mathrm{true} \mid a \mid \neg\phi \mid \phi \wedge \phi \mid \mathcal{P}_{\bowtie p}[\psi] \mid \mathcal{E}_{\bowtie c}[\phi], \quad \psi ::= \phi \mid \phi\,\mathcal{U}^{\leq k}\phi \mid \phi\,\mathcal{U}\phi$$

where a is an atomic proposition, $\bowtie \in \{\leq, <, \geq, >\}$, $p \in [0,1]$, $c \in \mathbb{R}^{\geq 0}$ and $k \in \mathbb{N}$. As before, the syntax is described in terms of state formulas ϕ and path formulas ψ, which are evaluated over states and paths, respectively, but properties of MDPs will always be expressed as state formulas.

The principal difference for the case of MDPs is the definition of the $\mathcal{P}_{\bowtie p}[\psi]$ and $\mathcal{E}_{\bowtie c}[\phi]$ operators. While the interpretation of path formulas ψ is identical to those for DTMCs, the meaning of a state s satisfying the formula $\mathcal{P}_{\bowtie p}[\psi]$, is that the probability of a path from s satisfying ψ is in the range $\bowtie p$ *for all adversaries*. Similarly, if $s \models \mathcal{E}_{\bowtie c}[\phi]$, then the expected cost of reaching a state satisfying ϕ is $\bowtie c$ for all adversaries.

DEFINITION 3.3.1. Let $\mathcal{M} = (S, \bar{s}, \mathrm{Act}, \mathrm{Steps}, L, \mathbf{c})$ be a labelled MDP. For any state $s \in S$, the satisfaction relation ϕ is defined inductively as follows:

$$\begin{array}{ll} s \models \mathrm{true} & \text{for all } s \in S, \\ s \models a & \Leftrightarrow a \in L(s), \\ s \models \neg\phi & \Leftrightarrow s \not\models \phi, \\ s \models \phi_1 \wedge \phi_2 & \Leftrightarrow s \models \phi_1 \wedge s \models \phi_2, \\ s \models \mathcal{P}_{\bowtie p}[\psi] & \Leftrightarrow p_s^A(\psi) \bowtie p \text{ for all } A \in \mathrm{Adv}_{\mathcal{M}}, \\ s \models \mathcal{E}_{\bowtie c}[\phi] & \Leftrightarrow e_s^A(\phi) \bowtie c \text{ for all } A \in \mathrm{Adv}_{\mathcal{M}}, \end{array}$$

where for any adversary $A \in \mathrm{Adv}_{\mathcal{M}}$:

$$p_s^A(\psi) \stackrel{\mathrm{def}}{=} \mathrm{Prob}_s^A(\{\omega \in \mathrm{Path}_s^A \mid \omega \models \psi\}), \quad e_s^A(\phi) \stackrel{\mathrm{def}}{=} E_s^A\big(\mathrm{cost}(\{s' \in S \mid s' \models \phi\}\big)$$

and for any path $\omega \in \mathrm{Path}$:

$$\begin{aligned} \omega \models \phi &\Leftrightarrow \omega(1) \models \phi, \\ \omega \models \phi_1 \mathcal{U}^{\leq k} \phi_2 &\Leftrightarrow \exists i \leq k \,.\, \big(\omega(i) \models \phi_2 \wedge \omega(j) \models \phi_1 \forall j < i\big), \\ \omega \models \phi_1 \mathcal{U} \phi_2 &\Leftrightarrow \exists k \geq 0 \,.\, \omega \models \phi_1 \mathcal{U}^{\leq k} \phi_2. \end{aligned}$$

Note that, as in the case for DTMCs, for any path formula ψ, state $s \in S$ and adversary A, the set of paths $\{\omega \in \mathrm{Path}_s^A \mid \omega \models \psi\}$ is measurable (see e.g. [**127**]).

As with PCTL for DTMCs, we can easily derive the standard logical operators false, $\vee$ and $\rightarrow$ and, since path formulas are interpreted identically, we can again define the path formula operators $\Diamond$ and $\Diamond^{\leq k}$ as:

$$\mathcal{P}_{\bowtie p}[\Diamond\phi] \equiv \mathcal{P}_{\bowtie p}[\mathrm{true}\,\mathcal{U}\phi], \quad \mathcal{P}_{\bowtie p}[\Diamond^{\leq k}\phi] \equiv \mathcal{P}_{\bowtie p}[\mathrm{true}\,\mathcal{U}^{\leq k}\phi].$$

Again, the case of $\Box$ and $\Box^{<k}$, is slightly more complex. We use the fact that, for a state s, adversary A and path formula ψ, $p_s^A(\neg\psi) = 1 - p_s^A(\psi)$. Then, for example, we can show the following:

$$\begin{aligned} \mathcal{P}_{\geq p}[\Box\phi] &\Leftrightarrow p_s(\Box\phi) \geq p \text{ for all adversaries } A \\ &\Leftrightarrow 1 - p_s(\Diamond\neg\phi) \geq p \text{ for all adversaries } A \\ &\Leftrightarrow p_s(\Diamond\neg\phi) \leq 1 - p \text{ for all adversaries } A \\ &\Leftrightarrow \mathcal{P}_{1-p}[\Diamond\neg\phi]. \end{aligned}$$

More generally:

$$\mathcal{P}_{\bowtie p}[\Box\phi] \equiv \mathcal{P}_{\overline{\bowtie}1-p}[\Diamond\neg\phi], \quad \mathcal{P}_{\bowtie p}[\Box^{\leq k}\phi] \equiv \mathcal{P}_{\overline{\bowtie}1-p}[\Diamond^{\leq k}\neg\phi]$$

where $\overline{\leq} \equiv \geq$, $\overline{<} \equiv >$, $\overline{\geq} \equiv \leq$ and $\overline{>} \equiv <$.

When writing specifications for MDPs in PCTL, it may sometimes be useful to reason about the *existence* of an adversary, rather than quantifying over *all* adversaries. This can be done via translation to a dual property, for example:

$$p_s^A(\psi) > p \text{ for some adversary } A \Leftrightarrow \neg\mathcal{P}_{\leq p}[\psi],$$
$$p_s^A(\psi) \leq p \text{ for some adversary } A \Leftrightarrow \neg\mathcal{P}_{>p}[\psi],$$
$$e_s^A(\psi) < p \text{ for some adversary } A \Leftrightarrow \neg\mathcal{E}_{\geq p}[\psi],$$
$$e_s^A(\psi) \geq p \text{ for some adversary } A \Leftrightarrow \neg\mathcal{E}_{<p}[\psi].$$

Using this idea, we can again construct an analogue of the CTL $\exists\Diamond\phi$ (EF ϕ) operator. The equivalent for an MDP is the existence of an adversary for which the probability of reaching a state satisfying ϕ is greater than zero. Hence:

$$\exists\Diamond\phi \equiv \neg\mathcal{P}_{\leq 0}[\Diamond\phi].$$

For the same reasons as in the DTMC case (see Fig. 2), we can only provide a weak analogue of the CTL $\forall\Diamond\phi$ (AF ϕ) operator and the following are not equivalent:

$$\forall\Diamond\phi \not\equiv \mathcal{P}_{\geq 1}[\Diamond\phi].$$

We conclude our discussion of PCTL for MDPs by pointing out that, as described in the previous chapter, the bound on the $\mathcal{P}_{\bowtie p}[\psi]$ and $\mathcal{E}_{\bowtie c}[\phi]$ operators can often be dropped in practice. In this case, we would compute either the maximum or minimum value, over all adversaries, of the probability or expected cost.

3.4. PCTL Model Checking

The model checking algorithm for PCTL over MDPs takes as input a labelled MDP $(S, \bar{s}, \text{Act}, \text{Steps}, L, \mathbf{c})$ and a PCTL formula ϕ and returns the set $\text{Sat}(\phi) \subseteq S$ of states satisfying the formula. The overall algorithm works exactly as for DTMCs (see page 107), constructing the parse tree for ϕ and treating each operator in the formula separately. Moreover, all operators are handled in identical fashion, with the exception of $\mathcal{P}_{\bowtie p}[\psi]$ and $\mathcal{E}_{\bowtie c}[\phi]$. For these, we need to determine whether $p_s(\psi)$ (or $e_s(\psi)$) satisfies the bound $\bowtie p$ (or $\bowtie c$) for all adversaries A. We do this by computing either the *minimum* or *maximum* probability or expected cost over all adversaries, depending on whether the relational operator $\bowtie$ defines an upper or lower bound. If $\bowtie$ is $\leq$ or $<$, then:

$$\text{Sat}(\mathcal{P}_{\bowtie p}[\psi]) = \{s \in S \mid p_s^{\max}(\psi) \bowtie p\}, \quad \text{Sat}(\mathcal{E}_{\bowtie c}[\phi]) = \{s \in S \mid e_s^{\max}(\phi) \bowtie c\}$$

and if $\bowtie$ is $\geq$ or $>$, then:

$$\text{Sat}(\mathcal{P}_{\bowtie p}[\psi]) = \{s \in S \mid p_s^{\min}(\psi) \bowtie p\}, \quad \text{Sat}(\mathcal{E}_{\bowtie c}[\phi]) = \{s \in S \mid e_s^{\min}(\phi) \bowtie c\}$$

where:

$$p_s^{\max}(\psi) \stackrel{\text{def}}{=} \sup_{A\in\text{Adv}_{\mathcal{M}}} [p_s^A(\psi)], \quad e_s^{\max}(\phi) \stackrel{\text{def}}{=} \sup_{A\in\text{Adv}_{\mathcal{M}}} [e_s^A(\phi)],$$
$$p_s^{\min}(\psi) \stackrel{\text{def}}{=} \inf_{A\in\text{Adv}_{\mathcal{M}}} [p_s^A(\psi)], \quad e_s^{\min}(\phi) \stackrel{\text{def}}{=} \inf_{A\in\text{Adv}_{\mathcal{M}}} [e_s^A(\phi)].$$

In the following sections we describe how to compute these values for the following four cases: $\mathcal{P}_{\bowtie p}[\phi]$, $\mathcal{P}_{\bowtie p}[\phi_1 \mathcal{U}^{\leq k} \phi_2]$, $\mathcal{P}_{\bowtie p}[\phi_1 \mathcal{U} \phi_2]$ and $\mathcal{E}_{\bowtie c}[\phi]$. The algorithms presented are those of [**25, 42, 44, 45, 51**].

3.4.1. The $\mathcal{P}_{\bowtie p}[\phi]$ operator. For this operator, we need to compute the probabilities $p_s^{\max}(\phi)$ or $p_s^{\min}(\phi)$ for all states $s \in S$. Like in the DTMC case, for each state s we only need to consider the immediate transitions from s:

$$p_s^{\max}(\phi) = \max_{(a,\mu)\in \mathrm{Steps}(s)} \left\{ \sum_{s'\in \mathrm{Sat}(\phi)} \mu(s') \right\},$$

$$p_s^{\min}(\phi) = \min_{(a,\mu)\in \mathrm{Steps}(s)} \left\{ \sum_{s'\in \mathrm{Sat}(\phi)} \mu(s') \right\}.$$

Let m be the total number of nondeterministic choices in all states of the MDP, i.e. $m = \sum_{s\in S} |\,\mathrm{Steps}(s)|$. We can represent the function Steps as an $m \times |S|$ matrix Steps, where each row of the matrix corresponds to a single nondeterministic choice. This means that there are $|\,\mathrm{Steps}(s)|$ rows corresponding to each state s, rather than just one, as in the DTMC case. Assume we also have a state-indexed vector $\underline{\phi}$ with $\underline{\phi}(s)$ equal to 1 if $s \models \phi$ and 0 otherwise. We can carry out the computation above in two steps: firstly, a matrix-vector multiplication $\mathrm{Steps} \cdot \underline{\phi}$ which computes the summation for each nondeterministic choice, producing a vector of length m; and secondly, an operation which selects, from this vector, the maximum or minimum value for each state, reducing the length from m to $|S|$.

EXAMPLE 3.4.1. Consider the MDP from (3.2.2) (see Fig. 2). The matrix Steps, described above, is given below. For clarity, we separate the rows for each state with a horizontal line

$$\mathrm{Steps} = \left(\begin{array}{cccc} 0 & 1 & 0 & 0 \\ \hline 0.7 & 0.3 & 0 & 0 \\ 0 & 0 & 0.5 & 0.5 \\ \hline 0 & 0 & 1 & 0 \\ \hline 0 & 0 & 0 & 1 \end{array}\right).$$

We will compute probabilities for the two PCTL formulas $\mathcal{P}_{\leq 0.5}[\mathrm{heads}]$ and $\mathcal{P}_{\geq 1}[\mathrm{heads}]$, i.e. we will determine both minimum and maximum probabilities for the path formula heads. The matrix-vector multiplication required for both cases is:

$$\mathrm{Steps} \cdot \underline{\phi} = \left(\begin{array}{cccc} 0 & 1 & 0 & 0 \\ \hline 0.7 & 0.3 & 0 & 0 \\ 0 & 0 & 0.5 & 0.5 \\ \hline 0 & 0 & 1 & 0 \\ \hline 0 & 0 & 0 & 1 \end{array}\right) \cdot \left(\begin{array}{c} 0 \\ 0 \\ 1 \\ 0 \end{array}\right) = \left(\begin{array}{c} 0 \\ \hline 0 \\ 0.5 \\ \hline 1 \\ \hline 0 \end{array}\right).$$

Extracting the minimum and maximum value for state 1 yields the following results: $\underline{p^{\max}(\mathrm{heads})} = [0, 0.5, 1, 0]$ and $\underline{p^{\min}(\mathrm{heads})} = [0, 0, 1, 0]$. Hence:

$$\mathrm{Sat}(\mathcal{P}_{\leq 0.5}[\mathrm{heads}]) = \{0, 1, 3\}, \quad \mathrm{Sat}(\mathcal{P}_{\geq 1}[\mathrm{heads}]) = \{2\}.$$

3.4.2. The $\mathcal{P}_{\bowtie p}[\phi_1 \mathcal{U}^{\leq k} \phi_2]$ operator. Here, we compute the probabilities $p_s^{\max}(\phi_1 \mathcal{U}^{\leq k} \phi_2)$ or $p_s^{\min}(\phi_1 \mathcal{U}^{\leq k} \phi_2)$. Again, the process is similar to the DTMC case. We first divide S into three disjoint subsets:

$$S^{\mathrm{no}} = S\backslash(\mathrm{Sat}(\phi_1) \cup \mathrm{Sat}(\phi_2)), \quad S^{\mathrm{yes}} = \mathrm{Sat}(\phi_2), \quad S^{?} = S\backslash(S^{\mathrm{no}} \cup S^{\mathrm{yes}})$$

where, for $s \in S^{\text{no}}$ and $s \in S^{\text{yes}}$, $p_s^{\max}(\phi_1 \mathcal{U}^{\leq k} \phi_2) = p_s^{\min}(\phi_1 \mathcal{U}^{\leq k} \phi_2)$ is trivially 0 or 1, respectively. Probabilities for the remaining states in $S^?$ are defined recursively. If $k = 0$, then $p_s^{\max}(\phi_1 \mathcal{U}^{\leq k} \phi_2) = p_s^{\min}(\phi_1 \mathcal{U}^{\leq k} \phi_2) = 0$. If $k > 0$:

$$p_s^{\max}(\phi_1 \mathcal{U}^{\leq k} \phi_2) = \max_{(a,\mu) \in \text{Steps}(s)} \left\{ \sum_{s' \in S} \mu(s') \cdot p_{s'}^{\max}(\phi_1 \ \mathcal{U}^{\leq k-1} \ \phi_2) \right\},$$

$$p_s^{\min}(\phi_1 \mathcal{U}^{\leq k} \phi_2) = \min_{(a,\mu) \in \text{Steps}(s)} \left\{ \sum_{s' \in S} \mu(s') \cdot p_{s'}^{\min}(\phi_1 \ \mathcal{U}^{\leq k-1} \ \phi_2) \right\}.$$

If we assume, as for "next" formulas above, that Steps is represented by the matrix $\mathbf{Steps}$, then the computation of $p_s^{\max}(\phi_1 \mathcal{U}^{\leq k} \phi_2)$ or $p_s^{\min}(\phi_1 \mathcal{U}^{\leq k} \phi_2)$ can be carried out in k iterations, each one similar to the process described for the PCTL "next" operator. Every iteration will comprise one matrix-vector multiplication and one maximum or minimum operation.

EXAMPLE 3.4.2. We again use the MDP from (3.2.2) (see Fig. 2) and model check the PCTL formula $\mathcal{P}_{<0.95}[\Diamond^{\leq 3}\text{init}]$. This is equivalent to the formula $\mathcal{P}_{<0.95}[\text{true } \mathcal{U}^{\leq 3}\text{init}]$, for which we must compute maximum probabilities. We have:

$$S^{\text{no}} = S \backslash \big(\text{Sat}(\text{true}) \cup \text{Sat}(\text{init})\big) = \emptyset, \quad S^{\text{yes}} = \text{Sat}(\text{init}) = \{0\}.$$

The vectors of probabilities are then computed successively as:

$$\underline{p^{\max}(\text{true}\mathcal{U}^{\leq 0}\text{init})} = [1, 0, 0, 0], \qquad \underline{p^{\max}(\text{true}\mathcal{U}^{\leq 1}\text{init})} = [1, 0.7, 0, 0],$$
$$\underline{p^{\max}(\text{true}\mathcal{U}^{\leq 2}\text{init})} = [1, 0.91, 0, 0], \quad \underline{p^{\max}(\text{true}\mathcal{U}^{\leq 3}\text{init})} = [1, 0.973, 0, 0].$$

Hence, $\text{Sat}(\mathcal{P}_{<0.95}[\Diamond^{\leq 3}\text{init}]) = \{2, 3\}$.

3.4.3. The $\mathcal{P}_{\bowtie p}[\phi_1 \mathcal{U} \phi_2]$ operator. For this operator, we are required to compute either the probabilities $p_s^{\max}(\phi_1 \mathcal{U} \phi_2)$ or $p_s^{\min}(\phi_1 \mathcal{U} \phi_2)$. In the remainder of this section, we will abbreviate these to $p_s^{\max}$ and $p_s^{\min}$, respectively. Similarly, we will use p_s^A to denote $p_s^A(\phi_1 \mathcal{U} \phi_2)$.

As for model checking the corresponding operator over DTMCs, we begin by computing the sets S^{no} and S^{yes} which contain all the states with probability equal to exactly 0 or 1, respectively. Again, this is done using graph-based precomputation algorithms. For MDPs, though, the algorithms used depend on whether we are computing minimum or maximum probabilities. For computation of $p_s^{\max}$ we have:

$$\begin{aligned} S^{\text{no}} &= \text{PROB0A}\big(\text{Sat}(\phi_1), \text{Sat}(\phi_2)\big), \\ S^{\text{yes}} &= \text{PROB1E}\big(\text{Sat}(\phi_1), \text{Sat}(\phi_2)\big), \\ S^? &= S \backslash (S^{\text{no}} \cup S^{\text{yes}}) \end{aligned}$$

and for $p_s^{\min}$:

$$\begin{aligned} S^{\text{no}} &= \text{PROB0E}\big(\text{Sat}(\phi_1), \text{Sat}(\phi_2)\big) \\ S^{\text{yes}} &= \text{Sat}(\phi_2), \\ S^? &= S \backslash (S^{\text{no}} \cup S^{\text{yes}}). \end{aligned}$$

When computing $p_s^{\max}$, S^{no} contains all the states for which $p_s^A = 0$ for *every* adversary A. We determine these with the precomputation algorithm PROB0A. Conversely, S^{yes} contains all the states for which $p_s^A = 1$ for *some* adversary A.

```
PROB0A(Sat(φ1), Sat(φ2))
R := Sat(φ2)
done := false
while (done = false)
      R' := R ∪ {s ∈ Sat(φ1) | ∃(a, μ) ∈ Steps(s) . ∃s' ∈ R . μ(s') > 0}
      if (R' = R) then done := true
      R := R'
endwhile
return S\R
```

ALGORITHM 3. The PROB0A algorithm.

```
PROB1E(Sat(φ1), Sat(φ2))
R := S
done := false
while (done = false)
      R' := Sat(φ2)
      done' := false
      while (done' = false)
            R'' := R' ∪ {s ∈ Sat(φ1) | ∃(a, μ) ∈ Steps(s).
                        (∀s' ∈ S . μ(s')>0 → s' ∈ R) ∧ (∃s' ∈ R' . μ(s')>0)}
            if (R'' = R') then done' := true
            R' := R''
      endwhile
      if (R' = R) then done := true
      R := R'
endwhile
return R
```

ALGORITHM 4. The PROB1E algorithm.

```
PROB0E(Sat(φ1), Sat(φ2))
R := Sat(φ2)
done := false
while (done = false)
      R' := R ∪ {s ∈ Sat(φ1) | ∀(a, μ) ∈ Steps(s) . ∃s' ∈ R . μ(s') > 0}
      if (R' = R) then done := true
      R := R'
endwhile
return S\R
```

ALGORITHM 5. The PROB0E algorithm.

For this, we use the precomputation algorithm PROB1E. The two algorithms are shown in Algorithms 3, 4, respectively.

PROB0A works in a similar fashion to the earlier algorithm PROB0. It first computes the states from which one can, under some adversary, with non-zero probability, reach a state satisfying ϕ_2 without leaving states satisfying ϕ_1. It then subtracts these from S to determine the states which have a zero probability for every adversary.

The PROB1E algorithm is more involved. It was first presented in [**51**] but is actually an extension of a related algorithm from [**44**]. In essence, it is similar to PROB1, given earlier, in that it works by identifying states for which $p_s^{\max}$ is less than 1. PROB1E is based on the computation of a double fixpoint and is hence implemented as two nested loops. The outer loop computes a set of states R. By the end of the algorithm, R will contain all the states s for which $p_s^A = 1$ for *some* adversary A, as required. Initially R is set to S, i.e. all states. In each iteration of the outer loop, states for which there is no adversary A with $p_s^A = 1$ are removed from R. The inner loop identifies the states to be removed as those from which one cannot reach a state in $\text{Sat}(\phi_2)$ without passing through either a state not in $\text{Sat}(\phi_1)$ or a state already removed by the algorithm (i.e. not in R).

When computing $p_s^{\min}$, S^{no} contains all states for which $p_s^A = 0$ for *some* adversary A. This is computed with the algorithm PROB0E, given in Algorithm 5, which works in an almost identical fashion to PROB0A. The approach taken in PROB1E, however, is not applicable to the problem of determining states where $p_s^A = 1$ for *every* adversary A. Hence, we take S^{yes} to be the set of states $\text{Sat}(\phi_2)$ for which $p_s^{\min}$ is trivially 1.

The computation of $p_s^{\max}$ or $p_s^{\min}$ for the remaining states in $S^?$ can then be performed by solving a linear optimisation problem over the set of variables $\{x_s \mid s \in S^?\}$. For the case of $p_s^{\max}$, the problem is:

$$\text{Minimise} \sum_{s \in S^?} x_s \text{ subject to the constraints:}$$
$$x_s \geq \sum_{s' \in S^?} \mu(s') \cdot x_{s'} + \sum_{s' \in S^{\text{yes}}} \mu(s')$$
$$\text{for all } s \in S^? \text{ and all } (a, \mu) \in \text{Steps}(s)$$

and for the case of $p_s^{\min}$:

$$\text{Maximise} \sum_{s \in S^?} x_s \text{ subject to the constraints:}$$
$$x_s \leq \sum_{s' \in S^?} \mu(s') \cdot x_{s'} + \sum_{s' \in S^{\text{yes}}} \mu(s')$$
$$\text{for all } s \in S^? \text{ and all } (a, \mu) \in \text{Steps}(s)$$

In either case, the problem admits a unique optimal solution and we can let $p_s^{\max} = x_s$ or $p_s^{\min} = x_s$, respectively. These results can be found in [**22, 42, 44**]. In fact, the computation required here is actually a simple case of a more general problem known as the *stochastic shortest path problem*, see [**22, 44**] for more details.

The linear optimisation problems above can be solved using classical techniques such as the Simplex or ellipsoid methods (see, for example, [**116**]). However, as with model checking of the PCTL "until" operator over DTMCs, these direct methods are not well suited to problems of the size we wish to deal with. Fortunately, there is an alternative method for computing the probabilities $p_s^{\max}$ and $p_s^{\min}$. As shown in [**11**], $p_s^{\max} = \lim_{n\to\infty} p_s^{\max(n)}$ where:

$$p_s^{\max(n)} = \begin{cases} 1 & \text{if } s \in S^{\text{yes}} \\ 0 & \text{if } s \in S^{\text{no}} \\ 0 & \text{if } s \in S^{?} \quad \text{and} \quad n = 0 \\ \max_{(a,\mu)\in \text{Steps}(s)} \left\{ \sum_{s'\in S} \mu(s') \cdot p_{s'}^{\max(n-1)} \right\} & \text{if } s \in S^{?} \quad \text{and} \quad n > 0 \end{cases}$$

and $p_s^{\min} = \lim_{n\to\infty} p_s^{\min(n)}$ where:

$$p_s^{\min(n)} = \begin{cases} 1 & \text{if } s \in S^{\text{yes}} \\ 0 & \text{if } s \in S^{\text{no}} \\ 0 & \text{if } s \in S^{?} \quad \text{and} \quad n = 0 \\ \min_{(a,\mu)\in \text{Steps}(s)} \left\{ \sum_{s'\in S} \mu(s') \cdot p_{s'}^{\min(n-1)} \right\} & \text{if } s \in S^{?} \quad \text{and} \quad n > 0. \end{cases}$$

The values of $p_s^{\max}$ and $p_s^{\min}$ can now be approximated by an iterative computation, which calculates $p_s^{\max(n)}$ and $p_s^{\min(n)}$ for successive n, stopping when some convergence criterion has been satisfied. Possible criteria can be found in Appendix B, which discusses iterative solution techniques for linear equation systems. The iterative process described here can actually be seen as a special case of *value iteration*, a common technique for the analysis of MDPs (see, for example, [**112**]).

Note the similarity between a single iteration of this method and one of those required for the "bounded until" operator. Hence, assuming again that Steps is represented by a matrix Steps, each iteration can be performed in the same way using one matrix-vector multiplication and one maximum or minimum operation.

EXAMPLE 3.4.3. Consider the MDP from (3.2.2) (see Fig. 2). We will illustrate model checking of the PCTL formula $\mathcal{P}_{\leq 0.5}[\neg\text{heads}\,\mathcal{U}\,\text{tails}]$. Since "$\leq 0.5$" defines an upper bound, we are required to compute the maximum probabilities, i.e. the values $p_s^{\max}(\neg\text{heads}\,\mathcal{U}\,\text{tails})$ for $s \in \{0,1,2,3\}$. The precomputation algorithms yield the following results:

$$\begin{aligned} S^{\text{no}} &= \text{PROB0A}(\text{Sat}(\neg\text{heads}), \text{Sat}(\text{tails})) \\ &= \text{PROB0A}(\{0,1,3\},\{3\}) \\ &= \{2\} \\ S^{\text{yes}} &= \text{PROB1E}(\text{Sat}(\neg\text{heads}), \text{Sat}(\text{tails})) \\ &= \text{PROB1E}(\{0,1,3\},\{3\}) \\ &= \{3\}. \end{aligned}$$

In fact, the probabilities for the states calculated above are clear anyway since state 3 satisfies tails and state 2 satisfies neither heads nor tails. The values of $p_s^{\max}(\neg\text{heads}\,\mathcal{U}\,\text{tails})$ for states in $S^{?} = \{0,1\}$ can be computed by solving the linear optimisation problem:

$$\begin{array}{ll} \text{Minimise:} & x_0 + x_1, \\ \text{such that:} & x_0 \geq x_1, \; x_1 \geq 0.5 \quad \text{and} \quad x_1 \geq 0.7 \cdot x_0 + 0.3 \cdot x_1 \end{array}$$

which has the unique, optimal solution $(x_0, x_1) = (0.5, 0.5)$. Hence, our required vector of probabilities is $(0.5, 0.5, 0, 1)$ and the formula $\mathcal{P}_{\leq 0.5}[\neg\text{heads}\,\mathcal{U}\,\text{tails}]$ is satisfied in states 0, 1 and 2.

Refer back to the adversary A_1, described in (3.2.2), and the associated DTMC $\mathcal{D}^{A_1}$ in Fig. 3. It is clear that this adversary would result in the maximum probabilities computed above.

EXAMPLE 3.4.4. We now consider the same MDP and the same PCTL formula, but with the bound reversed: $\mathcal{P}_{>0.5}[\neg\text{heads}\,\mathcal{U}\,\text{tails}]$. Since the formula now contains a lower bound, we need to compute the minimum probabilities, $p_s^{\min}(\neg\text{heads}\,\mathcal{U}\,\text{tails})$ for $s \in \{0,1,2,3\}$. The sets S^{no}, S^{yes} and $S^?$ are:

$$\begin{aligned} S^{\text{no}} &= \text{PROB0E}(\text{Sat}(\neg\text{heads}), \text{Sat}(\text{tails})) \\ &= \text{PROB0E}(\{0,1,3\},\{3\}) \\ &= \{0,1,2\} \\ S^{\text{yes}} &= \text{Sat}(\text{tails}) \\ &= \{3\} \\ S^? &= S\backslash(S^{\text{no}} \cup S^{\text{yes}}) \\ &= \emptyset. \end{aligned}$$

While the set S^{yes} is trivial to determine, note that the precomputation algorithm PROB0E has established that $p_s^{\min}(\neg\text{heads}\,\mathcal{U}\,\text{tails}) = 0$ for states 0, 1 and 2. Hence, the set $S^?$ is empty, no further numerical computation is required and we can conclude that the PCTL formula $\mathcal{P}_{>0.5}[\neg\text{heads}\,\mathcal{U}\,\text{tails}]$ is true only in state 3.

We can easily construct an adversary A which results in these minimum probabilities: the one which, in state 1, always selects the choice between states 0 and 1. Clearly, under this adversary, it is impossible to reach state 3 from either state 0 or state 1. Hence, $p_s^A(\neg\text{heads}\,\mathcal{U}\,\text{tails}) = 0$ for $s \in \{0,1\}$.

3.4.4. The $\mathcal{E}_{\bowtie c}[\phi]$ Operator. For the PCTL "expected cost" operator, we are required to compute either the cost $e_s^{\max}(\phi)$ or $e_s^{\min}(\phi)$, which we will abbreviate to $e_s^{\max}$ and $e_s^{\min}$, respectively, for the remainder of this section. Likewise, we will use e_s^A to denote $e_s^A(\phi)$.

Similarly to the DTMC case, we first compute the sets of states S^0 and S^∞, which have a maximum/minimum expected cost of 0 and ∞, respectively, of reaching a state satisfying ϕ. In either case, the set of states S^0 equals those states that satisfy ϕ, that is, the set of states $\text{Sat}(\phi)$. When computing $e_s^{\max}$, S^∞ contains all the states for which $p_s^A(\Diamond\phi) < 1$ for *some* adversary A. Whereas, when computing $e_s^{\min}$, S^∞ contains all the states for which $p_s^A(\Diamond\phi) < 1$ for *all* adversaries A. Hence, we can identify S^∞ for the maximum and minimum case by using the techniques of the previous section to model check the PCTL formulas $\neg\mathcal{P}_{\geq 1}[\Diamond\phi]$ and $\mathcal{P}_{<1}[\Diamond\phi]$, respectively.

The computation of $e_s^{\max}$ or $e_s^{\min}$ for the remaining states in $S^? = S\backslash(S^0 \cup S^\infty)$ is similar to the computation of $p_s^{\max}(\phi_1\mathcal{U}\phi_2)$ and $p_s^{\min}(\phi_1\mathcal{U}\phi_2)$, described in the previous section. There are two possible solution techniques. One alternative is to solve a linear optimisation problem over the set of variables $\{x_s \mid s \in S^?\}$. For the case of $e_s^{\max}$, the problem is:

$$\boxed{\begin{array}{c} \text{Minimise } \sum_{s\in S^?} x_s \text{ subject to the constraints:} \\ x_s \geq \mathbf{c}(a) + \sum_{s'\in S^?} \mu(s')\cdot x_{s'} \\ \text{for all } s \in S^? \text{ and all } (a,\mu) \in \text{Steps}(s) \end{array}}$$

and for the case of $e_s^{\min}$:

$$\boxed{\begin{array}{c}\text{Maximise } \sum\limits_{s \in S^?} x_s \text{ subject to the constraints:}\\ x_s \leq \mathbf{c}(a) + \sum\limits_{s' \in S^?} \mu(s') \cdot x_{s'}\\ \text{for all } s \in S^? \text{ and all } (a,\mu) \in \text{Steps}(s).\end{array}}$$

In either case, the problem admits a unique optimal solution and we can let $e_s^{\max} = x_s$ or $e_s^{\min} = x_s$, respectively. Alternatively, to avoid the inefficiency of direct linear programming techniques such as the Simplex method on large systems, we can use an iterative approach, computing $e_s^{\max} = \lim_{n\to\infty} e_s^{\max(n)}$ where:

$$e_s^{\max(n)} = \begin{cases} \infty & \text{if } s \in S^\infty \\ 0 & \text{if } s \in S^0 \\ 0 & \text{if } s \in S^? \text{ and } n = 0 \\ \max\limits_{(a,\mu)\in\text{Steps}(s)} \left\{ \mathbf{c}(a) + \sum_{s' \in S^?} \mu(s') \cdot e_s^{\max(n-1)} \right\} & \text{if } s \in S^? \text{ and } n > 0 \end{cases}$$

and $e_s^{\min} = \lim_{n\to\infty} e_s^{\min(n)}$ where:

$$e_s^{\min(n)} = \begin{cases} \infty & \text{if } s \in S^\infty \\ 0 & \text{if } s \in S^0 \\ 0 & \text{if } s \in S^? \text{ and } n = 0 \\ \min\limits_{(a,\mu)\in\text{Steps}(s)} \left\{ \mathbf{c}(a) + \sum_{s' \in S^?} \mu(s') \cdot e_s^{\min(n-1)} \right\} & \text{if } s \in S^? \text{ and } n > 0. \end{cases}$$

The values of $e_s^{\max}$ and $e_s^{\min}$ can thus be approximated iteratively, terminating when some desired level of precision has been achieved.

EXAMPLE 3.4.5. Consider the MDP of Fig. 2 and the PCTL "expected cost" formula $\mathcal{E}_{>1}[\text{heads} \vee \text{tails}]$. We will use the cost function $\mathbf{c}$ which assigns 1 to every state-action pair in the MDP. Hence, we essentially compute the minimum expected number of steps to reach either state 2 or 3. Clearly, $S^0 = \text{Sat}(\text{heads} \vee \text{tails}) = \{2,3\}$ and, since the precomputation algorithm PROB1E$(S, \{2,3\})$ returns the set $\{0,1,2,3\}$, we have $S^\infty = \emptyset$. Hence, $e_2^{\min}(\text{heads} \vee \text{tails}) = e_3^{\min}(\text{heads} \vee \text{tails}) = 0$ and, for the remaining two states, 0 and 1, we solve the linear optimisation problem:

$$\begin{array}{ll} \text{Maximise:} & x_0 + x_1, \\ \text{such that:} & x_0 \leq 1 + x_1, x_1 \leq 1 \quad \text{and} \quad x_1 \leq 1 + 0.7 \cdot x_0 + 0.3 \cdot x_1 \end{array}$$

which has the unique, optimal solution $(x_0, x_1) = (2,1)$. Hence, the resulting vector of expected costs is $\underline{e^{\min}(\text{heads} \vee \text{tails})} = (2,1,0,0)$ and the PCTL formula is only satisfied in state 0.

3.5. Fairness

When performing verification in the non-probabilistic setting, one may have to impose *fairness constraints* in order to ensure that liveness properties can be verified. In a distributed environment, fairness corresponds to a requirement that each concurrent component will progress whenever possible. Without fairness, certain liveness properties may trivially fail to hold in the presence of simultaneously enabled transitions of a concurrent component.

In the context of probabilistic systems which exhibit nondeterministic behaviour fairness constraints may also need to be imposed. For MDPs, this is done by distinguishing between *fair* and *unfair* adversaries. We can then, for example, reason about the minimum or maximum probability of certain behaviour being observed over all fair adversaries.

Several notions of fairness have been introduced in the literature, see for example [**17, 44, 127**]. The one we treat here is that of [**17**] which is loosely based on the notion of [**127**]. Intuitively, an adversary is fair if transitions enabled infinitely often along a computation path are selected by the adversary infinitely often. For an alternative approach, the interested reader is referred to [**44**]. We define fairness as follows.

DEFINITION 3.5.1. A path ω of an MDP $\mathcal{M}$ is *fair* if, for states s occurring infinitely often in ω, each choice $\mu \in \text{Steps}(s)$ is taken infinitely often. An adversary $A \in \text{Adv}_{\mathcal{M}}$ is *fair* if, for any state s in $\mathcal{M}$, $\text{Prob}_s^A(\{\omega \in \text{Path}_s^A \mid \omega \text{ is fair}\}) = 1$.

3.5.1. Model checking under fairness. We now discuss the problem of model checking PCTL formulas over MDPs where fairness is taken into consideration. For the non-probabilistic operators of PCTL the semantics, and hence model checking procedure, is identical. The $\mathcal{P}_{\bowtie p}[\psi]$ operator, however, now means that the probability $p_s^A(\psi)$ of the path formula ψ being satisfied in state s satisfies the bound $\bowtie p$ *for all fair adversaries* A. Similarly, the $\mathcal{E}_{\bowtie c}[\phi]$ operator reasons about the expected cost over fair adversaries only.

In fact, for the operators $\mathcal{P}_{\bowtie p}[\phi]$ and $\mathcal{P}_{\bowtie p}[\phi_1 \mathcal{U}^{\leq k} \phi_2]$ fairness can be ignored. Intuitively, this is because they relate to paths of bounded length and fairness only places restrictions on the long-run (infinite) behaviour of the model. For path formulas of the form $\psi = \phi_1 \mathcal{U} \phi_2$, though, the computation of $p_s^{\max}(\psi)$ or $p_s^{\min}(\psi)$ over fair adversaries is different. Algorithms to perform this were given in [**11, 17**] and we explain these in the next section. For the "expected cost" operator, $\mathcal{E}_{\bowtie c}[\phi]$, no such algorithms have been developed. It should be noted, though, that the case for computing minimum expected costs, $e_s^{\min}(\phi)$, in fact coincides with the case for all adversaries, i.e. fairness can be ignored. We also point out that, in [**44**], model checking for the "expected cost" operator using an alternative notion of fairness can be found.

3.5.1.1. *The* $\mathcal{P}_{\bowtie p}[\phi_1 \mathcal{U} \phi_2]$ *operator.* We now describe the case of computing $p_s^{\max}(\phi_1 \, \mathcal{U} \, \phi_2)$ and $p_s^{\min}(\phi_1 \, \mathcal{U} \, \phi_2)$ over fair adversaries only, as presented in [**11, 17**]. The case for computing $p_s^{\max}(\phi_1 \mathcal{U} \phi_2)$ actually remains unchanged, i.e. we can simply use the algorithm of Section 3.4.3. This can be explained as follows. Because we are now considering a more restricted class of adversaries, the maximum probability clearly cannot increase. More importantly, though, it cannot decrease either. This is because, for any adversary A, we can always construct a fair adversary A' for which $p_s^{A'}(\phi_1 \mathcal{U} \phi_2) \geq p_s^A(\phi_1 \mathcal{U} \phi_2)$. Intuitively, the reason for this is that fairness only places restrictions on infinite behaviour and, for a path to satisfy an "until" formula, only some finite, initial portion of it is relevant.

For the case of computing $p_s^{\min}(\phi_1 \mathcal{U} \phi_2)$, this argument cannot be applied. The minimum probability over all fair adversaries *can* be higher than the minimum for all adversaries. In (3.5.2), below, we will show a case where this is true. Fortunately, using the technique of [**11**], we can still compute $p_s^{\min}(\phi_1 \mathcal{U} \phi_2)$ without much additional effort. The basic idea is to solve a dual problem, calculating the probability that $\phi_1 \mathcal{U} \phi_2$ is *not* satisfied. In doing so, we convert the problem to one where the maximum, not the minimum, probability is needed. For this, as we saw above, we can ignore fairness. The desired probabilities can then be obtained from the computed probabilities by subtracting from 1.

The dual problem is constructed as follows. Let S^+ be the set of states for which $p_s^{\max}(\phi_1 \mathcal{U} \phi_2) > 0$. This can be computed using the PROB0A precomputation algorithm from Chapter 2. Then, let $S^{\#}$ be the set $S^+ \setminus \mathrm{Sat}(\phi_2)$. Using atomic propositions a^+ and $a^{\#}$ to label states in the sets S^+ and $S^{\#}$, respectively, it is shown in [**11**] that, for a fair adversary A, $p_s^A(\phi_1 \mathcal{U} \phi_2) = 1 - p_s^A(a^{\#} \mathcal{U} \neg a^+)$. Hence, as described above, the problem reduces to computing $p_s^{\max}(a^{\#} \mathcal{U} \neg a^+)$, for which fairness is irrelevant, and we can use the method of Section 3.4.3.

EXAMPLE 3.5.2. We now revisit (3.4.3), (3.4.4), which illustrated model checking of the formulas $\mathcal{P}_{\leq 0.5}[\neg \mathrm{heads}\ \mathcal{U}\ \mathrm{tails}]$ and $\mathcal{P}_{>0.5}[\neg \mathrm{heads}\ \mathcal{U}\ \mathrm{tails}]$ on the MDP from Fig. 2. This time, we model check over fair adversaries only. For the first of these two formulas, computation of maximum probabilities is required and hence the solution remains unchanged. Notice that the adversary A_1, which we cited in (3.4.3) as an example of an adversary resulting in the maximum probabilities, is fair.

For the second formula, we follow the model checking procedure outlined above. Reusing the computation of S^{no} in (3.4.3), we compute:

$$\begin{aligned} S^+ &= S \setminus \mathrm{PROB0A}\big(\mathrm{Sat}(\neg \mathrm{heads}), \mathrm{Sat}(\mathrm{tails})\big) = \{0, 1, 3\}, \\ S^{\#} &= S^+ \setminus \mathrm{Sat}(\mathrm{tails}) = \{0, 1\}. \end{aligned}$$

We now consider the dual problem of calculating $p_s^{\max}(a^{\#} \mathcal{U} \neg a^+)$ where $\mathrm{Sat}(a^{\#}) = \{0, 1\}$ and $\mathrm{Sat}(\neg a^+) = \{2\}$. The precomputation algorithms yield the following:

$$S^{\mathrm{no}} = \mathrm{PROB0A}(\{0,1\}, \{2\}) = \{3\}, \quad S^{\mathrm{yes}} = \mathrm{PROB1E}(\{0,1\}, \{2\}) = \{2\}.$$

For the states in $S^? = \{0, 1\}$, we solve the linear optimisation problem:

$$\begin{array}{ll} \text{Minimise:} & x_0 + x_1 \\ \text{such that:} & x_0 \geq x_1, x_1 \geq 0.5 \quad \text{and} \quad x_1 \geq 0.7 \cdot x_0 + 0.3 \cdot x_1 \end{array}$$

which has the unique, optimal solution $(x_0, x_1) = (0.5, 0.5)$, meaning that the probabilities for the dual problem are $(0.5, 0.5, 1, 0)$. Finally, we subtract from one yielding the minimum probabilities $(0.5, 0.5, 0, 1)$, implying that the PCTL formula $\mathcal{P}_{>0.5}[\neg \mathrm{heads}\ \mathcal{U}\ \mathrm{tails}]$ is satisfied only in state 3. These values are different to those computed for all adversaries. Observe that the adversary A, described in (3.4.4) as an example of an adversary which resulted in the minimum probabilities, is not fair since it always selects the same probability distribution from the two available in state 1.

3.5.2. Complexity of PCTL model checking. The overall time complexity for model checking a PCTL formula ϕ against an MDP $\mathcal{M} = (S, \bar{s}, \mathrm{Steps})$ is, as in the case for DTMCs, linear in the size of the formula and polynomial in the size of the model. For MDPs, we define the size of the model, $|\mathcal{M}|$, to be $\sum_{s \in S} |\mathrm{Steps}(s)|$, i.e. the total number of nondeterministic choices. Since we require $\mathrm{Steps}(s)$ to be non-empty for all states s, it is always the case that $|\mathcal{M}| \geq |S|$. We are required to model check each operator of ϕ individually. The most expensive cases are the "until" and "expected cost" operators, for which we must solve a linear optimisation problem of size $|\mathcal{M}|$. Using, for example, the ellipsoid method, this can be done in polynomial time.

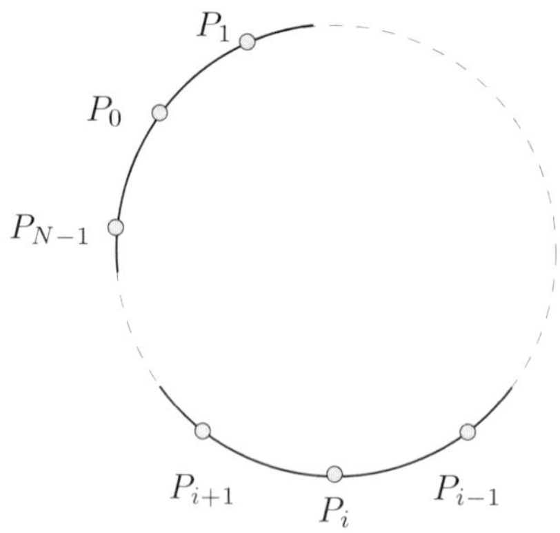

FIGURE 4. Ring of N processes.

3.6. Case Study: Self-Stabilisation Algorithms

In this section, we demonstrate how MDPs and PCTL model checking can be used to analyse two self-stabilisation protocols: those of Israeli and Jalfon [**75**] and Beauquier, Gradinariu and Johnen [**19**]. We begin by introducing the problem of self-stabilisation. Next, we outline the two protocols and explain how they can be represented as MDPs. Finally, we describe how PCTL can be used to both verify their correctness and study their performance.

A *self-stabilising protocol* for a network of processes is a protocol which, when initialised to some possibly illegal start state, returns to a legal state without any external intervention within some finite number of steps. We refer to the legal states of the system as *stable* states. In both protocols we study here, the network is a ring of N identical processes $P_0, P_1, \ldots, P_{N-1}$, as illustrated in Fig. 4, which are scheduled asynchronously. In this case, the stable states are those where exactly one process is designated as "privileged". A process is privileged if it is in possession of a *token*.

In the protocol of Israeli and Jalfon [**75**], each process P_i has a single boolean variable q_i which indicates whether or not the process has a token. When process P_i has a token, it is said to be *active*. A process can only be scheduled when it is active. When scheduled, it makes a (uniform) random choice as to whether to move the token to its left (to process $P_{i\ominus 1}$) or to its right (to process $P_{i\oplus 1}$), where $\ominus$ and $\oplus$ denote subtraction and addition modulo N. If a process acquires two tokens, they are merged into a single token.

In the protocol of Beauquier, Gradinariu and Johnen [**19**], the number of processes in the ring, N, must be odd. Each process has two boolean variables, d_i and p_i, which indicate the possession of two different types of token:

- if $d_i = d_{i-1}$, then process i is said to have a *deterministic token*;
- if $p_i = p_{i-1}$, then process i is said to have a *probabilistic token*.

A process is active and can be scheduled only if it has a deterministic token. When scheduled, a process sets d_i to be the negation of it current value (i.e. passes the deterministic token) and, if it also has a probabilistic token, makes a (uniform) random choice as to the next value of p_i (i.e. randomly selects whether to pass the probabilistic token or not). The stable states are those where there is only one probabilistic token.

We now explain how these protocols can be represented in terms of MDPs. For Israeli and Jalfon's protocol, the corresponding MDP $\mathcal{M} = (S, \bar{s}, \mathsf{Act}, \mathrm{Steps})$ is defined as follows:

- $S = \mathbb{B}^N$, i.e. each state is a boolean vector $\underline{q} = (q_0, q_1, \ldots, q_{N-1})$, the ith element of which indicates whether or not process P_i has a token;
- $\bar{s} = (1, \ldots, 1)$, the state where all processes have a token;
- $\mathsf{Act} = \{0, 1, \ldots, N-1\}$, where action i corresponds to the transitions where process i is scheduled;
- Steps: $S \to 2^{\mathsf{Act} \times \mathsf{Dist}(S)}$, defined as follows. For any $\underline{q} \in S$, we have $(i, \mu) \in \mathrm{Steps}(\underline{q})$ if and only if $q_i = 1$, where for any $\underline{q}'$:

$$\mu(\underline{q}') = \begin{cases} \frac{1}{2} & \text{if } \underline{q}'_{i\oplus 1} = 1, \underline{q}'_i = 0 \text{ and } \underline{q}'_j = \underline{q}_j \text{ for all } j \neq i, i \oplus 1 \\ \frac{1}{2} & \text{if } \underline{q}'_{i\ominus 1} = 1, \underline{q}'_i = 0 \text{ and } \underline{q}'_j = \underline{q}_j \text{ for all } j \neq i, i \ominus 1 \\ 0 & \text{otherwise.} \end{cases}$$

For the protocol of Beauquier, Gradinariu and Johnen [**19**], on the other hand, the MDP $\mathcal{M} = (S, \bar{s}, \mathsf{Act}, \mathrm{Steps})$ is defined as follows:

- $S = \mathbb{B}^N \times \mathbb{B}^N$, where each state is a pair boolean vectors $(\underline{d}, \underline{p})$, where $\underline{d} = (d_0, d_1, \ldots, d_{N-1})$ and $\underline{p} = (p_0, p_1, \ldots, p_{N-1})$, i.e. the ith elements of $\underline{d}$ and $\underline{p}$ indicate whether process P_i has a probabilistic and deterministic token, respectively;
- $\bar{s} = \big((1, \ldots, 1), (1, \ldots, 1)\big)$, the state where all processes have both a deterministic and a probabilistic token;
- $\mathsf{Act} = \{0, 1, \ldots, N-1\}$, where action i corresponds to the transitions where process i is scheduled;
- Steps $\to S \to 2^{\mathsf{Act} \times \mathsf{Dist}(S)}$, defined as follows. For any $(\underline{d}, \underline{p}) \in S$, we have $(i, \mu) \in \mathrm{Steps}(\underline{d}, \underline{p})$ if and only if $\underline{d}_i = \underline{d}_{i-1}$, where for any $(\underline{d}', \underline{p}')$:

$$\mu(\underline{d}', \underline{p}') = \begin{cases} 1 & \text{if } \underline{p}_i \neq \underline{p}_{i-1}, d'_i \neq d_i, \underline{p}' = \underline{p} \text{ and } \underline{d}'_j - \underline{d}_j \text{ for all } j \neq i \\ \frac{1}{2} & \text{if } \underline{p}_i = \underline{p}_{i-1}, \underline{d}'_i \neq \underline{d}_i, \underline{p}'_i = 0 \text{ and } \underline{d}'_j = \underline{d}_j, \underline{p}'_j = \underline{p}_j \text{ for all } j \neq i \\ \frac{1}{2} & \text{if } \underline{p}_i = \underline{p}_{i-1}, \underline{d}'_i \neq \underline{d}_i, \underline{p}'_i = 1 \text{ and } \underline{d}'_j = \underline{d}_j, \underline{p}'_j = \underline{p}_j \text{ for all } j \neq i \\ 0 & \text{otherwise.} \end{cases}$$

For the labelling of these MDPs, we let the atomic proposition stable denote the stable states of either protocol. Hence, for Israeli and Jalfon's protocol:

$$L(\text{stable}) = \big\{\underline{q} \in \mathbb{B}^N \mid \exists 0 \leq i \leq N \,.\, \big(\underline{q}_i = 1 \wedge \forall 0 \leq j \leq N \,.\, (j \neq i \to \underline{q}_j = 0)\big)\big\}$$

and in the case of the protocol of Beauquier, Gradinariu and Johnen, we set $L(\text{stable})$ equal to:

$$\big\{(\underline{d}, \underline{p}) \in \mathbb{B}^N \times \mathbb{B}^N \mid \exists 0 \leq i \leq N \,.\, \big(\underline{p}_i = \underline{p}_{i-1} \wedge \forall 0 \leq j \leq N \,.\, (j \neq i \to \underline{p}_j \neq \underline{p}_{j-1})\big)\big\}.$$

For both protocols, we define a cost structure which assigns a cost of 1 to all actions in all states. This allows us to reason about the total number of steps required by the protocols. Below, we formalise in PCTL the properties of the protocols we wish to study. Note that, since the processes in both protocols operate asynchronously under arbitrary schedulers, no fairness conditions are required for model checking.

- $\mathcal{P}_{\geq 1}[\Diamond \text{stable}]$—under any adversary a stable state is reached with probability 1;

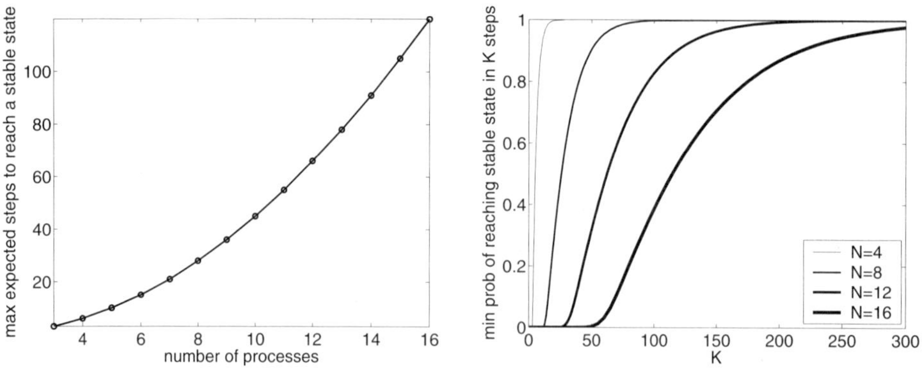

FIGURE 5. Results for the protocol of Israeli and Jalfon.

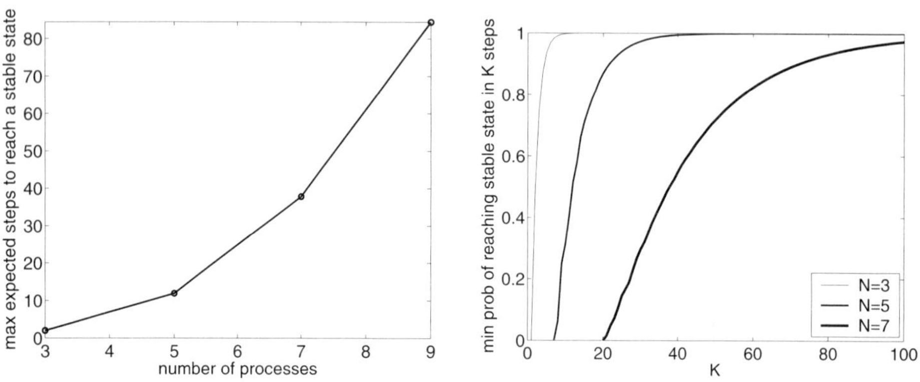

FIGURE 6. Results for the protocol of Beauquier, Gradinariu and Johnen.

- $\mathcal{P}_{\geq p}[\Diamond^{\leq k}\text{stable}]$—under any adversary the probability that a stable state is reached within k steps is at least p;
- $\mathcal{E}_{\leq r}[\text{stable}]$—under any adversary the expected number of steps to reach a stable state is at most r.

In order to consider these properties with respect to all possible initial configurations of the protocols, we extend the MDP models described to include an initial configuration step (with cost zero) in which the values of the variables of each process are set nondeterministically. It then follows, for example, that if $\mathcal{P}_{\geq 1}[\Diamond\text{stable}]$ holds, then for all initial configurations and adversaries a stable state is reached with probability 1.

The results of model checking results can be summarised as follows. Firstly, $\mathcal{P}_{\geq 1}[\Diamond\text{stable}]$ holds for both protocols, confirming that they are self-stabilisation algorithms. For the second and third properties, Figs. 5 and 6 illustrate the probabilities and expected costs obtained for the initial state of each of the two protocols. More specifically, the values plotted are, over all adversaries and initial configurations, the minimum probability of reaching a stable state within k steps and the maximum expected number of steps for this to occur. In the graphs, we vary both the number of processes, N, and the bound, k. We see that, as the number of processes in the ring increases, the probability of reaching a stable state within k steps decreases and the expected number of steps to reach a stable state increases. See the PRISM website [**111**] for more information.

3.7. Discussion

MDPs, as presented in this chapter, are essentially the same as the "concurrent probabilistic systems" of [**17**] and the "simple probabilistic automata" of [**117**]. The model can be also seen as a generalisation of the "concurrent Markov chains" of [**127**]. MDPs naturally arise as models of randomised distributed algorithms. Several such algorithms have been verified in addition to the case study mentioned here, including [**84, 90**].

The logic PCTL discussed in this chapter is due to [**25**], extended with the expectation operator based on [**44**]. More expressive logics include LTL and PCTL* [**9, 25**], a probabilistic extension of CTL*. Model checking algorithms for these logics can be found in [**9, 17, 25, 41, 127**]. These involve a transformation of the MDP to a larger one. Work on quantitative extensions of the μ-calculus can be found in [**73, 102**].

One of the principal challenges for probabilistic model checking of MDPs is to increase the size and complexity of model which can feasibly be analysed. Promising approaches are those based on abstraction and compositionality [**46, 50, 101**].

CHAPTER 4

Continuous-Time Markov Chains

4.1. Introduction

The third type of model we consider is the *continuous-time Markov chain* (CTMC). This model also extends discrete-time Markov chains (DTMCs) but in a different way. While each transition between states in a DTMC corresponds to a discrete time-step, in a CTMC transitions can occur in real time. Formally, a CTMC is a family of random variables $\{X(t) \mid t \geq 0\}$, where $X(t)$ is an observation made at time instant t and t varies over the non-negative reals. The state space, the set of all possible values taken by $X(t)$, is discrete and, for the purposes of this book, will in fact be assumed to be finite. Like a DTMC, a CTMC must satisfy the Markov (memoryless) property, which in this case can be expressed as follows: for any integer $k \geq 0$, sequence of time instances $t_0 < t_1 < \cdots < t_k$ and states $s_0, \ldots, s_k$:

$$P[X(t_k)=s_k \mid X(t_{k-1})=s_{k-1}, \ldots, X(t_0)=s_0]=P[X(t_k)=s_k \mid X(t_{k-1})=s_{k-1}].$$

Intuitively, this means that the probability of making a transition to a particular state depends only on the current state, not the previous history of states passed through. Moreover, the probability is also independent of the time which has elapsed so far in the current state. There is only one continuous probability distribution which exhibits this memoryless property: the exponential distribution (sometimes referred to as the negative exponential distribution). Rather than associating each pair of states with a probability, as in a DTMC, we assign it a *rate* λ, which is used as the parameter of the exponential distribution. The probability of making a transition between the pair of states within t time-units is $1 - e^{-\lambda \cdot t}$. Intuitively, the rate λ represents the average number of times that the transition would occur per unit of time. Formally, we define a CTMC as follows.

DEFINITION 4.1.1. A CTMC is a tuple $(S, \bar{s}, \mathbf{R})$ where:

- S is a finite set of *states*;
- $\bar{s} \in S$ is the *initial state*;
- $\mathbf{R}\colon S \times S \to \mathbb{R}_{\geq 0}$ is the *transition rate matrix*.

The transition rate matrix $\mathbf{R}$ assigns rates to each pair of states in the CTMC. A transition can only occur between states s and s' if $\mathbf{R}(s, s') > 0$. The likelihood of this transition occurring is determined by the value of $\mathbf{R}(s, s')$, as described above. Typically, in a state s, there is more than one state s' for which $\mathbf{R}(s, s') > 0$. This is known as a *race condition*. The first transition to occur determines the next state of the CTMC. The time spent in state s, before any such transition occurs, is exponentially distributed with rate $E(s)$, where:

$$E(s) \stackrel{\text{def}}{=} \sum_{s' \in S} \mathbf{R}(s, s').$$

$E(s)$ is known as the *exit rate* of state s. We can also determine the actual probability of each state s' being the next state to which a transition is made from state s, independent of the time at which this occurs. This is defined by the following DTMC.

DEFINITION 4.1.2. The *embedded Markov chain* $\text{emb}(\mathcal{C})$ of a CTMC $\mathcal{C} = (S, \bar{s}, \mathbf{R})$ is the DTMC $(S, \bar{s}, \mathbf{P})$ where for $s, s' \in S$:

$$\mathbf{P}(s, s') = \begin{cases} \mathbf{R}(s, s')/E(s) & \text{if } E(s) \neq 0 \\ 1 & \text{if } E(s) = 0 \text{ and } s = s' \\ 0 & \text{otherwise.} \end{cases}$$

Notice that, unlike for DTMCs and MDPs, we allow states of a CTMC to have no outgoing transitions, i.e. the case where $E(s) = 0$. Such a state is called an *absorbing state*.

Using the above definitions, we can consider the behaviour of the CTMC in an alternative way. It will remain in a state s for a delay which is exponentially distributed with rate $E(s)$ and then make a transition. The probability that this transition is to state s' is given by $\mathbf{P}(s, s')$.

We also define the following matrix, which will be used when we perform analysis of the CTMC.

DEFINITION 4.1.3. The *infinitesimal generator matrix* for the CTMC $(S, \bar{s}, \mathbf{R})$ is the matrix $\mathbf{Q}\colon S \times S \to \mathbb{R}$ defined as:

$$\mathbf{Q}(s, s') = \begin{cases} \mathbf{R}(s, s') & \text{if } s \neq s' \\ -\sum_{s'' \neq s} \mathbf{R}(s, s'') & \text{otherwise.} \end{cases}$$

As for DTMCs and MDPs, in order to perform analysis we annotate CTMCs with additional information. As before, we use a labelling function L, which maps states to sets of atomic propositions from a set AP. In addition, we assign costs to the model but, for CTMCs, there are two distinct types of cost: *instantaneous* costs, which are assigned to transitions, and *cumulative* costs, which are assigned to states. These are defined using the cost functions $\mathbf{C}$ and $\mathbf{c}$, respectively. The instantaneous cost $\mathbf{C}(s, s')$ is incurred each time a transition between states s and s' occurs. The cumulative cost $\mathbf{c}(s)$ is the rate at which cost is incurred in state s, i.e. a cost of $t \cdot \mathbf{c}(s)$ is incurred if the CTMC remains in state s for t time units.

DEFINITION 4.1.4. A *labelled* CTMC is a tuple $(\mathcal{C}, L, \mathbf{C}, \mathbf{c})$ where:

- $\mathcal{C} = (S, \bar{s}, \mathbf{R})$ is a CTMC;
- $L\colon S \to 2^{\text{AP}}$ is a *labelling function*;
- $\mathbf{C}\colon S \times S \to \mathbb{R}^{\geq 0}$ is an *instantaneous cost function*;
- $\mathbf{c}\colon S \to \mathbb{R}^{\geq 0}$ is a *cumulative cost function*.

EXAMPLE 4.1.5. Figure 1 shows a simple example of a CTMC with four states $\{0, 1, 2, 3\}$. The graphical notation is identical to that of DTMCs, except that transitions are now labelled with rates rather than probabilities. The CTMC models a queue of jobs, where state i indicates that there are i jobs in the queue. Initially, the queue is empty and the maximum size is 3. Jobs arrive with rate 1.5 and and

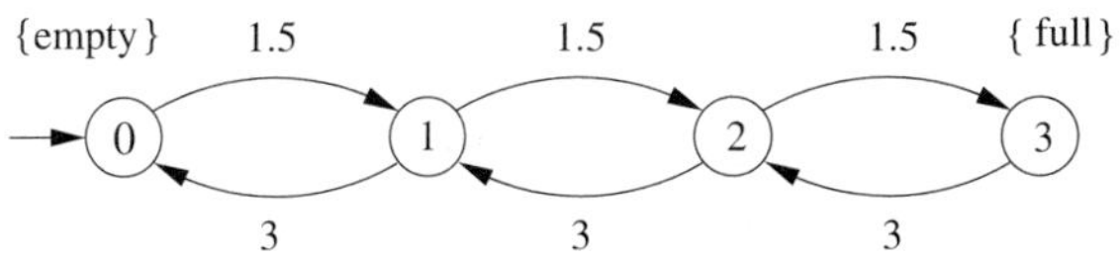

FIGURE 1. A four state CTMC.

are removed from the queue with rate 3. The associated transition rate matrix **R**, infinitesimal generator matrix **Q** and transition probability matrix **P** for the embedded Markov chain are as follows:

$$\mathbf{R} = \begin{pmatrix} 0 & 1.5 & 0 & 0 \\ 3 & 0 & 1.5 & 0 \\ 0 & 3 & 0 & 1.5 \\ 0 & 0 & 3 & 0 \end{pmatrix},$$

$$\mathbf{Q} = \begin{pmatrix} -1.5 & 1.5 & 0 & 0 \\ 3 & -4.5 & 1.5 & 0 \\ 0 & 3 & -4.5 & 1.5 \\ 0 & 0 & 3 & -3 \end{pmatrix}, \quad \mathbf{P} = \begin{pmatrix} 0 & 1 & 0 & 0 \\ \frac{2}{3} & 0 & \frac{1}{3} & 0 \\ 0 & \frac{2}{3} & 0 & \frac{1}{3} \\ 0 & 0 & 1 & 0 \end{pmatrix}.$$

We have labelled the state where the queue contains no jobs with the atomic proposition empty and the state where it has the maximum number of jobs with full. These are also illustrated in Fig. 1.

4.2. Paths and Probability Measures

A path in a CTMC is a non-empty sequence $s_0 t_0 s_1 t_1 s_2 \ldots$ where $\mathbf{R}(s_i, s_{i+1}) > 0$ and $t_i \in \mathbb{R}_{>0}$ for all $i \geq 0$. The value t_i represents the amount of time spent in the state s_i. As with DTMCs and MDPs, we denote by $\omega(i)$ the ith state of a path ω, namely s_i. In addition, we denote by $\mathrm{time}(\omega, i)$ the amount of time spent in state s_i, namely t_i, and by $\omega@t$ the state occupied at time t, i.e. $\omega(k)$ where k is the smallest index for which $\sum_{i=0}^{k} t_i \geq t$.

We again denote by Path_s the set of all infinite paths starting in state s. The probability measure Prob_s over Path_s, taken from [**16**], can be defined as follows. If the states $s_0, \ldots, s_n \in S$ satisfy $\mathbf{R}(s_i, s_{i+1}) > 0$ for all $0 \leq i < n$ and $I_0, \ldots, I_{n-1}$ are non-empty intervals in $\mathbb{R}_{\geq 0}$, then the *cylinder set* $C(s_0, I_0, \ldots, I_{n-1}, s_n)$ is defined to be the set containing infinite all paths $s'_0 t_0 s'_1 t_1 s'_2 \ldots$ where $s_i = s'_i$ for $i \leq n$ and $t_i \in I_i$ for $i < n$.

We then let Σ_s be the smallest σ-algebra (see Appendix A) on Path_s which contains all the cylinder sets $C(s_0, I_0, \ldots, I_{n-1}, s_n)$, where $s_0, \ldots, s_n \in S$ range over all sequences of states with $s_0 = s$ and $\mathbf{R}(s_i, s_{i+1}) > 0$ for $0 \leq i < n$, and $I_0, \ldots, I_{n-1}$ range over all sequences of non-empty intervals in $\mathbb{R}_{\geq 0}$. The probability measure Prob_s on Σ_s is then the unique measure defined inductively by $\mathrm{Prob}_s(C(s_0)) = 1$ and $\mathrm{Prob}_s(C(s_0, \ldots, s_n, I_n, s_{n+1}))$ equal to:

$$\mathrm{Prob}_s(C(s_0, \ldots, s_n)) \cdot \mathbf{P}(s_n, s_{n+1}) \cdot (e^{-\inf I_n \cdot E(s_n)} - e^{-\sup I_n \cdot E(s_n)})$$

where **P** is the transition probability matrix of the CTMC's embedded Markov chain.

EXAMPLE 4.2.1. Consider the CTMC from Fig. 1 and the sequence of states and intervals $0, [0,2], 1$ (i.e. taking $s_0 = 0$, $I_0 = [0,2]$ and $s_1 = 1$ in the notation of the previous paragraph). Using the probability measure for the initial state, 0, of the CTMC, for the cylinder set $C(0,[0,2],1)$, we have:

$$\mathrm{Prob}_0\big(C(0,[0,2],1)\big) = 1\cdot\mathbf{P}(0,1)\cdot(e^{-0\cdot E(0)} - e^{-2\cdot E(0)}) = 1\cdot 1\cdot(e^0 - e^{-3}) = 1 - e^{-3}.$$

Intuitively, this means that the probability of leaving the initial state 0 and passing to state 1 within the first 2 time units is $1 - e^{-3} \simeq 0.950213$.

Using this probability measure we can now define, as we did for DTMCs and MDPs, the expected cost of reaching a set of target states. For a state s, set of target states $F \subseteq S$ and path $\omega \in \mathrm{Path}_s$, we define the cost $\mathrm{cost}(F)(\omega)$ of reaching a state in F along the path ω as follows:

$$\mathrm{cost}(F)(\omega) \stackrel{\mathrm{def}}{=} \begin{cases} \displaystyle\sum_{i=1}^{\min\{j|\omega(j)\in F\}} \mathbf{C}\big(\omega(i-1),\omega(i)\big)+\mathbf{c}\big(\omega(i-1)\big)\cdot \mathrm{time}(\omega,i-1) & \text{if } \exists j\in\mathbb{N} \,.\, \omega(j)\in F \\ \infty & \text{otherwise.} \end{cases}$$

Letting E_s be the usual expectation with respect to the measure Prob_s (see [**24**]), the expected cost of reaching F from s is then $E_s\big(\mathrm{cost}(F)\big)$. As previously, we always define the cost of a path which does not reach F to be ∞, and hence the expected cost of reaching F from s is finite if and only if a state in F is reached from s with probability 1.

4.3. Steady-State and Transient Behaviour

In addition to path probabilities, we consider two more traditional properties of CTMCs: *transient* behaviour, which relates to the state of the model at a particular time instant; and *steady-state* behaviour, which describes the state of the CTMC in the long run. The transient probability $\pi_{s,t}(s')$ is defined as the probability, having started in state s, of being in state s' at time instant t. Using the definitions of the previous section:

$$\pi_{s,t}(s') \stackrel{\mathrm{def}}{=} \mathrm{Prob}_s(\{\omega \in \mathrm{Path}_s \mid \omega@t = s'\}).$$

The steady-state probability $\pi_s(s')$, the probability of, having started in state s, being in state s' in the long run, can now be defined as:

$$\pi_s(s') \stackrel{\mathrm{def}}{=} \lim_{t\to\infty} \pi_{s,t}(s').$$

The steady-state probability distribution, i.e. the values $\pi_s(s')$ for all $s' \in S$, can be used to infer the percentage of time, in the long run, that the CTMC spends in each state. For the class of CTMCs which we consider here, i.e. those that are homogeneous and finite-state, the limit in the above definition always exists [**122**]. Furthermore, for CTMCs which are *irreducible*, that is, those for which there exists a finite path from each of its states to every other state, the steady-state probabilities $\pi_s(s')$ are independent of the initial state s.

4.4. Continuous Stochastic Logic (CSL)

We write specifications of CTMCs using the logic CSL (Continuous Stochastic Logic), an extension of the logic PCTL used for DTMCs and MDPs. CSL was first introduced in [**8**] and later extended in [**16**]. It allows reasoning about both path-based properties as well as steady-state and transient behaviour.

DEFINITION 4.4.1. The syntax of CSL is as follows:

$$\phi ::= \text{true} \mid a \mid \neg\phi \mid \phi \wedge \phi \mid \mathcal{P}_{\bowtie p}[\psi] \mid \mathcal{S}_{\bowtie p}[\phi] \mid \mathcal{E}_{\bowtie c}[\phi]$$
$$\psi ::= \phi \mid \phi\,\mathcal{U}^{I}\,\phi \mid \phi\mathcal{U}\phi$$

where a is an atomic proposition, $\bowtie \in \{\leq, <, \geq, >\}$, $p \in [0,1]$, $c \in \mathbb{R}^{\geq 0}$ and I is an interval of $\mathbb{R}^{\geq 0}$.

As for PCTL, $\mathcal{P}_{\bowtie p}[\psi]$ indicates that the probability of the path formula ψ being satisfied from a given state meets the bound $\bowtie p$. Path formulas are the same for CSL as for PCTL, except that the parameter of the bounded version of the "until" operator is an interval I of the non-negative reals, rather than simply an integer upper bound. The path formula $\phi\,\mathcal{U}^{I}\,\phi$ holds if ϕ_2 is satisfied at some time instant in the interval I and ϕ_1 holds at all preceding time instants. To avoid confusion, we will refer to this as the "time-bounded until" operator.

The $\mathcal{S}$ operator describes the steady-state behaviour of the CTMC. The formula $\mathcal{S}_{\bowtie p}[\phi]$ asserts that the steady-state probability of being in a state satisfying ϕ meets the bound $\bowtie p$. As in the previous two chapters, the formula $\mathcal{E}_{\bowtie c}[\phi]$ states that the expected cost to reach a state satisfying ϕ meets the bound $\bowtie c$. The only difference is that, in this case, this cost is actually a combination of instantaneous and cumulative costs.

As with PCTL, we write $s \models \phi$ to indicate that a CSL formula ϕ is satisfied in a state s of a CTMC and denote by $\text{Sat}(\phi)$ the set $\{s \in S \mid s \models \phi\}$. Similarly, for a path formula ψ satisfied by path ω, we write $\omega \models \psi$. The semantics of CSL over CTMCs are as follows.

DEFINITION 4.4.2. Let $\mathcal{C} = (S, \bar{s}, \mathbf{R}, L, \mathbf{C}, \mathbf{c})$ be a labelled CTMC. For any state $s \in S$ the relation $s \models \phi$ is defined inductively by:

$$\begin{aligned}
&s \models \text{true} && \text{for all } s \in S,\\
&s \models a && \Leftrightarrow a \in L(s),\\
&s \models \neg\phi && \Leftrightarrow s \not\models \phi,\\
&s \models \phi_1 \wedge \phi_2 && \Leftrightarrow s \models \phi_1 \wedge s \models \phi_2,\\
&s \models \mathcal{P}_{\bowtie p}[\psi] && \Leftrightarrow p_s(\psi) \bowtie p,\\
&s \models \mathcal{S}_{\bowtie p}[\phi] && \Leftrightarrow \sum_{s' \models \phi} \pi_s(s') \bowtie p,\\
&s \models \mathcal{E}_{\bowtie c}[\phi] && \Leftrightarrow e_s(\phi) \bowtie c,
\end{aligned}$$

where:

$$p_s(\psi) \stackrel{\text{def}}{=} \text{Prob}_s(\{\omega \in \text{Path}_s \mid \omega \models \psi\}),$$
$$e_s(\phi) \stackrel{\text{def}}{=} E_s\big(\text{cost}(\{t \in S \mid t \models \phi\})\big)$$

and for any path $\omega \in \text{Path}_s$:

$$\omega \models \phi \Leftrightarrow \omega(1) \text{ is defined and } \omega(1) \models \phi,$$
$$\omega \models \phi_1 \mathcal{U}^I \phi_2 \Leftrightarrow \exists t \in I \,.\, (\omega@t \models \phi_2 \wedge \omega@x \models \phi_1 \;\forall x \in [0,t)),$$
$$\omega \models \phi_1 \mathcal{U} \phi_2 \Leftrightarrow \exists k \geq 0 \,.\, \big(\omega(k) \models \phi_2 \wedge \omega(j) \models \phi_1 \;\forall j < k\big).$$

As discussed in [**16**], for any path formula ψ, the set $\{\omega \in \text{Path}_s \mid \omega \models \psi\}$ is a measurable set of Path_s, and hence Prob_s is well defined over this set.

As with PCTL, we can easily derive CSL operators for false, $\vee$ and $\rightarrow$. Similarly, we can use the $\Diamond$ and $\Box$ temporal operators:

$$\mathcal{P}_{\bowtie p}[\Diamond\phi] \equiv \mathcal{P}_{\bowtie p}[\text{true } \mathcal{U}\phi],$$
$$\mathcal{P}_{\bowtie p}[\Diamond^I\phi] \equiv \mathcal{P}_{\bowtie p}[\text{true } \mathcal{U}^I\phi],$$
$$\mathcal{P}_{\bowtie p}[\Box\phi] \equiv \mathcal{P}_{\overline{\bowtie} 1-p}[\Diamond\neg\phi],$$
$$\mathcal{P}_{\bowtie p}[\Box^I\phi] \equiv \mathcal{P}_{\overline{\bowtie} 1-p}[\Diamond^I\neg\phi].$$

It is also worth noting that, despite the fact that CSL does not explicitly include operators to reason about transient probabilities, the following can be used to reason about the probability of satisfying a formula ϕ at time instant t:

$$\mathcal{P}_{\bowtie p}[\Diamond^{[t,t]}\phi].$$

Again, similarly to PCTL, when the outermost operator of a CSL formula is $\mathcal{P}_{\bowtie p}$, $\mathcal{S}_{\bowtie p}$ or $\mathcal{E}_{\bowtie c}$, we can omit the bound ($\bowtie p$ or $\bowtie c$) and return the actual value of the associated probability or expected cost.

4.5. CSL Model Checking

In this section, we consider the computation of probabilities for the CSL "next", "time-bounded", "until", "steady-state" and "expected cost" operators. CSL model checking was shown to be decidable (for rational time bounds) in [**8**] and a model checking algorithm first presented in [**16**]. We use these techniques plus the subsequent improvements made in [**14, 78**].

4.5.1. The $\mathcal{P}_{\bowtie p}[\phi]$ and $\mathcal{P}_{\bowtie p}[\phi_1 \mathcal{U} \phi_2]$ Operators. The CSL "next" and "unbounded until" operators are defined as for the PCTL equivalents. Furthermore, their definition does not relate to any of the real-time aspects of CTMCs: the "next" operator depends only on the probability of moving to the next immediate state; and the "until" operator depends on the probability of eventually reaching some states. Hence, these two operators can actually be model checked by using the PCTL algorithms of Section 2.4.1 and Section 2.4.3, respectively, on the embedded Markov chain of the CTMC.

EXAMPLE 4.5.1. Consider the CTMC from Fig. 1 and the CSL "next" formula $\mathcal{P}_{\geq 0.5}[\text{full}]$. Working with the embedded Markov chain and following the algorithm of Section 2.4.1, we multiply the matrix $\mathbf{P}$, given in (4.1.5) by the vector $(0,0,0,1)$, yielding the probabilities $\underline{p(\text{full})} = (0,0,\frac{1}{3},0)$. Hence, the formula is not true in any state of the CTMC.

4.5.2. The $\mathcal{P}_{\bowtie p}[\phi_1 \mathcal{U}^I \phi_2]$ operator. For this operator, we need to determine the probabilities $p_s(\phi_1 \mathcal{U}^I \phi_2)$ for all states s where I is an arbitrary interval of the non-negative real numbers. Noting that:

$$p_s(\phi_1 \mathcal{U}^I \phi_2) = p_s(\phi_1 \mathcal{U}^{\mathrm{cl}(I)} \phi_2)$$

where $\mathrm{cl}(I)$ is the closure of the interval I, and that:

$$p_s(\phi_1 \mathcal{U}^{[0,\infty)} \phi_2) = p_s(\phi_1 \mathcal{U} \phi_2)$$

we need only deal with the following three cases for the interval I:

- $I = [0, t]$ for $t \geq 0$;
- $I = [t, t']$ for $t > 0$ and $t' \geq t$;
- $I = [t, \infty)$ for $t > 0$.

Following the methods proposed in [**14, 78**], we will see that these probabilities can be computed using variants of standard techniques for transient analysis. Hence, we begin with a slight detour into this problem.

4.5.2.1. *Transient analysis.* To compute the transient probabilities $\pi_{s,t}(s')$ for all states $s, s' \in S$, we will use a technique called *uniformisation* (sometimes also known as "randomisation" or "Jensen's method"). We denote by $\mathbf{\Pi}_t$ the matrix of all such probabilities, i.e. $\mathbf{\Pi}_t(s, s') = \pi_{s,t}(s')$. It can be shown (see for example [**122**]) that $\mathbf{\Pi}_t$ can be expressed as a matrix exponential, and hence evaluated as a power series:

$$\mathbf{\Pi}_t = e^{\mathbf{Q}\cdot t} = \sum_{i=0}^{\infty} \frac{(\mathbf{Q}\cdot t)^i}{i!}.$$

Unfortunately, this computation tends to be unstable. As an alternative, we construct the *uniformised* DTMC corresponding to the CTMC, whose transition probability matrix $\mathbf{P}$ is defined as:

$$\mathbf{P} = \mathbf{I} + \mathbf{Q}/q \text{ where } q \geq \max_{s\in S} |\mathbf{Q}(s, s)|.$$

The matrix of transient probabilities can therefore be expressed as:

$$\mathbf{\Pi}_t = \sum_{i=0}^{\infty} \gamma_{i,q\cdot t} \cdot \mathbf{P}^i \text{ where } \gamma_{i,q\cdot t} = e^{-q\cdot t} \cdot (q \cdot t)^i / i!.$$

In fact, this reformulation has a fairly intuitive explanation. Each step of the uniformised DTMC corresponds to one exponentially distributed delay, with parameter q, in the CTMC. The matrix power $\mathbf{P}^i$ gives the probability of jumping between each pair of states in the DTMC in i steps and $\gamma_{i,q\cdot t}$ is the ith Poisson probability with parameter $q \cdot t$, the probability of i such steps occurring in time t, given that the delay is exponentially distributed with rate q.

This approach has a number of important advantages. Firstly, unlike $\mathbf{Q}$, the matrix $\mathbf{P}$ is *stochastic*, meaning that all entries are in the range $[0, 1]$ and all rows sum to one. Computations using $\mathbf{P}$ are therefore more numerically stable. In particular, $\mathbf{Q}$ contains both positive and negative values which can cause severe round-off errors.

Secondly, the infinite sum is now easy to truncate. For example, the techniques of Fox and Glynn [**57**], which allow efficient computation of the Poisson probabilities $\gamma_{i,q\cdot t}$, also produce an upper and lower bound $(L_\varepsilon, R_\varepsilon)$, for some desired precision ε, below and above which the probabilities are insignificant. Hence, the sum can be computed only over this range.

Lastly, the computation can be carried out efficiently using vector-matrix multiplications, rather than more costly matrix-matrix multiplications. Consider the problem of computing $\pi_{s,t}(s')$ for a fixed state s. These values can be obtained by pre-multiplying the matrix $\mathbf{\Pi}_t$ by the initial probability distribution, in this case the vector $\underline{\pi_{s,0}}$ where $\underline{\pi_{s,0}}(s')$ is equal to 1 if $s' = s$ and 0 otherwise:

$$\underline{\pi_{s,t}} = \underline{\pi_{s,0}} \cdot \mathbf{\Pi}_t = \underline{\pi_{s,0}} \cdot \sum_{i=0}^{\infty} \gamma_{i,q\cdot t} \cdot \mathbf{P}^i.$$

Rearranging, this can be expressed as a sum of vectors, rather than a sum of matrix powers:

$$\underline{\pi_{s,t}} = \sum_{i=0}^{\infty} (\gamma_{i,q\cdot t} \cdot \underline{\pi_{s,0}} \cdot \mathbf{P}^i)$$

where the vector required in each element of the summation is computed by a single vector-matrix multiplication, using the vector from the previous iteration:

$$\underline{\pi_{s,0}} \cdot \mathbf{P}^i = (\underline{\pi_{s,0}} \cdot \mathbf{P}^{i-1}) \cdot \mathbf{P}.$$

Hence, the total work required is R_ε vector-matrix multiplications.

4.5.2.2. *The case $I = [0, t]$.* We now return to the problem of computing probabilities for the $\mathcal{P}_{\bowtie p}[\phi_1 \mathcal{U}^I \phi_2]$ operator. We begin with the case where $I = [0, t]$ for some $t \geq 0$, i.e. we compute $p_s(\phi_1 \mathcal{U}^{[0,t]} \phi_2)$ for all states $s \in S$. The probability for states which satisfy $\neg\phi_1 \wedge \neg\phi_2$ or ϕ_2 is trivially 0 or 1, respectively. The probabilities for the remaining states must be computed numerically. Originally, [**16**] proposed to do this via approximate solution of Volterra integral equation systems. Experiments in [**68**] showed that this method was generally slow and, in [**14**], a simpler alternative was presented which reduces the problem to transient analysis. The basic idea is to modify the CTMC, making all states which satisfy $\neg\phi_1 \wedge \neg\phi_2$ or ϕ_2 absorbing (i.e. removing all their outgoing transitions). For this purpose, we introduce the following notation. For a CTMC $\mathcal{C} = (S, \bar{s}, \mathbf{R})$ and CSL formula ϕ:

$$\mathcal{C}[\phi] \stackrel{\text{def}}{=} (S, \bar{s}, \mathbf{R}[\phi]) \text{ where } \mathbf{R}[\phi](s, s') = \begin{cases} \mathbf{R}(s, s') & \text{if } s \not\models \phi \\ 0 & \text{otherwise} \end{cases}$$

i.e. $\mathcal{C}[\phi]$ is $\mathcal{C}$ with states satisfying ϕ made absorbing.

Consider now the CTMC $C[(\neg\phi_1 \wedge \neg\phi_2) \vee \phi_2] = C[\neg\phi_1 \vee \phi_2]$. Since a path in this CTMC cannot exit a state satisfying ϕ_2 once it reaches one, and will never be able to reach a state satisfying ϕ_2 if it enters one satisfying $\neg\phi_1 \wedge \neg\phi_2$, the probability of the path formula $\phi_1 \mathcal{U}^{[0,t]} \phi_2$ being satisfied in CTMC $\mathcal{C}$ is equivalent to the transient probability of being in a state satisfying ϕ at time t in CTMC $C[\neg\phi_1 \vee \phi_2]$. Using superscripts to differentiate notation for different CTMCs:

$$\begin{aligned} p_s^{\mathcal{C}}(\phi_1 \mathcal{U}^{[0,t]} \phi_2) &= \sum_{s' \models \phi_2} \pi_{s,t}^{\mathcal{C}[\neg\phi_1 \vee \phi_2]}(s') \\ &= \underline{\pi_{s,t}^{\mathcal{C}[\neg\phi_1 \vee \phi_2]}} \cdot \underline{\phi_2} \end{aligned}$$

where $\underline{\pi_{s,t}^{\mathcal{C}[\neg\phi_1 \vee \phi_2]}}$ is the vector of probabilities $\pi_{s,t}^{\mathcal{C}[\neg\phi_1 \vee \phi_2]}(s')$ for all $s' \in S$ and $\underline{\phi_2}$ is a column vector with $\underline{\phi_2}(s)$ equal to 1 if $s \models \phi_2$ and 0 otherwise.

As shown in [**78**], uniformisation can be adapted to compute the probabilities for the CSL "time-bounded until" operator. Let $\underline{p(\phi_1 \mathcal{U}^{[0,t]} \phi_2)}$ be the required vector of probabilities $p_s(\phi_1 \mathcal{U}^{[0,t]} \phi_2)$ for each state s. We have:

$$\begin{aligned} \underline{p^{\mathcal{C}}(\phi_1 \mathcal{U}^{[0,t]} \phi_2)} &= \mathbf{\Pi}_t^{\mathcal{C}[\neg\phi_1 \vee \phi_2]} \cdot \underline{\phi_2} \\ &= \left(\sum_{i=0}^{\infty} \gamma_{i,q\cdot t} \cdot \mathbf{P}^i \right) \cdot \underline{\phi_2} \\ &= \sum_{i=0}^{\infty} (\gamma_{i,q\cdot t} \cdot \mathbf{P}^i \cdot \underline{\phi_2}) \end{aligned}$$

where $\mathbf{P}$ is the transition probability matrix of the uniformised DTMC of the CTMC $\mathcal{C}[\neg\phi_1 \vee \phi_2]$. Note that the inclusion of the vector $\underline{\phi_2}$ within the brackets is vital since, like with uniformisation, it allows explicit computation of the matrix powers $\mathbf{P}^i$ to be avoided. Instead, each product $\mathbf{P}^i \cdot \underline{\phi_2}$ is calculated as $\mathbf{P} \cdot (\mathbf{P}^{i-1} \cdot \underline{\phi_2})$, reusing the computation from the previous iteration. Again, the infinite summation can be truncated using the techniques of Fox and Glynn [**57**]. Hence, the bulk of the work required thus reduces to R_ε matrix-vector multiplications. In fact the summation can be truncated even sooner if the vector converges. As an additional optimisation, we can reuse the PROB0 algorithm, from Algorithm 1 in Chapter 2, to initially identify additional states for which the probability is actually 0.

EXAMPLE 4.5.2. Consider the CTMC in Fig. 1 and the CSL "time-bounded until" formula $\mathcal{P}_{>0.65}[\Diamond^{[0,7.5]}\text{full}]$. This is equivalent to $\mathcal{P}_{>0.65}[\text{true } \mathcal{U}^{[0,7.5]}\text{full}]$ so we in fact compute the probabilities $p_s(\text{true } \mathcal{U}^{[0,7.5]} \text{ full})$, i.e. the probability from each $s \in \{0, 1, 2, 3\}$ that a state satisfying atomic proposition full is reached within 7.5 time units. Following the procedure outlined above, we begin by observing that state 3 satisfies full and no states satisfy $\neg\,\text{true} \wedge \neg\text{full}$. Hence, we will actually analyse the modified CTMC with state 3 made absorbing, i.e. with the transition between states 3 and 2 removed. The transition probability matrix for the uniformised DTMC of this modified CTMC is:

$$\mathbf{P} = \begin{pmatrix} \frac{2}{3} & \frac{1}{3} & 0 & 0 \\ \frac{2}{3} & 0 & \frac{1}{3} & 0 \\ 0 & \frac{2}{3} & 0 & \frac{1}{3} \\ 0 & 0 & 0 & 1 \end{pmatrix}$$

using the value $q = 4.5$ for uniformisation. Computing the summation of matrix-vector multiplications described above yields the solution:

$$\underline{p(\text{true } \mathcal{U}^{[0,7.5]}\text{full})} = (0.6482, 0.6823, 0.7811, 1)$$

and we conclude that the CSL property is satisfied in states 1, 2 and 3.

4.5.2.3. *The case* $I = [t, t']$. For this case, we split the computation into two parts. As shown in [**14**], we can consider separately the probabilities of:

(a) staying in ϕ_1 states for time t;
(b) reaching a ϕ_2 state within time $t' - t$.

For the former, we use a similar idea to that of the previous section, computing transient probabilities in a CTMC for which ϕ_1 states have been made absorbing.

We have:

$$\begin{aligned} p_s^{\mathcal{C}}(\phi_1 \mathcal{U}^{[t,t']} \phi_2) &= \sum_{s' \models \phi_1} \pi_{s,t}^{\mathcal{C}[\neg\phi_1]}(s') \cdot p_{s'}^{\mathcal{C}}(\phi_1 \mathcal{U}^{[0,t'-t]} \phi_2) \\ &= \underline{\pi_{s,t}^{\mathcal{C}[\neg\phi_1]}} \cdot \underline{p_{\phi_1}^{\mathcal{C}}(\phi_1 \mathcal{U}^{[0,t'-t]} \phi_2)} \end{aligned}$$

where $\underline{p_{\phi_1}^{\mathcal{C}}(\phi_1 \mathcal{U}^{[0,t'-t]} \phi_2)}$ is a vector with:

$$\underline{p_{\phi_1}^{\mathcal{C}}(\phi_1 \mathcal{U}^{[0,t'-t]} \phi_2)}(s) = \begin{cases} p_s^{\mathcal{C}}(\phi_1 \mathcal{U}^{[0,t'-t]} \phi_2) & \text{if } s \models \phi_1 \\ 0 & \text{otherwise} \end{cases}$$

which can be computed as in the previous section. We can perform the overall computation as a summation, in the style of uniformisation, to determine the probability for all states at once:

$$\begin{aligned} \underline{p^{\mathcal{C}}(\phi_1 \mathcal{U}^{[t,t']} \phi_2)} &= \mathbf{\Pi}_t^{\mathcal{C}[\neg\phi_1]} \cdot \underline{p_{\phi_1}^{\mathcal{C}}(\phi_1 \mathcal{U}^{[0,t'-t]} \phi_2)} \\ &= \left(\sum_{i=0}^{\infty} \gamma_{i,q \cdot t} \cdot \mathbf{P}^i \right) \cdot \underline{p_{\phi_1}^{\mathcal{C}}(\phi_1 \ \mathcal{U}^{[0,t'-t]} \phi_2)} \\ &= \sum_{i=0}^{\infty} \left(\gamma_{i,q \cdot t} \cdot \mathbf{P}^i \cdot \underline{p_{\phi_1}^{\mathcal{C}}(\phi_1 \mathcal{U}^{[0,t'-t]} \phi_2)} \right) \end{aligned}$$

where $\mathbf{P}$ is the transition probability matrix of the uniformised DTMC of the CTMC $\mathcal{C}[\neg\phi_1]$. Again, this summation can be truncated and performed using matrix-vector multiplication.

4.5.2.4. *The case* $I = [t, \infty)$. This case is, in fact almost identical to the previous one. We again split the computation into two parts. Here however, the second part is an unbounded, rather than time-bounded, "until" formula:

$$p_s^{\mathcal{C}}(\phi_1 \mathcal{U}^{[t,\infty)} \phi_2) = \underline{\pi_{s,t}^{\mathcal{C}[\neg\phi_1]}} \cdot \underline{p_{\phi_1}^{\mathcal{C}}(\phi_1 \mathcal{U} \phi_2)}.$$

Similarly, we compute for all states as:

$$\underline{p^{\mathcal{C}}(\phi_1 \mathcal{U}^{[t,t']} \phi_2)} = \sum_{i=0}^{\infty} \left(\gamma_{i,q \cdot t} \cdot \mathbf{P}^i \cdot \underline{p_{\phi_1}^{\mathcal{C}}(\phi_1 \mathcal{U} \phi_2)} \right)$$

truncating as above and using matrix vector multiplications.

4.5.3. The $\mathcal{S}_{\bowtie p}[\phi]$ operator. A state s satisfies the steady-state formula $\mathcal{S}_{\bowtie p}[\phi]$ if $\sum_{s' \models \phi} \pi_s(s') \bowtie p$. Therefore, to model check the $\mathcal{S}_{\bowtie p}[\phi]$ operator, we must compute the steady-state probabilities $\pi_s(s')$ for all states s and s'. As described in Section 4.3, for CTMCs which are irreducible, these probabilities are independent of the initial state s. In this case, we denote the steady-state probability of being in state s' as $\pi(s')$ and the vector of all such values for $s' \in S$ as $\underline{\pi}$. The latter can be computed by solving the linear equation system:

$$\underline{\pi} \cdot \mathbf{Q} = \underline{0} \quad \text{and} \quad \sum_{s' \in S} \pi(s') = 1.$$

By transposing, we can convert this problem to one of the standard form $\mathbf{A} \cdot \underline{x} = \underline{b}$, as required for PCTL model checking of DTMCs and as discussed further in Appendix B. The satisfaction of the CSL formula, which in this case will be the same for all states $s \in S$ of the CTMC, can be determined by summing the steady-state probabilities for states satisfying ϕ and comparing to the bound in the formula:

$$s \models \mathcal{S}_{\bowtie p}[\phi] \Leftrightarrow \sum_{s' \models \phi} \pi(s') \bowtie p.$$

For the case where the CTMC is *reducible* (i.e. not irreducible), the procedure is more complex. We define a *strongly connected component* (SCC) of the CTMC as a set of states for which there exists a finite path between every pair of the states. We define a *bottom strongly connected component* (BSCC) as an SCC from which there are no transitions in the CTMC which leave the component. Each individual BSCC B can now be treated as an irreducible CTMC, for which we can compute the steady-state probabilities as described above. These probabilities are independent of any initial state and we denote them $\pi^B(s')$ for $s' \in B$. In the special case where B comprises a single state s', we have $\pi^B(s') = 1$.

Next, we determine the probability of reaching each BSCC B from state s. In fact, this is simply $p_s(\Diamond a_B)$, where a_B is an atomic proposition true only in the states $B \subseteq S$. This can be computed exactly as would be done for model checking the CSL formula $\mathcal{P}_{\bowtie p}[\Diamond a_B] \equiv \mathcal{P}_{\bowtie p}[\text{true}\ \mathcal{U}\ a_B]$, as is described in Section 4.5.1. Essentially, this reduces to the solution of a linear equation system. Then, for states $s, s' \in S$:

$$\pi_s(s') = \begin{cases} p_s(\Diamond a_B) \cdot \pi^B(s') & \text{if } s' \text{ is in a BSCC } B \\ 0 & \text{otherwise.} \end{cases}$$

Note that, since the steady-state probabilities $\pi^B(s')$ are independent of s and the procedure for computing $p_s(\Diamond a_B)$ does so for all $s \in S$ at once, the total work required to compute $\pi_s(s')$ for all $s, s' \in S$ is the solution of two linear equation systems for each BSCC in the CTMC. Computation of the BSCCs in the CTMC requires an analysis of its underlying graph structure and can be performed using classic algorithms based on depth-first search [**125**].

EXAMPLE 4.5.3. Consider the CTMC in Figure 1 and the CSL "steady-state" formula $\mathcal{S}_{<0.1}[\text{full}]$. From inspection, we see that the CTMC comprises a single BSCC containing all 4 states. Hence, the steady-state probabilities are computed by solving the linear equation system:

$$\begin{aligned} -1.5x_0 + 3x_1 &= 0, \\ 1.5x_0 - 4.5x_1 + 3x_2 &= 0, \\ 1.5x_1 - 4.5x_2 + 3x_3 &= 0, \\ 1.5x_2 - 3x_3 &= 0, \\ x_0 + x_1 + x_2 + x_3 &= 1 \end{aligned}$$

which has the solution $\underline{\pi} \simeq (0.4923, 0.3077, 0.1231, 0.0769)$. State 3 is the only state satisfying atomic proposition full, and thus the CSL formula is true in all states.

4.5.4. The $\mathcal{E}_{\bowtie c}[\phi]$ operator. In this case, we compute the expected cost $e_s(\phi)$ for all states $s \in S$. Similarly to the "until" operator, we can use the embedded Markov chain of the CTMC. However, since, unlike in the DTMC case, we have cumulative as well as instantaneous costs, there are differences compared to the DTMC algorithm given in Section 2.4.4.

First, similarly to the DTMC case, for the embedded DTMC $\text{emb}(\mathcal{C}) = (S, \bar{s}, \mathbf{P})$ of the CTMC, we identify the sets of states for which $e_s(\phi)$ is either 0 or ∞. We denote these S^0 and S^∞. The set S^0 is simply those states that satisfy ϕ, while the set S^∞ is those states from which the probability of reaching a ϕ state is less than 1. It remains to compute $e_s(\phi)$ for the states in $S^? = S \setminus (S^0 \cup S^\infty)$. This is achieved by solving the linear equation system in variables $\{x_s \mid s \in S^?\}$:

$$x_s = \left(\sum_{s' \in S^?} \mathbf{P}(s, s') \cdot \left(\mathbf{C}(s, s') + x_{s'}\right) \right) + \frac{1}{E(s)} \cdot \mathbf{c}(s)$$

and then letting $e_s(\phi) = x_s$ for $s \in S^?$. Recall that, for any $s \in S$, $E(s)$ denotes the exit rate of state s and hence $1/E(s)$ is the expected time the system remains in s before a transition is taken. As before, this can be solved using any standard direct or iterative method.

4.6. Complexity of CSL Model Checking

The overall time complexity for model checking a CSL formula ϕ against a CTMC $\mathcal{C} = (S, \bar{s}, \mathbf{R})$ is linear in $|\phi|$, polynomial in $|S|$ and linear in $q \cdot t_{\max}$, where $q = \max_{s \in S} |\mathbf{Q}(s, s)|$ and $t_{\max}$ is the maximum value found in the parameter of a "time-bounded until" operator. For the $\mathcal{P}_{\bowtie p}[\phi_1 \mathcal{U} \phi_2]$ and $\mathcal{E}_{\bowtie c}[\phi]$ operators, solution of a linear equation system of size $|S|$ is required. This can be done with Gaussian elimination, the complexity of which is cubic in the size of the system. For the $\mathcal{P}_{\bowtie p}[\phi_1 \mathcal{U}^I \phi_2]$ operator, we must perform at most two iterative summations, each step of which requires a vector-matrix multiplication. This operation is quadratic in the size of the matrix, i.e. $|S|$. The total number of iterations required is determined by the upper bound supplied by the algorithm of Fox and Glynn [**57**], which for large $q \cdot t$ is linear in $q \cdot t$.

4.7. Case Study: Dynamic Power Management

Dynamic Power Management(DPM) is a technique for saving energy in devices which can be turned on and off under operating system control. DPM has gained considerable attention over the last few years, both in the research literature and in the industrial setting, with schemes such as OnNow and ACPI becoming prevalent. One of the main reasons for this interest is the continuing growth in the use of mobile, hand-held and embedded devices, for which minimisation of power consumption is a key issue.

DPM-enabled devices typically have several *power states* with different power consumption rates. A DPM *policy* is used to decide when commands to transition between these states should be issued, based on the current state of the system. We consider stochastic approaches to DPM which use randomised policies, i.e. where the choice between power states in a given instance is probabilistic rather than deterministic. These approaches are based on the formulation of a stochastic model of the DPM system. This model makes probabilistic assumptions (based on empirical observations) about characteristics such as the usage patterns and response times of devices. From this stochastic model, and for a given performance constraint, an optimisation problem is constructed, the solution to which is the optimal DPM policy. An example of a constraint would be that the queue of jobs to be served by a device does not exceed a specified size. The policy will be optimal in the sense that it minimises system power consumption under this constraint.

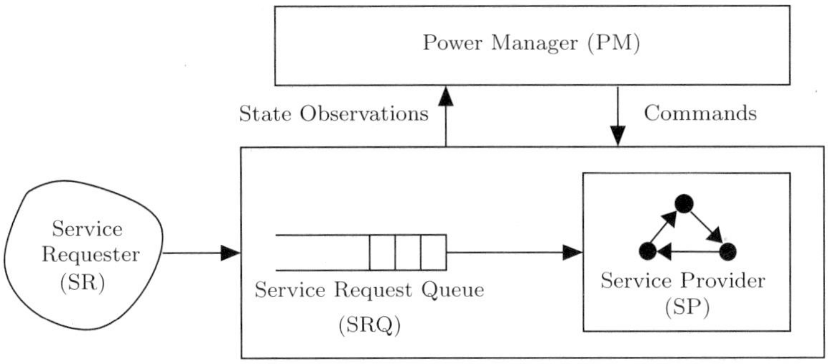

FIGURE 2. The system model.

	sleep	*standby*	*idle*	*busy*
sleep	0	0.6	1.6	-
standby	0.3	0	1.2	-
idle	0.67	0.4	0	0
busy	-	-	0	0

The stochastic models which have been used in the DPM literature include both DTMCs [**21, 108**] and CTMCs [**113, 114**]. The basic structure of the DPM model, in each case, can be seen in Fig. 2. The model consists of:

- a Service Provider (SP), which represents the device under power management control;
- a Service Requester (SR), which issues requests to the device;
- a Service Request Queue (SRQ), which stores requests that are not serviced immediately;

and the Power Manager (PM), which issues commands to the SP, based on observations of the system and a stochastic DPM policy.

This case study is based on a CTMC model of a Fujitsu disk drive with four power states taken from [**114**]. The SP has 4 states: *sleep*, *standby*, *idle* and *busy*. When the SP is in states *sleep* and *standby*, it is inactive and no requests can be served. When the SP is in *idle*, the SP is active but not working on any requests, which means that the SRQ is empty. The SP makes transitions from *idle* to *busy* whenever a request arrives for service. Similarly, it makes a transition from *busy* to *idle* whenever it finishes the service of a request. Transitions between *sleep*, *standby* and *idle* are controlled by the PM (that is, by the DPM policy). Figure 3 shows the SP and the average times for transitions between power states. Tables 1 and 2 show the energy used for these transitions and the average power consumption in each state. These will translate into cumulative and instantaneous costs, respectively, in our CTMC model.

With regards to the other components of the system, the SR represents the fact that the inter-arrival distribution of requests is an exponential distribution (with rate 100/72); the SRQ models requests arriving in the queue from the SR and being served by the SP. The maximum size of the SRQ is 20. Note that, if a request arrives from the SR and the queue is full (20 requests are already awaiting service), then they are presumed lost. For further details, see [**114**].

TABLE 1. Average energy consumption values for state transitions of the SP.

	sleep	*standby*	*idle*	*busy*
sleep	0	5.1	7	-
standby	0.006	0	2	-
idle	0.067	0.001	0	0
busy	-	-	0	0

TABLE 2. Average power consumption for states of the SP.

	sleep	*standby*	*idle*	*busy*
av. power	0.13	0.35	0.95	2.15
av service time	0	0	0	0.008

TABLE 3. Optimum policies under varying constraints.

Constraint : (**average queue size**)	**Optimum policy:**

≤ 10	if the SP is in *standby* and there are 19 requests awaiting service:
	remain in *standby* with probability 0.879121
	go to *idle* with probability 0.120879
	if the SP is in *standby* and there are 20 requests awaiting service:
	go to *idle*
	if the SP is in *standby* and there are more than 9 requests awaiting service:
	go to *idle*
	if the SP is in *idle* and there are no requests awaiting service:
	go to *standby*

≤ 5	if the SP is in *standby* and there are 9 requests awaiting service:
	remain in *standby* with probability 0.392547
	go to *idle* with probability 0.607453
	if the SP is in *standby* and there are more than 9 requests awaiting service:
	go to *idle*
	if the SP is in *idle* and there are no requests awaiting service:
	go to *standby*

≤ 2	if the SP is in *standby* and there are more than 7 requests awaiting service:
	go to *idle*
	if the SP is in *idle* and there are no requests awaiting service:
	remain in *idle* with probability 0.884841
	go to *standby* with probability 0.115159

≤ 1	if the SP is in *standby* and there are more than 7 requests awaiting service:
	go to *idle*
	if the SP is in *idle* and there are no requests awaiting service:
	remain in *idle* with probability 0.956711
	go to *standby* with probability 0.043289

≤ 0.1	if the SP is in *standby* and there are more than 7 requests awaiting service:
	go to *idle*
	if the SP is in *idle* and there are no requests awaiting service:
	remain in *idle* with probability 0.996813
	go to *standby* with probability 0.003187

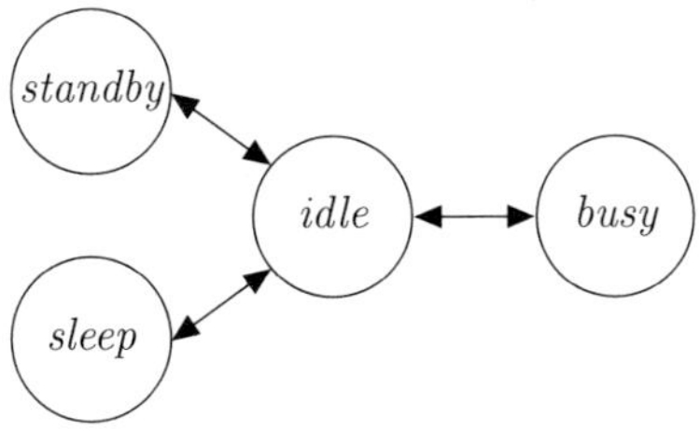

FIGURE 3. Average transition times for state transitions of the SP.

The PM, which represents a randomised DPM policy is constructed automatically for a given performance constraint. These constraints take the form of an upper bound on the average size of the queue of jobs for the disk drive. In Table 3, we illustrate a range of policies which have been constructed in this way. Each policy's performance can then be analysed using probabilistic model checking. Below we formalise in CSL the properties we wish to study

- $\mathcal{P}_{\bowtie p}[\Diamond^{\leq t}(q \geq N)]$ the probability that the queue size becomes greater than or equal to N by time t;
- $\mathcal{P}_{\bowtie p}[\Diamond^{\leq t}(\text{served})]$—the probability that a request is served by time t;
- $\mathcal{P}_{\bowtie p}[\Diamond^{\leq t}(\text{lost} \geq N)]$—the probability that at least N requests get lost by time t;
- $\mathcal{E}_{\bowtie c}[(\text{served} = N)]$—the expected power consumption until N requests are served.

Note that, for the third and fourth properties, we must change the model to include variables which count the number of lost customers and the number of served customers. This is similar to the approach taken in the DTMC case study (see Section 2.5). The instantaneous and cumulative costs used for the final property are taken from values in Tables 1 and 2.

Figure 4 presents results for the probability that, when starting from the SP in *standby* and the queue empty, the number of requests awaiting service is greater than or equal to N by time t, for the cases $N = 10$ and 15. The results show that this probability becomes larger as the constraint on the queue size increases. This is to be expected since, for larger constraints on the queue, the PM lets more requests await service before switching the SP on.

Figure 5 reports results for the probability that a request is served by time t, both when it arrives into the 5th and the 10th position of the queue. Note that each request eventually gets served with probability 1. For larger constraints, the PM does not switch the SP on as often, and hence the probability of serving the request within any time bound is smaller.

Figure 6 plots the probability that at least N requests get lost by time t for $N = 1$ and 2. This probability is almost 1 for large time bounds; this is because, unlike for the other constraints (see Table 3), under this constraint the SP is not switched on until the queue is nearly full, meaning that there is a high chance that more requests arrive before the SP has time to switch itself on and start serving requests. For the remaining cases, as the constraint on the average queue size decreases, the probability of requests getting lost also decreases; this is because, as the constraints get smaller, the corresponding PM switches the SP on earlier.

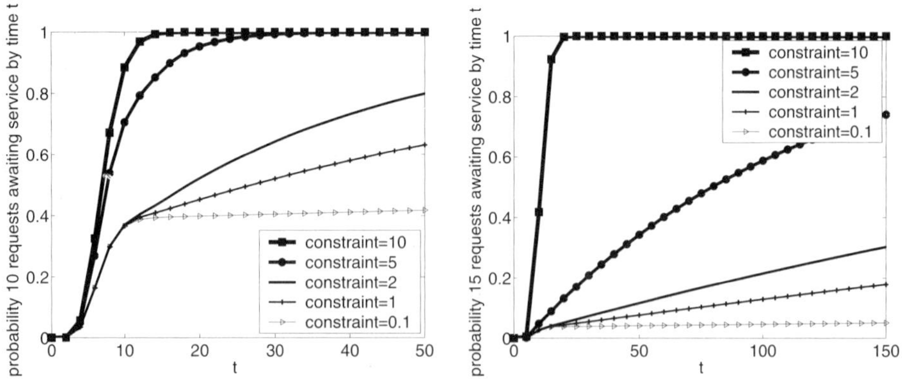

FIGURE 4. Probability that the queue becomes greater than or equal to N by time t.

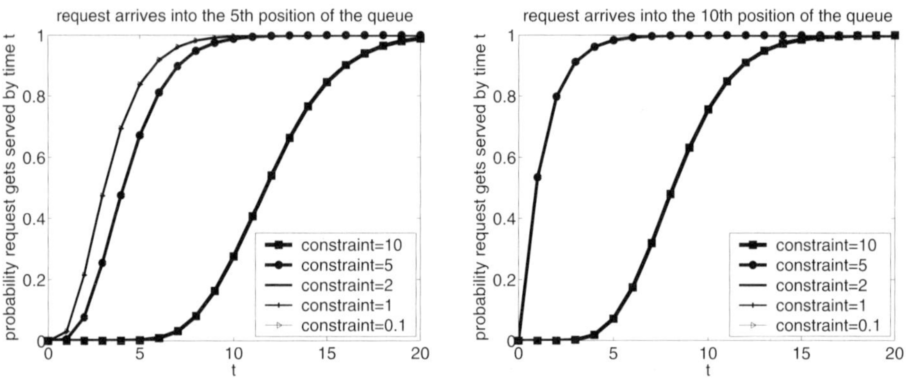

FIGURE 5. Probability that a request is served by time t.

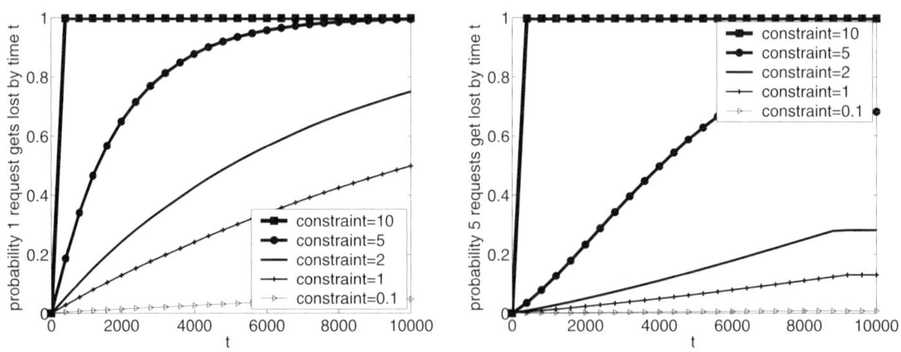

FIGURE 6. Probability that N requests gets lost by time t.

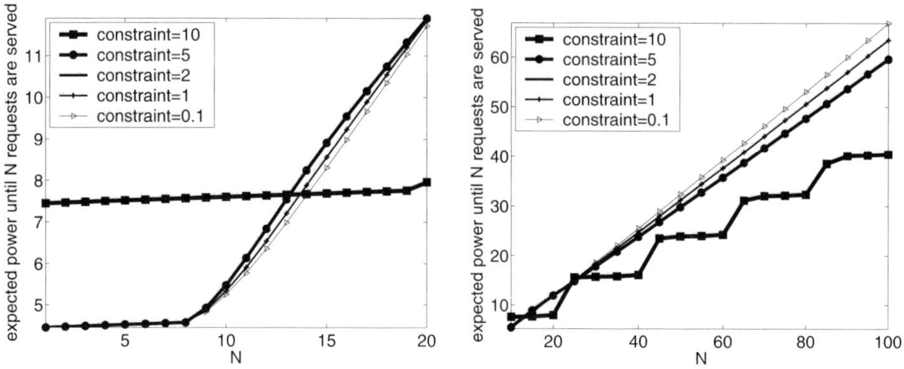

FIGURE 7. Expected power consumption until N requests get served.

Finally, Fig. 7 reports on the expected power consumption until N requests are served, both as N varies between 1 and 20 and between 10 and 100. For larger values of N we see that the expected power consumption is smaller when the constraint is larger, which is to be expected since for larger constraints the PM allows the SP to spend more time in standby preserving energy. An interesting observation in this case corresponds to the optimal policy derived when the constraint on the queue is 10. Recall that this policy (see Table 3) does not do anything until the queue is full, and then once it has served all pending requests, returns the SP to standby. This can be seen in the graphs: the SP waiting until the queue is full then serving all requests justifies why the power consumption is almost the same for $N \in \{0, \ldots, 19\}$, the fact that the SP is then switched to standby until the queue is almost full again explains the jumps at $N = 40$, 60 and 80. For further details about probabilistic model checking of DPM, see [**104, 105**] or the PRISM website [**111**].

4.8. Discussion

CTMC models are well established in performance modelling, and can be derived from a variety of high-level languages, for example, stochastic process algebras such as PEPA [**71**]. In fact, the probabilistic model checker PRISM [**87**] now supports CTMC specifications in PEPA.

A number of extensions of the logic CSL have been presented in the literature, along with associated model checking algorithms: [**15, 64**] introduce the logics CRL and CSRL which add support for reward-based properties; [**85, 86**] augment CSL with random time-bounded until and random expected-time operators, respectively.

A challenging direction of research is into the verification of models which allow more general probability distributions. While the restriction to exponential distributions imposed by CTMCs is important for the tractability of their model checking, it may prove too simplistic for some modelling applications. See [**58**] for an introduction to this area.

CHAPTER 5

Probabilistic Timed Automata

The final class of models we consider are *probabilistic timed automata* which exhibit dense time, nondeterminism and probabilistic choice, a combination of features used in randomised distributed protocols with timing delays, for example random back-off schemes. To derive an appropriate model we extend the formalism of *timed automata* [**5**] with probabilistic choice. A timed automaton is an automaton extended with *clocks*, positive real-valued variables that increase uniformly with time, and whose nodes and edges are labelled with *clock constraints*, e.g. $(x \geq 5) \wedge (y < 1)$, respectively called *invariants* and *guards*. The invariants dictate when the automaton may remain in a node, letting time pass, and guards when the corresponding edge can be taken. Transitions are instantaneous, and can involve *clock resets* of the form $x := 0$.

The denseness of the time domain means that the state space of a timed automaton is infinite. Therefore, conventional model checking techniques for timed automata are based on constructing a finite-state quotient representation. By suitably adapting such quotients to the probabilistic case, model checking for probabilistic timed automata can be reduced to an application of probabilistic model checking techniques introduced earlier, namely for Markov decision processes as described in Chapter 3. Here we report on two such quotient constructions, the first based on the *region graph* [**4**] and the second on *digital clocks* [**66**]. Alternative approaches are briefly discussed in Section 5.9.

We begin by reviewing preliminary concepts necessary to define probabilistic timed automata.

5.1. Time, Clocks and Zones

Let $\mathbb{T} \in \{\mathbb{R}^{\geq 0}, \mathbb{N}\}$ be the *time domain* of either the non-negative reals or naturals. (By abuse of notation, for the remainder of this chapter we use $\mathbb{R}$ to denote the non-negative reals.) We assume the existence of a finite set $\mathcal{X}$ of *clocks*, variables $x \in \mathcal{X}$ that take values from the time domain $\mathbb{T}$. A point $\nu \in \mathbb{T}^{|\mathcal{X}|}$ is referred to as a *clock valuation*, with $\mathbf{0} \in \mathbb{T}^{|\mathcal{X}|}$ the valuation that assigns 0 to all clocks in $\mathcal{X}$. For any $\nu \in \mathbb{T}^{|\mathcal{X}|}$ and $t \in \mathbb{T}$, the clock valuation $\nu \oplus t$ denotes the *time increment* for ν and t. We use $\nu[X := 0]$ to denote the clock valuation obtained from ν by resetting all of the clocks in $X \subseteq \mathcal{X}$ to 0, with the value of the remaining clocks unaffected.

A *zone* of $\mathcal{X}$, written ζ, is a subset of the valuation space $\mathbb{T}^{|\mathcal{X}|}$ described by a conjunction of constraints. Formally, for a given set of clocks $\mathcal{X}$, the set of zones of $\mathcal{X}$, denoted $\text{Zones}(\mathcal{X})$ is defined by the following syntax:

$$\zeta ::= x \sim c \mid x - y \sim c \mid \zeta \wedge \zeta$$

where $x, y \in \mathcal{X}$, $c \in \mathbb{N}$ and $\sim \in \{\leq, <, >, \geq\}$. We only consider zones in canonical form (where constraints are as tight as possible), and can thus freely interchange the semantic and syntactic interpretations, respectively, as sets of valuations and conjunction of constraints. A zone is called:

closed: if it does not feature atomic constraints of the form $x > c$ or $x < c$ for $x \in \mathcal{X}$ and $c \in \mathbb{N}$;

diagonal-free: if it does not feature atomic constraints of the form $x - y \sim c$ for $x, y \in \mathcal{X}$ and $c \in \mathbb{N}$.

Let $k_{\max}(\zeta)$ be the largest constant appearing in zone ζ.

The clock valuation ν *satisfies* the zone ζ, written $\nu \triangleleft \zeta$, if and only if ζ resolves to true after substituting each clock $x \in \mathcal{X}$ with the corresponding clock value from ν. If $\nu \triangleleft \zeta$ then we also denote this by $\nu \in \zeta$. For any zones $\zeta, \zeta' \in \text{Zones}(\mathcal{X})$, and a set of clocks $X \subseteq \mathcal{X}$, let $\zeta[X := 0] = \{\nu[X := 0] \mid \nu \in \zeta\}$.

As in previous chapters, we let $\mathsf{Dist}(Y)$ denote the family of probability distributions over subsets of the set Y. For $y \in Y$ the point distribution is denoted by δ_y.

5.2. Syntax of Probabilistic Timed Automata

In this section we give the definition of probabilistic timed automata [**98**] extended with urgent events, a well-established concept for classical timed automata [**49**].

DEFINITION 5.2.1. A *probabilistic timed automaton* (PTA) is a tuple $\mathcal{T} = (\text{Loc}, \bar{l}, \mathcal{X}, \Sigma, \text{inv}, \text{prob})$ where:

- Loc is a finite set of *locations*;
- $\bar{l} \in \text{Loc}$ is the *initial location*;
- Σ is a finite set of *events* disjoint from $\mathbb{T}$, of which $\Sigma_u \subseteq \Sigma$ are declared as being *urgent*;
- the function inv: $\text{Loc} \to \text{Zones}(\mathcal{X})$ is the *invariant condition*;
- the finite set $\text{prob} \subseteq \text{Loc} \times \text{Zones}(\mathcal{X}) \times \Sigma \times \mathsf{Dist}(2^{\mathcal{X}} \times \text{Loc})$ is the *probabilistic edge relation.*

Unless otherwise stated, we will usually consider probabilistic timed automata over the time domain $\mathbb{R}$ of non-negative reals. A *state* of a probabilistic timed automaton is a pair $\langle l, \nu \rangle$ where $l \in \text{Loc}$ and $\nu \in \mathbb{T}^{|\mathcal{X}|}$ are such that $\nu \triangleleft \text{inv}(l)$. The invariant condition describes the set of admissible states. Transitions are *probabilistic edges* $(l, g, \sigma, p) \in \text{prob}$ where l is the source location, g is an enabling guard, σ an event label and p a probability distribution on target locations, together with a set of clocks to be reset as the edge is taken. No time passage is allowed if there are enabled transitions labelled with urgent events, and hence they must be taken immediately.

The probabilistic timed automaton executes as follows. The model starts in the initial location $\bar{l}$ with all clocks set to 0, and hence the initial state is $\langle \bar{l}, \mathbf{0} \rangle$. In any state $\langle l, \nu \rangle$, a nondeterministic choice is made between:

- making a *discrete transition* according to any $(l, g, \sigma, p) \in \text{prob}$ with source location l such that the guard g is *enabled* ($\nu \triangleleft g$), which results in moving to the location l' and resetting all of the clocks in X to 0 with probability $p(X, l')$;
- letting *time pass*, as long as the invariant condition $\text{inv}(l)$ is satisfied while time elapses and there does not exist an enabled probabilistic edge with an urgent event.

Note that a *timed automaton* [**5**] is a probabilistic timed automaton over the time domain $\mathbb{R}$ for which every probabilistic edge (l, g, σ, p) is a point distribution $p = \delta_{(X,l')}$ for some target location l' and set of resetting clocks $X \subseteq \mathcal{X}$.

We use $\mathrm{Zones}(\mathcal{T})$ to denote the set of all zones appearing as an invariant or enabling condition of $\mathcal{T}$. Formally, let:

$$\mathrm{Zones}(\mathcal{T}) \stackrel{\mathrm{def}}{=} \{\mathrm{inv}(l) \mid l \in \mathrm{Loc}\} \cup \{\zeta \mid (-, \zeta, -, -) \in \mathrm{prob}\}.$$

We let $k_{\max}(\mathcal{T})$ denote the maximum constant used in the description of $\mathcal{T}$, i.e. $k_{\max}(\mathcal{T}) = \max\{k_{\max}(\zeta) \mid \zeta \in \mathrm{Zones}(\mathcal{T})\}$.

DEFINITION 5.2.2. A *closed diagonal-free* probabilistic timed automaton is a timed automaton $\mathcal{T} = (\mathrm{Loc}, \bar{l}, \mathcal{X}, \Sigma, \mathrm{inv}, \mathrm{prob})$ whose zones $\mathrm{Zones}(\mathcal{T})$ are closed and diagonal-free.

Similarly to the models considered previously, we annotate probabilistic timed automata with additional information. As in the case of MDPs, see Definition 3.1.2, we introduce a labelling function, assigning atomic propositions from a set AP to locations, and a cost function over the events. However, in addition, we allow the cost function to range over the *time domain* which represents the costs associated with letting time elapse; more precisely, for any $t \in \mathbb{T}$, the value of $\mathbf{c}(t)$ is the cost associated with letting t time units elapse in any location.

DEFINITION 5.2.3. A *labelled probabilistic timed automaton* is a tuple $(\mathcal{T}, L, \mathbf{c})$ where:

- $\mathcal{T}$ is a probabilistic timed automaton;
- $L\colon \mathrm{Loc} \to 2^{\mathrm{AP}}$ is a *labelling function*;
- $\mathbf{c}\colon \Sigma \cup \mathbb{T} \to \mathbb{R}$ is the *cost function*.

Note that, by allowing costs to range over both events and time, we will be able to consider properties relating to the functional and timing behaviour, as well as combinations of the two. For example, if $r \in \mathbb{R}$ gives the rate at which cost is accumulated as time passes, and $\mathbf{e}\colon \Sigma \to \mathbb{R}$ is a function assigning the cost of executing each event in Σ, then one can define an associated cost function $c_{r,\mathbf{e}}$ by $\mathbf{c}_{r,\mathbf{e}}(t) = t \cdot r$ for all $t \in \mathbb{T}$, and $\mathbf{c}_{r,\mathbf{e}}(a) = \mathbf{e}(a)$ for all $a \in \Sigma$. The use of pairs such as $(r, \mathbf{e})$ induces a model over the time domain $\mathbb{R}$ which is a probabilistic generalisation of uniformly priced timed automata [**20**].

EXAMPLE 5.2.4. Consider the PTA modelling a simple probabilistic communication protocol given in Fig. 1. The protocol attempts to send data packets which may be lost by the communication medium, and features timing delays. There are two clocks, x and y. Clock y counts the total transmission time, while clock x captures the time delay until resending a data packet that was lost. The nodes represent the locations:

- II—sender and receiver both idle (initial location);
- DI—sender has data and receiver idle;
- SI—sender sent data and receiver idle;
- SR—sender sent data and receiver received.

The protocol starts in location II. As soon as data has been received by the sender, the protocol transfers to the location DI with probability 1. In DI, after between 1 and 2 time units, it makes a transition either to SR with probability 0.9 (data received), or to SI with probability 0.1 (data lost). In SI, the protocol will attempt

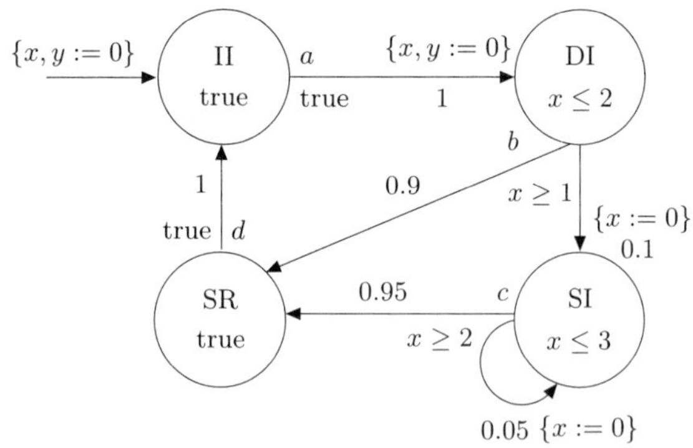

FIGURE 1. A probabilistic timed automaton modelling a simple protocol.

to resend the data after 2 to 3 time units, which again can be lost, this time with probability 0.05.

5.3. Parallel Composition

It is often useful to describe models in terms of a parallel composition of components. The *parallel composition* $\|$ of probabilistic timed automata derives from the corresponding composition operators over (untimed) probabilistic systems [**118**] and classical timed automata [**5**]. It allows the automata to either perform independent moves through interleaving, or synchronise on common events, in which case the resulting probability distribution is a product of the component probability distributions.

DEFINITION 5.3.1. Let $\mathcal{T}_i = (\mathrm{Loc}_i, \bar{l}_i, \mathcal{X}_i, \Sigma_i, \mathrm{inv}_i, \mathrm{prob}_i)$ for $i \in \{1,2\}$ such that $\mathcal{X}_1 \cap \mathcal{X}_2 = \emptyset$. The *parallel composition* of two probabilistic timed automata $\mathcal{T}_1$ and $\mathcal{T}_2$ is the probabilistic timed automaton $\mathcal{T}_1 \| \mathcal{T}_2 = \big(\mathrm{Loc}_1 \times \mathrm{Loc}_2, (\bar{l}_1, \bar{l}_2), \mathcal{X}_1 \cup \mathcal{X}_2, \Sigma_1 \cup \Sigma_2, \mathrm{inv}, \mathrm{prob}\big)$ where $\mathrm{inv}(l, l') = \mathrm{inv}_1(l) \wedge \mathrm{inv}_2(l')$ for all $(l, l') \in \mathrm{Loc}_1 \times \mathrm{Loc}_2$ and $\big((l_1, l_2), g, \sigma, p\big) \in \mathrm{prob}$ if and only if one of the following conditions holds:

- $\sigma \in \Sigma_1 \setminus \Sigma_2$ and there exists $(l_1, g, \sigma, p_1) \in \mathrm{prob}_1$ such that $p = p_1 \otimes \delta_{(\emptyset, l_2)}$;
- $\sigma \in \Sigma_2 \setminus \Sigma_1$ and there exists $(l_2, g, \sigma, p_2) \in \mathrm{prob}_2$ such that $p = \delta_{(\emptyset, l_1)} \otimes p_2$;
- $\sigma \in \Sigma_1 \cap \Sigma_2$ and there exist $(l_1, g_1, \sigma, p_1) \in \mathrm{prob}_1$ and $(l_2, g_2, \sigma, p_2) \in \mathrm{prob}_2$ such that $g = g_1 \wedge g_2$ and $p = p_1 \otimes p_2$

where for any $l_1 \in \mathrm{Loc}_1$, $l_2 \in \mathrm{Loc}_2$, $X_1 \subseteq \mathcal{X}_1$ and $X_2 \subseteq \mathcal{X}_2$:

$$p_1 \otimes p_2\big(X_1 \cup X_2, (l_1, l_2)\big) = p_1(X_1, l_1) \cdot p_2(X_2, l_2).$$

Note that parallel composition can introduce nondeterminism even if there was no nondeterminism present in the individual probabilistic timed automata.

5.4. Semantics of Probabilistic Timed Automata

The semantics of a probabilistic timed automaton is defined in terms of an infinite-state Markov decision process over an abstract time domain of durations. We refer to the semantic models arising from probabilistic timed automata *timed probabilistic systems*, to distinguish them from the untimed counterparts discussed in Chapter 3.

DEFINITION 5.4.1. A *timed probabilistic system* is a tuple $\mathsf{PS}=(S,\bar{s},\mathrm{Act},\mathbb{T},\mathrm{Steps})$ where:

- S is the (possible infinite) set of *states*;
- $\bar{s}\in S$ is the *initial state*;
- Act is the set of *actions*;
- $\mathbb{T}$ is the set of *durations*;
- $\mathrm{Steps}\subseteq S\times(\mathrm{Act}\cup\mathbb{T})\times\mathsf{Dist}(S)$ is the *probabilistic transition relation*, such that, if $(s,t,\mu)\in\mathrm{Steps}$ for any $t\in\mathbb{T}$, then μ is a point distribution.

The reader should note close correspondence to Definition 3.1.1 of Markov decision processes. Observe that transitions are labelled either by actions from Act or durations from $\mathbb{T}$; we will often let a range over $\mathrm{Act}\cup\mathbb{T}$. A *probabilistic transition* $s\xrightarrow{a,\mu}s'$ is made from a state $s\in S$ by first nondeterministically selecting an action-distribution or duration-distribution pair (a,μ) such that $(s,a,\mu)\in\mathrm{Steps}$, and second by making a probabilistic choice of target state s' according to the distribution μ, such that $\mu(s')>0$.

As before, we introduce a labelling function, mapping states to atomic propositions from a set AP, and a cost function over the actions.

DEFINITION 5.4.2. A *labelled timed probabilistic system* is a tuple $(\mathsf{PS},L,\mathbf{c})$ where:

- PS is a timed probabilistic system;
- $L\colon S\to 2^{\mathrm{AP}}$ is a *labelling function*;
- $\mathbf{c}\colon \mathrm{Act}\cup\mathbb{T}\to\mathbb{R}$ is the *cost function*.

As for MDPs, we represent the computation of a timed probabilistic system by unfolding it into paths. A *path* of a timed probabilistic system is a non-empty finite or infinite sequence of probabilistic transitions

$$\omega = s_0\xrightarrow{a_0,\mu_0}s_1\xrightarrow{a_1,\mu_1}\cdots.$$

Again, we denote by $\omega(i)$ the $(i+1)$th state of ω, $\mathrm{last}(\omega)$ is the last state of ω if ω is finite, and $\mathrm{step}(\omega,i)$ the action or duration associated with the i-th step, i.e. $\mathrm{step}(\omega,i)=a_k$. For any infinite path $\omega=s_0\xrightarrow{a_0,\mu_0}s_1\xrightarrow{a_1,\mu_1}\cdots$, the duration up to the $(n+1)$th state of ω is defined by:

$$\mathcal{D}_\omega(n+1)\stackrel{\mathrm{def}}{=}\sum\{|a_i\mid 0\le i\le n\wedge a_i\in\mathbb{T}|\}.$$

Since the underlying semantics of a probabilistic timed automaton is a Markov decision process, we utilise *adversaries* as a means to resolve nondeterminism. These are an adaptation of those introduced for MDPs in Definition 3.2.1. Formally, an adversary of a timed probabilistic system is a function A mapping every finite path ω to a pair (a,μ) such that $\big(\mathrm{last}(\omega),a,\mu\big)\in\mathrm{Steps}$. For any adversary A, let Path_s^A denote the set of infinite paths associated with A which have initial state s. Then, we define the probability measure Prob_s^A over Path_s^A as for MDPs.

We restrict our attention to *time-divergent adversaries*, conventionally imposed in real-time systems to rule out unrealisable behaviour (i.e. corresponding to time not advancing beyond a time bound, also called *zeno* behaviour). We say that an infinite path ω is *divergent* if for any $t\in\mathbb{T}$, there exists $j\in\mathbb{N}$ such that $\mathcal{D}_\omega(j)>t$. It can be shown that the set of divergent paths is measurable under any adversary.

DEFINITION 5.4.3. An adversary A of a timed probabilistic system PS is *divergent* if and only if, for each state s, the probability Prob_s^A of the divergent paths of Path_s^A is 1. Furthermore, let $\mathrm{Adv}_{\mathsf{PS}}$ be the set of divergent adversaries of PS.

Such *probabilistic* divergent adversaries, as opposed to a stronger notion in which an adversary is divergent if and only if all its paths are divergent, are essential, as will be explained in (5.4.7).

We now give the semantics of probabilistic timed automata in terms of timed probabilistic systems. As already informally described in Section 5.2, transitions consist of two types. In Definition 5.4.4 below, this is reflected by distinguishing between time and discrete transitions. Since we model urgency of events, we must ensure there is no delay if an urgent transition can be taken. Hence, *time transitions* correspond to the passage of time in the current location, provided that no urgent transitions are available and that the invariant condition for the location is satisfied. *Discrete transitions* correspond to a probabilistic edge being taken, with summation required for the cases in which multiple clock resets result in the same target state $\langle l', \nu' \rangle$. We disallow the situation in which an automaton moves to a state where an invariant is not satisfied.

Observe that the definition is parameterised both by the time domain $\mathbb{T}$ and the time increment $\oplus$.

DEFINITION 5.4.4. Let $\mathcal{T} = (\mathrm{Loc}, \bar{l}, \mathcal{X}, \Sigma, \mathrm{inv}, \mathrm{prob})$ be a probabilistic timed automaton. The *semantics of $\mathcal{T}$ with respect to the time domain $\mathbb{T}$ and the time increment* $\oplus$ is the timed probabilistic system $[\![\mathcal{T}]\!]_{\mathbb{T}}^{\oplus} = (S, \bar{s}, \Sigma, \mathbb{T}, \mathrm{Steps})$ where:

- $S \subseteq \mathrm{Loc} \times \mathbb{T}^{|\mathcal{X}|}$ such that $\langle l, \nu \rangle \in S$ if and only if $\nu \triangleleft \mathrm{inv}(l)$;
- $\bar{s} = \langle \bar{l}, \mathbf{0} \rangle$;
- $(\langle l, \nu \rangle, a, \mu) \in \mathrm{Steps}$ if and only if one of the following conditions holds:

  Time transitions: $a \in \mathbb{T}$ and $\mu = \delta_{\langle l, \nu \oplus a \rangle}$ is a point distribution such that:
  1. $\nu \oplus t' \triangleleft \mathrm{inv}(l)$ for all $0 \leq t' \leq a$;
  2. for all probabilistic edges of the form $(l, g, \sigma, -) \in \mathrm{prob}$, if $\nu \oplus t' \triangleleft g$ for any $0 \leq t' \leq a$, then $\sigma \notin \Sigma_u$;

  Discrete transitions: $a \in \Sigma$ and there exists $(l, g, a, p) \in \mathrm{prob}$ such that $\nu \triangleleft g$ and for any $\langle l', \nu' \rangle \in S$:

$$\mu(\langle l', \nu' \rangle) = \sum_{X \subseteq \mathcal{X} \,\&\, \nu' = \nu[X:=0]} p(X, l')$$

  and for each $(X, l') \in 2^{\mathcal{X}} \times \mathrm{Loc}$, if $p(X, l') > 0$, then $\nu[X := 0] \triangleleft \mathrm{inv}(l')$.

For a labelled probabilistic timed automaton $(\mathcal{T}, L, \mathbf{c})$, the corresponding labelled timed probabilistic system is given by $([\![\mathcal{T}]\!]_{\mathbb{T}}^{\oplus}, L', \mathbf{c})$ where $L'(\langle l, \nu \rangle) = L(l)$ for all $\langle l, \nu \rangle \in S$.

EXAMPLE 5.4.5. Consider the probabilistic timed automaton of (5.2.4) (see Fig. 1). An example of an adversary for this automaton is shown in Fig. 2.

Note that we have chosen time to be *weakly monotonic* [**67**], and thus time need not advance between consecutive transitions. In particular, time does not advance for discrete transitions. This approach allows us, for example, to model simultaneous transitions arising through the interleaving of probabilistic timed automata.

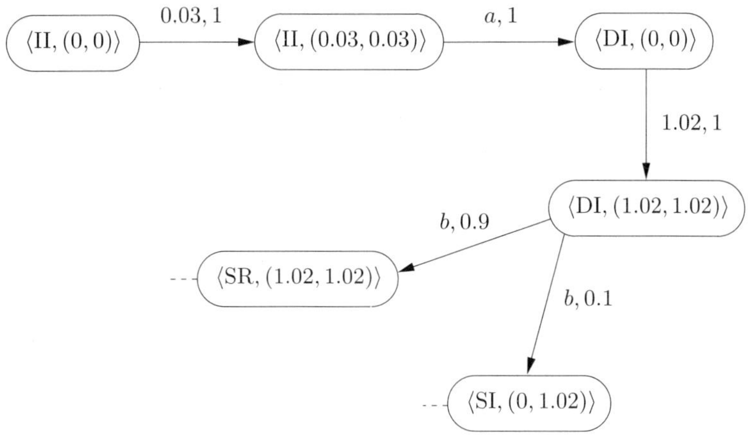

FIGURE 2. Illustration of an adversary for the probabilistic timed automaton of Fig. 1.

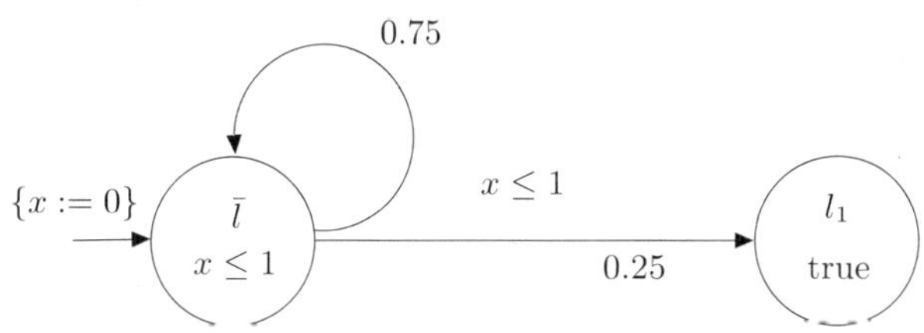

FIGURE 3. Example demonstrating time-divergent adversaries.

Traditionally, the semantics of probabilistic timed automata assumes that the reals $\mathbb{R}$ form the underlying model of time, implying uncountable models. We refer to an adaptation of this model to probabilistic timed automata as the dense-time semantics.

DEFINITION 5.4.6. The *dense-time semantics* of a probabilistic timed automaton $\mathcal{T}$ is defined as the timed probabilistic system $[\![\mathcal{T}]\!]_{\mathbb{R}}^{+}$; that is, $\mathbb{T} = \mathbb{R}$ and $\oplus = +$.

The dense-time semantics of the parallel composition of two probabilistic timed automata corresponds, via isomorphism, to the dense-time semantics of the parallel composition of their individual semantic probabilistic systems. Formally, for the probabilistic timed automata $\mathcal{T}_1$ and $\mathcal{T}_2$ with disjoint clock sets, $[\![\mathcal{T}_1 \| \mathcal{T}_2]\!]_{\mathbb{R}}^{+}$ and $[\![\mathcal{T}_1]\!]_{\mathbb{R}}^{+} \| [\![\mathcal{T}_2]\!]_{\mathbb{R}}^{+}$ are isomorphic. In the composed automaton, time t passes only if both components can delay by t, and common actions happen simultaneously.

EXAMPLE 5.4.7. The following example motivates our choice of probabilistic time-divergent adversaries, as given in Definition 5.4.3. Consider the probabilistic timed automaton in Fig. 3. Take any adversary A of the underlying timed probabilistic system $[\![\mathcal{T}]\!]_{\mathbb{R}}^{+}$; then there always exists a unique infinite path of the adversary A which remains in $\bar{l}$, resulting in a path of the form (we omit the action label for simplicity):

$$\omega = \langle \bar{l}, \nu_0 \rangle \xrightarrow{t_0, \mu_0} \langle \bar{l}, \nu_1 \rangle \xrightarrow{0.75} \langle \bar{l}, \nu_1 \rangle \xrightarrow{t_1, \mu_1} \langle \bar{l}, \nu_2 \rangle \xrightarrow{0.75} \langle \bar{l}, \nu_2 \rangle \xrightarrow{t_2, \mu_2} \cdots$$

where t_i are such that $\mathcal{D}(\omega) = \sum_{i=0}^{\infty} t_i \leq 1$, $\nu_i = \nu_{i-1} + t_i$, and $\mu_i = \delta_{\langle \bar{l}, \nu_{i+1} \rangle}$ are point distributions. This path ω is not time divergent, and so requiring that *all* of the paths of the adversary are divergent in Definition 5.4.3 would mean no divergent adversaries of this model would exist. Therefore, Definition 5.4.3 admits unrealisable, non-divergent paths, provided that the probability measure of these paths is 0.

5.5. Probabilistic Timed Computation Tree Logic (PTCTL)

We now describe the probabilistic real-time logic PTCTL (Probabilistic Timed Computation Tree Logic) which can be used to specify properties of probabilistic timed automata. Unlike in the previous chapters, we do not enhance PTCTL with an expectation operator. We defer the discussion of how to specify and verify expected cost properties to Section 5.7.

PTCTL is a combination of two extensions of the branching temporal logic CTL, the real-time temporal logic TCTL [**67**] and the probabilistic temporal logic PCTL discussed in Section 3.3. The logic TCTL can express timing constraints and includes the reset quantifier $z.\phi$, used to reset the clock z so that ϕ is evaluated from a state at which $z = 0$. PTCTL is obtained by enhancing TCTL with the probabilistic operator $\mathcal{P}_{\bowtie p}[\cdot]$ from PCTL. In the derived logic we can express properties such as "with probability 0.95 or greater, the value of the system clock x does not exceed 3 before 8 time units have elapsed", which is represented as the PTCTL formula $z \,.\, \mathcal{P}_{\geq 0.95}[(x \leq 3)\mathcal{U}(z = 8)]$.

We define PTCTL for the time domain $\mathbb{R}$ of reals only. As in TCTL, PTCTL employs a set of *formula clocks*, $\mathcal{Z}$, disjoint from the clocks $\mathcal{X}$ of the PTA. Formula clocks are assigned values by a *formula clock valuation* $\mathcal{E} \colon \mathcal{Z} \to \mathbb{R}$, which uses the notation for clock valuations in the standard way.

DEFINITION 5.5.1. The syntax of PTCTL is defined as follows:

$$\phi ::= a \mid \zeta \mid \neg\phi \mid \phi \wedge \phi \mid z \,.\, \phi \mid \mathcal{P}_{\bowtie p}[\phi \mathcal{U} \phi]$$

where $a \in AP$, $\zeta \in \text{Zones}(\mathcal{X} \cup \mathcal{Z})$, $z \in \mathcal{Z}$, $\bowtie \in \{\leq, <, >, \geq\}$ and $p \in [0, 1]$.

Note that the values of system clocks in $\mathcal{X}$ and formula clocks in $\mathcal{Z}$ can be obtained from a state and a formula clock valuation, respectively. Take a particular $\zeta \in \text{Zones}(\mathcal{X} \cup \mathcal{Z})$, and consider the syntactic interpretation of ζ as a conjunction of constraints. Then, given a state $\langle l, \nu \rangle$ and a formula clock valuation $\mathcal{E}$, we denote by $\zeta[\langle l, \nu \rangle, \mathcal{E}]$ the boolean value obtained by replacing each occurrence of a system clock $x \in \mathcal{X}$ in ζ by $\nu(x)$, and of a formula clock $z \in \mathcal{Z}$ in ζ by $\mathcal{E}(z)$.

Similarly to the case of probabilistic timed automata models, it is useful to identify the maximal integer constant that is referred to in a zone appearing in a PTCTL formula ϕ. Let $\text{Zones}(\phi) \stackrel{\text{def}}{=} \{\zeta \mid \zeta \text{ is a subformula of } \phi\}$, i.e. the set of zones occurring in ϕ, and define $k_{\max}(\phi) = \max\{k_{\max}(\zeta) \mid \zeta \in \text{Zones}(\phi)\}$ be the required maximal constant.

DEFINITION 5.5.2. Let $\mathcal{M} = (\mathsf{PS}, L, \mathbf{c})$ be a labelled timed probabilistic system. For any state s of $\mathcal{M}$, formula clock valuation $\mathcal{E}$, and PTCTL formula ϕ, the satisfaction relation $s, \mathcal{E} \models \phi$ is defined inductively as follows:

$$\begin{aligned}
s, \mathcal{E} &\models \text{true} && \text{for all } s \text{ and } \mathcal{E}, \\
s, \mathcal{E} &\models a && \Leftrightarrow a \in L(s), \\
s, \mathcal{E} &\models \zeta && \Leftrightarrow \zeta[s, \mathcal{E}] = \text{true}, \\
s, \mathcal{E} &\models \phi_1 \wedge \phi_2 && \Leftrightarrow s, \mathcal{E} \models \phi_1 \quad \text{and} \quad s, \mathcal{E} \models \phi_2, \\
s, \mathcal{E} &\models \neg\phi && \Leftrightarrow s, \mathcal{E} \not\models \phi, \\
s, \mathcal{E} &\models z.\phi && \Leftrightarrow s, \mathcal{E}[z := 0] \models \phi, \\
s, \mathcal{E} &\models \mathcal{P}_{\bowtie p}[\phi_1 \mathcal{U} \phi_2] && \Leftrightarrow p^A_{s,\mathcal{E}}(\phi_1 \mathcal{U} \phi_2) \bowtie p \text{ for all } A \in \text{Adv}_{\mathcal{M}}
\end{aligned}$$

where for any adversary $A \in \text{Adv}_{\mathcal{M}}$:

$$p^A_{s,\mathcal{E}}(\phi_1 \mathcal{U} \phi_2) = \text{Prob}^A_s(\{\omega \in \text{Path}^A_s \mid \omega, \mathcal{E} \models \phi_1 \mathcal{U} \phi_2\})$$

and for any path $\omega \in \text{Path}$:

$$\begin{aligned}
\omega, \mathcal{E} \models \phi_1 \mathcal{U} \phi_2 \Leftrightarrow\ & \text{there exist } i \in \mathbb{N}, t \in \mathbb{R} \text{ such that } \omega(i) + t, \mathcal{E} + \mathcal{D}_\omega(i) + t \models \phi_2, \\
& \text{and} \quad \text{for all } j \leq i, t' \in \mathbb{R} \text{ such that } j = i \text{ implies } t' < t \text{ we have:} \\
& \omega(j) + t', \mathcal{E} + \mathcal{D}_\omega(j) + t' \models \phi_1 \vee \phi_2.
\end{aligned}$$

The temporal logic PTCTL extends PCTL with the means to express real-time constraints. Recall that PCTL could only express discrete time. Although the logic CSL can express real-time bounded "until", it is interpreted over continuous time Markov chain models which do not feature nondeterminism exhibited by probabilistic timed automata.

Below we give two examples of timed properties of probabilistic timed automata that can be expressed by means of the temporal logic PTCTL. Untimed properties are also expressible, in the same way as for PCTL.

Time-bounded reachability: The system can reach a certain set of states within a certain time deadline with a given probability. For example, "with probability 0.975 or greater, it is possible to correctly deliver a data packet within 5 time units", which can be expressed by:

$$z \,.\, \neg\big(\mathcal{P}_{\leq 0.975}[\text{true}\ \mathcal{U}\ (\text{delivered} \wedge z \leq 5)]\big).$$

Bounded response: The system inevitably reaches a certain set of states within a certain time deadline with a given probability. For example, the property "with probability 0.99 or greater, a data packet will always be delivered within 5 time units" is stated in PTCTL as follows

$$z \,.\, \big(\mathcal{P}_{\geq 0.99}[\text{true}\ \mathcal{U}\ (\text{delivered} \wedge z \leq 5)]\big).$$

Similarly to PCTL, the logic PTCTL can be extended with the "diamond" ("eventually") and "box" ("always") temporal operators and "time-bounded until".

Because all clocks are real-valued, the state space of the dense semantics of a probabilistic timed automaton is uncountably infinite. However, it has been shown in [**5**] that one can partition the clock valuations of a timed automaton into a finite set of zones called *clock regions*, each satisfying the same TCTL formulas [**4**], and derive a structure, called a *region graph*, which can be used for model checking using conventional methods. In the next section we will show that an adaptation of this construction to the probabilistic case can be used for model checking probabilistic timed automata against PTCTL formulas. This result first appeared in [**96, 98**].

5.6. Model Checking: The Region Graph Approach

The region graph construction is an equivalence quotient of the state space that crucially relies on equivalence of clock valuations originally given in [**4, 5**]. Consider the probabilistic timed automaton $\mathcal{T} = (\text{Loc}, \bar{l}, \mathcal{X}, \Sigma, \text{inv}, \text{prob})$ over the time domain $\mathbb{R}$, and let $\mathbf{k} = k_{\max}(\mathcal{T})$. For any $t \in \mathbb{R}$, let $\lfloor t \rfloor$ denotes the integral part of t. Then, for any $t, t' \in \mathbb{R}$, t and t' *agree on their integral parts* if and only if:

- $\lfloor t \rfloor = \lfloor t' \rfloor$,
- both t and t' are integers or neither is an integer.

The valuations $\nu, \nu' \in \mathbb{R}^{|\mathcal{X}|}$ are *clock equivalent*, denoted by $\nu \cong \nu'$, if and only if they satisfy the following conditions:

- for any $x \in \mathcal{X}$, either $\nu(x)$ and $\nu'(x)$ agree on their integral parts, or both $\nu(x) > \mathbf{k}$ and $\nu'(x) > \mathbf{k}$,
- for any $x, x' \in \mathcal{X}$, either $\nu(x) - \nu(x')$ and $\nu'(x) - \nu'(x')$ agree on their integral parts, or both $\nu(x) - \nu(x') > \mathbf{k}$ and $\nu'(x) - \nu'(x') > \mathbf{k}$.

Let the valuations $\nu, \nu' \in \mathbb{R}^{|\mathcal{X}|}$ be such that $\nu \cong \nu'$, then the following are key properties that ensure soundness of the quotient construction:

(1) $\nu[X := 0] \cong \nu'[X := 0]$ for all $X \subseteq \mathcal{X}$,
(2) for any $t \in \mathbb{R}$, there exists $t' \in \mathbb{R}$ such that $\nu + t \cong \nu' + t'$,
(3) for any zone $\zeta \in \text{Zones}(\mathcal{T})$, $\nu \triangleleft \zeta$ if and only if $\nu' \triangleleft \zeta$.

Clock equivalence classes $[\nu]$ for $\nu \in \mathbb{R}^{|\mathcal{X}|}$ are a special type of zones. We refer to pairs of the form $\langle l, [\nu] \rangle$ where l is a location as *regions*, sometimes omitting the location component. Observe that, for finite $\mathbf{k}$, clock equivalence has a finite number of classes. Thus, the set of regions of the probabilistic timed automaton $\mathcal{T}$ is finite since $k_{\max}(\mathcal{T})$ is finite and the set of locations is finite.

EXAMPLE 5.6.1. Consider the set of clocks $\mathcal{X} = \{x, y\}$ and suppose the maximal constant $\mathbf{k}$ equals 1. The set of equivalence classes of clock valuations is given in Fig. 4, where each vertical, horizonal or diagonal line segment, open area and corner point is a region. The regions $\alpha_1, \alpha_2, \ldots, \alpha_5$ given in Fig. 4 are illustrated below

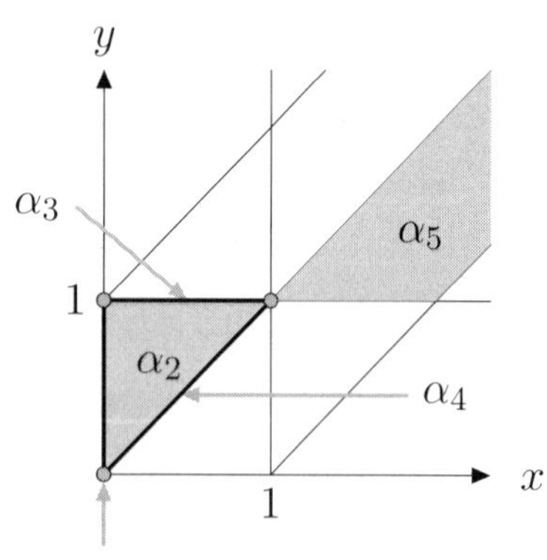

FIGURE 4. Equivalence classes for the set of clocks $\mathcal{X} = \{x, y\}$ when $\mathbf{k} = 1$.

with the help of a clock valuation contained in each region and the corresponding representation as a conjunction of constraints

$$\begin{aligned}\alpha_1 &= [x=0, y=0] = (x=0) \wedge (y=0),\\ \alpha_2 &= [x=0.4, y=0.67] = (0<x) \wedge (x-y<0) \wedge (y<1),\\ \alpha_3 &= [x=0.78, y=1] = (0<x) \wedge (x-y<0) \wedge (y=1),\\ \alpha_4 &= [x=0.666, y=0.666] = (0<x) \wedge (x<1) \wedge (x=y),\\ \alpha_5 &= [x=24.3, y=23.7] = (y>1) \wedge (x>y) \wedge (x-y>1).\end{aligned}$$

Clock equivalence naturally extends to formula clocks and the set of all valuations of a probabilistic timed automaton. Let $(\nu,\xi)\colon \mathcal{X}\cup\mathcal{Z} \to \mathbb{R}$ be the clock valuation that assigns a real value to each of the system and formula clocks, and let $\mathbb{R}^{|\mathcal{X}\cup\mathcal{Z}|}$ be the set of all such valuations for PTA. For a valuation $(\nu,\xi) \in \mathbb{R}^{|\mathcal{X}\cup\mathcal{Z}|}$, and $X \subseteq \mathcal{X}\cup\mathcal{Z}$, we use the notation $(\nu,\xi)[X:=0]$ in the usual way. For some $t \in \mathbb{R}$, $(\nu+t, \xi+t)$ denotes the clock valuation for which all clocks x in $\mathcal{X}\cup\mathcal{Z}$ take the value $(\nu,\xi)(x)+t$.

The equivalence relation for such valuations is defined with respect to a particular PTCTL formula ϕ. Let ξ' be the restriction of ξ over the clocks of $\mathcal{Z}$ that are referred to in ϕ. We can then extend the equivalence relation from $\cong$ to $\cong^*$ simply by taking (ν,ξ') instead of ν and $\mathcal{X}\cup\mathcal{Z}$ instead of $\mathcal{X}$, requiring that $\mathbf{k} = \max\{k_{\max}(\mathcal{T}), k_{\max}(\phi)\}$ and using the obvious definition of equivalence classes of the form $[\nu,\xi']$. Because our construction of the equivalence classes will always be with respect to a particular ϕ, we henceforth write ξ for ξ'. An element of the form $\langle l, [\nu,\xi]\rangle$ is called an *augmented region*.

5.6.1. The region graph. We now adapt the region graph construction to probabilistic timed automata and obtain a Markov decision process over regions, henceforth called *region graph*, instead of the usual transition system in the case of conventional timed automata [4]. The finiteness of the resulting representation permits the use of probabilistic model checking techniques for MDPs as described in Chapter 3.

First, we classify regions into three types. Let α be an equivalence class of the form $[\nu,\xi]$ and $\zeta \in \mathrm{Zones}(\mathcal{T}) \cup \mathrm{Zones}(\phi)$ be a zone either appearing in the description of $\mathcal{T}$ or as a subformula of ϕ. We extend the satisfaction relation in the obvious way: α *satisfies* ζ (denoted $\alpha \lhd \zeta$) if and only if, for any $(\nu,\xi) \in \alpha$, the value of ζ after substituting each occurrence of $x \in \mathcal{X}$ with $\nu(x)$, and each occurrence of $z \in \mathcal{Z}$ with $\xi(z)$, is true.

DEFINITION 5.6.2. Let α and β be distinct clock equivalence classes of $\mathbb{R}^{|\mathcal{X}\cup\mathcal{Z}|}$.

- **Successor class:** The equivalence class β is said to be the *successor* of α if and only if, for each $(\nu,\xi) \in \alpha$, there exists a positive $t \in \mathbb{R}$ such that $(\nu+t, \xi+t) \in \beta$, and $(\nu+t', \xi+t') \in \alpha \cup \beta$ for all $t' \leq t$.
- **x-zero class:** For a clock $x \in \mathcal{X}\cup\mathcal{Z}$, the equivalence class α is said to be *x-zero* if and only if, for each $(\nu,\xi) \in \alpha$, $(\nu,\xi)(x) = 0$.
- **x-unbounded class:** For a clock $x \in \mathcal{X}\cup\mathcal{Z}$, the equivalence class α is said to be *x-unbounded* if and only if, for each $(\nu,\xi) \in \alpha$, $\big(\nu,\xi)(x) > \max(k_{\max}(\mathcal{T}), k_{\max}(\phi)\big)$.

The successor relation can be extended to regions in the following way: $\langle l', \beta \rangle$ is the *successor region* of $\langle l, \alpha \rangle$ if $l' = l$ and $\beta = \mathrm{succ}(\alpha)$. Similarly, the region $\langle l, \alpha \rangle$ is x-zero (x-unbounded) if α is x-zero (x-unbounded).

EXAMPLE 5.6.3. Consider again the equivalence classes of clocks given in Fig. 4; we have that α_1 is an x-zero class, α_5 is an x-unbounded class, and α_4 is the successor of α_1 while α_3 the successor of α_2.

Next, we define a Markov decision process over regions. In contrast to timed probabilistic systems, the transition relation of the region graph abstracts from exact timing information.

DEFINITION 5.6.4. For a PTA $\mathcal{T} = (\mathrm{Loc}, \bar{l}, \mathcal{X}, \Sigma, \mathrm{inv}, \mathrm{prob})$ and PTCTL formula ϕ, the *region graph* $R(\mathcal{T}, \phi)$ is defined to be the Markov decision process $(V^*, \bar{v}^*, \mathrm{Act}^*, \mathrm{Steps}^*)$ where:

- the set of augmented regions V^* is the set of states;
- the initial state $\bar{v}^*$ equals $\langle \bar{l}, [\mathbf{0}, \mathbf{0}] \rangle$;
- the set of actions Act^* equals $\{\mathrm{succ}, \mathrm{loop}\} \cup \Sigma$;
- the probabilistic transition function $\mathrm{Steps}^* : V^* \to 2^{\mathrm{Act}^* \times \mathsf{Dist}(V^*)}$ includes three classes of transitions for each augmented region $\langle l, \alpha \rangle \in V^*$:

  Passage of time: If
  (1) the invariant condition $\mathrm{inv}(l)$ is satisfied by $\mathrm{succ}(\alpha)$;
  (2) for all probabilistic edges of the form $(l, g, \sigma, -) \in \mathrm{prob}$, $\alpha \lhd g$ implies $\sigma \notin \Sigma_u$;

  then $(\mathrm{succ}, \mu^*) \in \mathrm{Steps}^* \langle l, \alpha \rangle$ where $\mu^* = \delta_{\langle l, \mathrm{succ}(\alpha) \rangle}$ is a point distribution.

  Discrete transitions: If $(l, g, \sigma, p) \in \mathrm{prob}$ and $\alpha \lhd g$, then $(\sigma, \mu^*) \in \mathrm{Steps}^* \langle l, \alpha \rangle$ such that for any $l' \in \mathrm{Loc}$ and equivalence class β:

$$\mu^* \langle l', \beta \rangle = \sum_{X \subseteq \mathcal{X} \,\&\, \alpha[X:=0] = \beta} p(l', X).$$

  Self-loops: If for all probabilistic edges of the form $(l, g, \sigma, -) \in \mathrm{prob}$, $\alpha \lhd g$ implies $\sigma \notin \Sigma_u$, then $(\mathrm{loop}, \mu^*) \in \mathrm{Steps}^* \langle l, \alpha \rangle$, where $\mu^* = \delta_{\langle l, \alpha \rangle}$ is a point distribution.

Self-loops are included purely for technical convenience, and could be removed in certain contexts (for example, in an implementation of the model checking method).

Next we endow the region graph $R(\mathcal{T}, \phi)$ with labelling with atomic propositions (we omit costs since they are not considered in this section). For every zone ζ appearing in the given PTCTL formula ϕ, we extend the set AP with the atomic proposition a_ζ, by setting:

$$\mathrm{AP}^* \stackrel{\mathrm{def}}{=} \mathrm{AP} \cup \{a_\zeta \mid \zeta \in \mathrm{Zones}(\phi)\}.$$

The labelling function $L^* \colon V^* \to 2^{\mathrm{AP}^*}$ is then defined as follows. For each augmented region $\langle l, \alpha \rangle \in V^*$:

$$L^* \langle l, [\nu, \xi] \rangle \stackrel{\mathrm{def}}{=} L(l) \cup \{a_\zeta \mid \zeta \in \mathrm{Zones}(\phi) \text{ and } [\nu, \xi] \lhd \zeta\}.$$

The set of paths and adversaries of the region graph $R(\mathcal{T}, \phi)$ can now be defined in the standard way following the approach in Chapter 3.

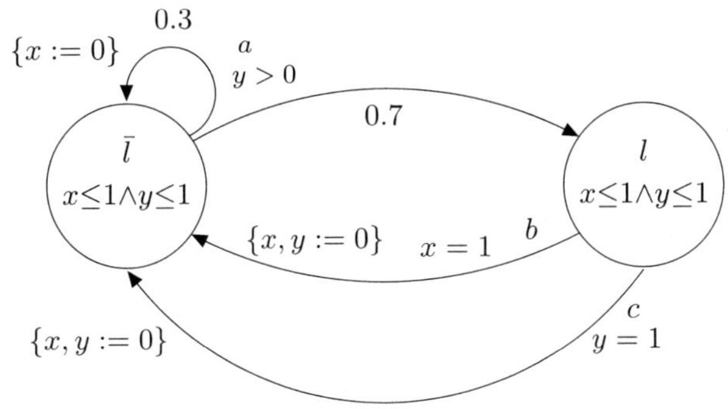

FIGURE 5. The probabilistic timed automaton $\mathcal{T}'$.

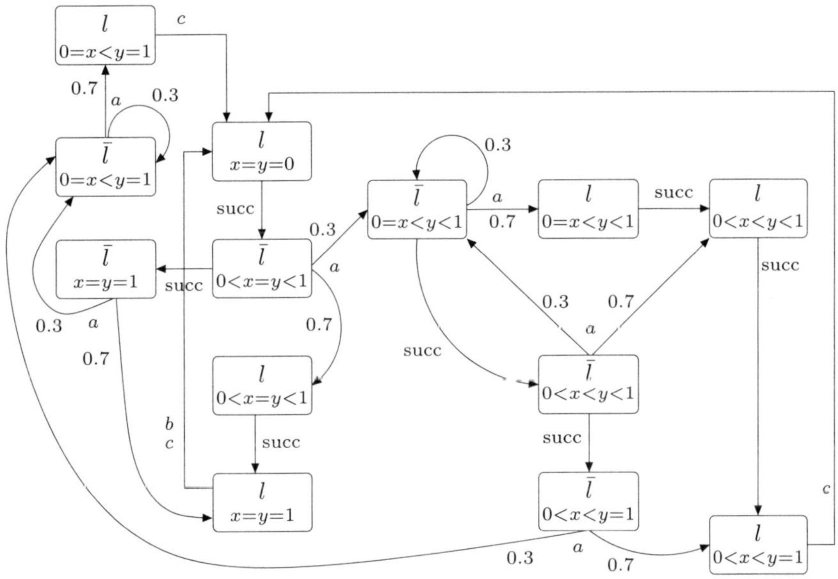

FIGURE 6. The region graph of the probabilistic timed automaton $\mathcal{T}'$.

Note that an adversary on the region graph $R(\mathcal{T}, \phi)$ corresponds to an *infinite* number of adversaries on the underlying timed probabilistic system. This is because a time transition in the dense-time semantics of a probabilistic timed automaton (see Definition 5.4.6) allows any non-negative real to be chosen, whereas the passage of time in any state $\langle l, \alpha \rangle$ of the region graph is given by the single action-distribution pair (succ, μ^*).

EXAMPLE 5.6.5. Consider the probabilistic timed automaton $\mathcal{T}'$ given in Fig. 5 and the PTCTL formula

$$\phi_1 = \neg\mathcal{P}_{<1}\big[(y=0)\,\mathcal{U}\,(x>0) \wedge \neg\mathcal{P}_{\leq 0.7}[(x>0)\,\mathcal{U}\,(y=0)]\big].$$

The corresponding region graph $R(\mathcal{T}', \phi_1)$ is given in Fig. 6 (note that we omit self-loops of the form (succ, μ^*)).

5.6.1.1. *Divergence on the region graph.* Next we introduce the notion of path divergence directly on the region graph, following the classical construction of [**29**]. As we allow for the possibility of an infinite number of discrete transitions in zero time, our construction differs from that in [**4**].

DEFINITION 5.6.6. Let $\omega^* = \langle l_0, \alpha_0 \rangle \xrightarrow{a_0, \mu_0^*} \langle l_1, \alpha_1 \rangle \xrightarrow{a_1, \mu_1^*} \cdots$ be an infinite path of the region graph. Then ω^* is

Possibly progressive: If and only if, for each clock $x \in \mathcal{X} \cup \mathcal{Z}$, one of the following conditions holds:
(1) for every $i \in \mathbb{N}$, there exists $j \geq i$ such that α_j is an x-zero class;
(2) there exists $i \in \mathbb{N}$ such that, for all $j \geq i$, α_j is an x-unbounded class.

Zero: If and only if there exists a clock $x \in \mathcal{X}$ and $i \in \mathbb{N}$ such that, for all $j > i$, α_j is x-zero.

Divergent: If and only if it is possibly progressive and not zero.

Possibly progressive paths are those in which, for each clock, either the clock is reset infinitely often, or eventually the value of the clock could become arbitrarily large. Possibly progressive paths, though, do not capture the concept of the divergence of time, as it is possible for a path to have a suffix of zero-regions (called zero-path). Thus, divergent paths are possibly progressive and not zero.

We can now define adversaries of the region graph.

DEFINITION 5.6.7 (Divergent adversaries on the region graph). An adversary A^* is *divergent* if and only if for all $v^* \in V^*$:

$$\mathrm{Prob}_{v^*}^{A*}\{\omega^* \in \mathrm{Path}_{v^*}^{A^*} \mid \omega^* \text{ is divergent}\} = 1.$$

Let $\mathrm{Adv}_{R(\mathcal{T},\phi)}$ be the set of divergent adversaries on the region graph.

Observe that divergent adversaries on the region graph $R(\mathcal{T}, \phi)$ correspond to an infinite number of adversaries on the underlying timed probabilistic system $[\![\mathcal{T}]\!]_{\mathbb{R}}^+$, some of which will be divergent in the sense of Definition 5.4.3. Conversely, for any divergent adversary of $[\![\mathcal{T}]\!]_{\mathbb{R}}^+$, the corresponding adversary on $R(\mathcal{T}, \phi)$ is divergent.

5.6.2. Model checking for PTCTL using the region graph. The region graph construction enables model checking for probabilistic timed automata against PTCTL formulas as follows. First, the finite-state region graph is built for the probabilistic timed automaton in question, then the PTCTL formula is translated to PCTL enhanced with clock resets, and finally model checking is performed for this obtained formula on the region graph using standard MDP model checking methods.

To extend PCTL with a reset quantifier expression, for a PCTL formula ϕ and $z \in \mathcal{Z}$ we add a new PCTL formula $z \,.\, \phi$ and extend the satisfaction relation by setting:

$$\langle l, [\nu, \xi] \rangle \models z.\Phi \Leftrightarrow \langle l, [\nu, \xi[z := 0]] \rangle \models \Phi.$$

This allows for a PCTL formula, Φ, to be *derived from* a PTCTL formula, ϕ, by just replacing a zone ζ appearing in ϕ by the atomic proposition a_ζ.

EXAMPLE 5.6.8. Consider again the PTA of (5.6.5), then the PCTL formula derived from the formula ϕ_1 equals:

$$\Phi_1 = \neg \mathcal{P}_{<1}\left[a_{(y=0)} \mathcal{U} a_{(x>0)} \wedge \neg \mathcal{P}_{<0.7}[a_{(x>0)} \mathcal{U} a_{(y=0)}]\right].$$

It is straightforward to show that Φ_1 is satisfied by this region graph, and therefore we conclude that the probabilistic timed automaton satisfies ϕ_1. Note that the following PTCTL formula, ϕ_2, is *not* satisfied by the region graph:

$$\phi_2 = \neg\mathcal{P}_{<1}\big[(y=0)\mathcal{U}(x>0) \wedge \neg\mathcal{P}_{\leq 0.7}[(x>0)\mathcal{U}(y=0)]\big].$$

The following proposition establishes the feasibility of model checking probabilistic timed automata against PTCTL using the region graph approach. It relies on the fact that probabilities and time are orthogonal, and so deriving the time-abstract quotient generalises to the probabilistic case without the need to restrict the logic.

PROPOSITION 5.6.9. [**98**] *Given a probabilistic timed automaton $\mathcal{T}$, we say that the state $\langle l,\nu\rangle$ of $[\![\mathcal{T}]\!]_{\mathbb{R}}^{+}$ and formula clock valuation ξ satisfy the* PTCTL *formula ϕ if and only if state $\langle l,[\nu,\xi]\rangle$ of $R(\mathcal{T},\phi)$ satisfies the* PCTL *formula Φ, where Φ is derived from ϕ.*

The proof involves first constructing, for a path ω of the dense-time semantics $[\![\mathcal{T}]\!]_{\mathbb{R}}^{+}$, a corresponding path $[\omega]$ of $R(\mathcal{T},\phi)$ such that ω is divergent if and only if $[\omega]$ is divergent, and then extending this construction to adversaries and showing that the basic cylinders corresponding to the finite paths ω and $[\omega]$ are associated with the same probability value for the respective adversaries. Finally, it can be shown that $\omega,\xi \models \phi_1\mathcal{U}\phi_2$ if and only if $[\omega] \models \Phi_1\mathcal{U}\Phi_2$, where the initial augmented state of $[\omega]$ comprises of ξ. For the complete proof see [**98**].

The main drawback of the region graph approach is that the number of states of the region graph is exponential in the number of clocks and size of constants, which makes verification prohibitively expensive. The current practice in real-time verification is to employ coarser partitioning into zones; some of these have been adapted to probabilistic timed automata in [**47, 98**] but will not be discussed here. Instead, in the next section we will consider an alternative, coarser, partitioning based on *digital clocks* of [**66**], which is more amenable to model checking.

5.7. Model Checking: The Digital Clocks Approach

As we have seen, timed automata and their probabilistic counterparts give rise to uncountable semantic descriptions and potentially undecidable verification problems. In the previous section we demonstrated the feasibility of model checking probabilistic timed automata against PTCTL properties. Unfortunately, this region graph approach suffers from the well known state space explosion problem. Therefore, in this section we describe an alternative approach which, although limited to a subclass of PTCTL properties, is more feasible in practice.

Our approach will be based on an *integer*-time model, instead of the more expressive but less tractable *dense*-time model. In an integer-time model, clocks assume integer values only, and are known as *digital clocks*. Such clocks were studied in the non-probabilistic context in [**66**], where *ε-digitisation* was introduced, and applied to timed automata in [**28, 107**], where it is shown that the subclass of *closed*, *diagonal-free* timed automata (automata whose constraints do not compare clocks or use strict comparison operators) admit integer-time verification techniques. Other related work includes [**6, 23**], and the investigations of the power of digital clocks of [**59, 106**].

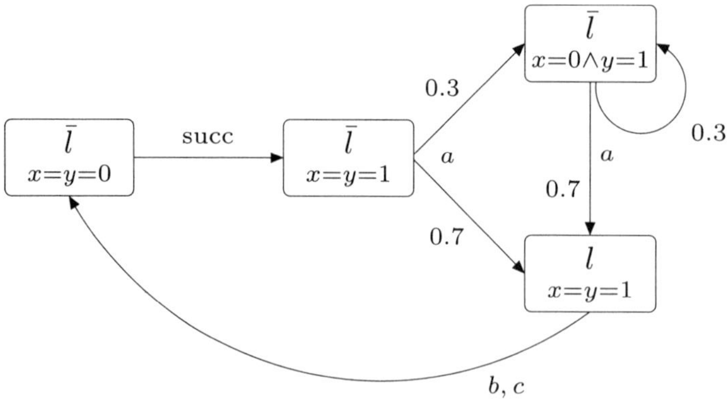

FIGURE 7. Integer-time semantics of the PTA of Fig. 6.

5.7.1. The integer-time semantics. We define the integer-time semantics for probabilistic timed automata by instantiating the semantics introduced in Definition 5.4.4 for an abstract time domain $\mathbb{T}$.

DEFINITION 5.7.1. Let $\mathcal{T} = (\text{Loc}, \bar{l}, \mathcal{X}, \Sigma, \text{inv}, \text{prob})$ be a probabilistic timed automaton. For any $x \in \mathcal{X}$, let $\mathbf{k}_x$ denote the greatest constant the clock x is compared to in the zones of $\mathcal{T}$. Define $\oplus_{\mathbb{N}}$ such that, for any clock valuation $\nu \in \mathbb{N}^{|\mathcal{X}|}$ and time duration $t \in \mathbb{N}$, the clock valuation $\nu \oplus_{\mathbb{N}} t$ assigns the value $\min\{\nu(x) + t, \mathbf{k}_x + 1\}$ to all clocks $x \in \mathcal{X}$. Then the *integer-time semantics* of $\mathcal{T}$ is defined as $[\![\mathcal{T}]\!]_{\mathbb{N}}^{\oplus_{\mathbb{N}}}$; that is, $\mathbb{T} = \mathbb{N}$ and $\oplus = \oplus_{\mathbb{N}}$.

For clarity, we omit the superscript indicating the operators $+$ and $\oplus_{\mathbb{N}}$, so that the dense-time semantics of a PTA $\mathcal{T}$ is denoted by $[\![\mathcal{T}]\!]_{\mathbb{R}}$ and the integer-time semantics by $[\![\mathcal{T}]\!]_{\mathbb{N}}$.

The integer-time semantics for probabilistic timed automata is a probabilistic analogue of the definition presented in [**23**]. The fact that the integer-time semantics of a probabilistic timed automaton is finite can be derived from the definitions. The integer-time semantics of the parallel composition of two probabilistic timed automata corresponds to the parallel composition of their integer-time semantics [**94**], in the sense that, for $\mathcal{T}_1$ and $\mathcal{T}_2$ with disjoint clock sets, $[\![\mathcal{T}_1 \| \mathcal{T}_2]\!]_{\mathbb{N}}$ and $[\![\mathcal{T}_1]\!]_{\mathbb{N}} \| [\![\mathcal{T}_2]\!]_{\mathbb{N}}$ are isomorphic.

EXAMPLE 5.7.2. Consider the probabilistic timed automaton given in Fig. 5 (p. 167). The corresponding integer-time semantic model of this probabilistic timed automaton is given in Fig. 7. Notice that the graph is smaller than the region graph given in Fig. 6 (p. 167).

5.7.2. Probabilistic and expected reachability properties. In previous chapters, we were able to present verification frameworks against temporal logics extended with probabilistic and expectation operators (namely, PCTL and CSL). Model checking for probabilistic timed automata is much more complex, which often necessitates restrictions on the logic and/or models. A typical restriction, in the case of classical timed automata, is to verify reachability properties only. In this section we consider a generalisation of reachability for probabilistic timed automata, namely *probabilistic* and *expected reachability* properties. We show that

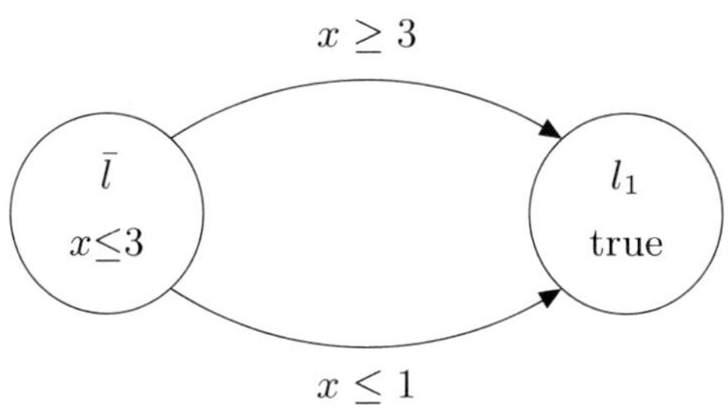

FIGURE 8. Counter-example for nested formulas.

the integer-time semantics suffices for the verification of probabilistic and expected reachability properties for the subclass of probabilistic timed automata that are closed and diagonal-free, see Definition 5.2.2. This method, however, does not extend to full PTCTL with nested formulas, as demonstrated below.

EXAMPLE 5.7.3. We illustrate why the integer-time semantics does not suffice for the verification of full PTCTL. Consider the (probabilistic) timed automaton given in Fig. 8. In the integer-time semantic model, no state satisfies the formula $\phi = z \,.\, (\mathcal{P}_{<1}[\text{true}\ \mathcal{U}\, a_{l_1} \wedge z \leq 1])$. More precisely, if the automaton is in location $\bar{l}$ and the clock x has an integer value then there exists an adversary such that (with probability 1) location l_1 is reached. On the other hand, for the dense-time semantics of Section 5.4, when the automaton is in location $\bar{l}$ and the clock x is in the interval $(1,2)$, then l_1 cannot be reached without letting time elapse by a value exceeding 1. Therefore, since no state of the integer-time semantic model can reach a state satisfying ϕ, the formula $\mathcal{P}_{<1}[\text{true}\ \mathcal{U}\phi]$ is trivially true in the initial state $\langle \bar{l}, \mathbf{0}\rangle$. However, in the dense-time model, by starting from the initial state $\langle \bar{l}, \mathbf{0}\rangle$ and letting time pass until $x \in (1,2)$ the formula ϕ becomes true (with probability 1), and hence $\mathcal{P}_{<1}[\text{true}\ \mathcal{U}\phi]$ is not satisfied in the initial state.

Probabilistic and expected reachability properties can be thought of as performance measures and have been studied in the context of probabilistic timed automata in [**93, 94, 98**]. We define *probabilistic reachability* [**94, 98**] as the maximal and minimal probability of reaching, from the initial state, a certain set of goal or target states. For a timed probabilistic system $\mathsf{PS} = (S, \bar{s}, \text{Act}, \mathbb{T}, \text{Steps})$, target set $F \subseteq S$ of target states and adversary $A \in \text{Adv}_{\mathsf{PS}}$, let:

$$\text{ProbReach}^A(F) \stackrel{\text{def}}{=} \text{Prob}^A_{\bar{s}}\{\omega \in \text{Path}^A_{\bar{s}} \mid \exists i \in \mathbb{N} \,.\, \omega(i) \in F\}.$$

DEFINITION 5.7.4. The *maximal and minimal reachability probabilities* of reaching the set of states F of the timed probabilistic system PS are defined as follows:

$$\text{MaxProbReach}_{\mathsf{PS}}(F) = \sup_{A \in \text{Adv}_{\mathsf{PS}}} \text{ProbReach}^A_{\bar{s}}(F),$$

$$\text{MinProbReach}_{\mathsf{PS}}(F) = \inf_{A \in \text{Adv}_{\mathsf{PS}}} \text{ProbReach}^A_{\bar{s}}(F).$$

Expected reachability [**93**] is defined with respect to a cost function, mapping actions to real values, and a set $F \subseteq S$ of target states. It corresponds to the *expected cost* (with respect to the given cost function) of reaching the set of target states F. For a labelled timed probabilistic system $\mathcal{M} = (\mathsf{PS}, L, \mathbf{c})$, where $\mathbf{c} : \text{Act} \cup \mathbb{T} \to \mathbb{R}$ is a cost function and $F \subseteq S$ is a set of target states, and adversary $A \in \text{Adv}_{\mathsf{PS}}$, let $\text{cost}(F)$ be defined for paths ω of $\mathcal{M}$ as in Section 3.2 and $\mathrm{E}_{\bar{s}}^{A}(\text{cost}(F))$ denote the usual expectation with respect to the measure $\text{Prob}_{\bar{s}}^{A}$ over $\text{Path}_{\bar{s}}^{A}$. *Expected time reachability* (the expected time with which a given set of states can be reached from the initial state) corresponds to the case when $\mathbf{c}(a) = 0$ for all $a \in \text{Act}$ and $\mathbf{c}(t) = t$ for all $t \in \mathbb{T}$.

DEFINITION 5.7.5. The *maximal and minimal expected costs* of reaching a set of states F in the labelled timed probabilistic system $\mathcal{M} = (\mathsf{PS}, L, \mathbf{c})$ are defined as follows:

$$\text{MaxExpReach}_{\mathcal{M}}(F) = \sup_{A \in \text{Adv}_{\mathsf{PS}}} \mathrm{E}_{\bar{s}}^{A}\big(\text{cost}(F)\big),$$
$$\text{MinExpReach}_{\mathcal{M}}(F) = \inf_{A \in \text{Adv}_{\mathsf{PS}}} \mathrm{E}_{\bar{s}}^{A}\big(\text{cost}(F)\big).$$

The verification of expected cost properties typically involves a modification of the underlying probabilistic timed automaton by the addition of a distinct clock (to represent the elapsed time) and a new location such that the automaton moves to this new location, from all locations, once the clock has reached a specified value.

Probabilistic reachability can be used to express the following properties:

Probabilistic reachability: "with probability at least 0.999, a data packet is correctly delivered";
Time-bounded reachability: "with probability 0.01 or less, a data packet is lost within 5 time units";
Cost-bounded reachability: "with probability at least 0.75, a data packet is correctly delivered with at most 4 retransmissions";
Invariance: "with probability 0.875 or greater, the system never aborts"; and
Bounded response: "with probability 0.99 or greater, a data packet will always be delivered within 5 time units".

On the other hand, expected reachability allows us to express:

Expected time: "the expected time until a data packet is delivered is at most 20ms" and "the expected time until a packet collision occurs is at least 100 seconds"; and
Expected cost: "the expected number of retransmissions before the message is correctly delivered is less than 3", "the expected number of packets sent before failure is at least 300" and "the expected number of lost messages within the first 200 seconds is at most 10".

We adapt the notion of digitisability, which was introduced for real-time systems in [**65, 66**], to timed probabilistic systems. A verification problem is said to be digitisable if it is invariant under the time model (dense or integer), and hence verification can be performed directly on the (often simpler) integer-time model. Digitisability is based on the notion of ε-*digitisation*, which discretises the real-times t by shifting the value to the preceding or following integer depending on the fractional value of t being less than some $\varepsilon \in [0, 1]$. In the remainder of this section we give a flavour of this construction and outline the main results.

Let $t \in \mathbb{R}$ be a (precise) real-time value and $\varepsilon \in [0,1]$. Define the digital value of t by rounding it with respect to ε:

$$[t]_\varepsilon = \begin{cases} \lfloor t \rfloor & \text{if } t \leq \lfloor t \rfloor + \varepsilon \\ \lceil t \rceil & \text{otherwise.} \end{cases}$$

The following property can be shown:

$$[t]_1 \leq t \leq [t]_0 \text{ for all } t \in \mathbb{R}.$$

Let $\mathcal{T} = (\mathrm{Loc}, \bar{l}, \mathcal{X}, \Sigma, \mathrm{inv}, \mathrm{prob})$ be a closed, diagonal-free probabilistic timed automaton. Firstly, observe that closed, diagonal-free constraints are preserved under ε-digitisation [**65**].

LEMMA 5.7.6. *For any $t, t' \in \mathbb{R}$, $c \in \mathbb{N}$ and $\sim \in \{\leq, =, \geq\}$, if $t - t' \sim c$ then $[t]_\varepsilon - [t']_\varepsilon \sim c$ for all $\varepsilon \in [0,1]$.*

Next, we digitise paths containing precise real-valued times. We shall require the following property, which can be shown for a path by choosing j as the most recent state at which clock x was reset.

LEMMA 5.7.7. *For any path $\omega = \langle l_0, \nu_0 \rangle \xrightarrow{a_0, \mu_0} \langle l_1, \nu_1 \rangle \xrightarrow{a_1, \mu_1} \cdots$, $x \in \mathcal{X}$ and $i \in \mathbb{N}$, there exists $j \leq i$ such that $\nu_i(x) = \mathcal{D}_\omega(i) - \mathcal{D}_\omega(j)$.*

We can now define the ε-digitisation of a path of the dense-time semantics $[\![\mathcal{T}]\!]_\mathbb{R}$ of a probabilistic timed automaton $\mathcal{T}$, which yields a path of the integer-time semantics $[\![\mathcal{T}]\!]_\mathbb{N}$ of $\mathcal{T}$.

DEFINITION 5.7.8 (ε-digitisation). For any path:

$$\omega = \langle \bar{l}, \mathbf{0} \rangle \xrightarrow{a_0, \mu_0} \langle l_1, \nu_1 \rangle \xrightarrow{a_1, \mu_1} \cdots$$

of $[\![\mathcal{T}]\!]_\mathbb{R}$, its ε-digitisation is the path:

$$[\omega]_\varepsilon = \langle \bar{l}, \mathbf{0} \rangle \xrightarrow{a'_0, \mu_0} \langle l_1, [\nu_1]_\varepsilon \rangle \xrightarrow{a'_1, \mu_1} \cdots$$

of $[\![\mathcal{T}]\!]_\mathbb{N}$ where for any $i \in \mathbb{N}$ and $x \in \mathcal{X}$:

- $[\nu_i]_\varepsilon(x) = \min([\mathcal{D}_\omega(i)]_\varepsilon - [\mathcal{D}_\omega(j)]_\varepsilon, \mathbf{k}_x + 1)$ and $j \leq i$ such that $\nu_i(x) = \mathcal{D}_\omega(i) - \mathcal{D}_\omega(j)$ which exists by Lemma 5.7.7;
- if $a_i \in \Sigma$, then $a'_i = a_i$;
- if $a_i \in \mathbb{R}$, then $a'_i = [\mathcal{D}_\omega(i+1)]_\varepsilon - [\mathcal{D}_\omega(i)]_\varepsilon$.

That this is well-defined, that is, $[\omega]_\varepsilon$ is a path of $[\![\mathcal{T}]\!]_\mathbb{N}$, follows from Lemmas 5.7.7 and 5.7.6, and the fact that $\mathcal{T}$ is diagonal-free and closed.

We can show that the total duration of ε-digitisation of a path in the dense-time semantics corresponds to the ε-digitisation of the duration of the original path.

LEMMA 5.7.9. *For any path $\omega \in \mathrm{Path}_{\bar{s}}$, $\varepsilon \in [0,1]$ and $i \in \mathbb{N}$: $\mathcal{D}_{[\omega]_\varepsilon}(i) = [\mathcal{D}_\omega(i)]_\varepsilon$.*

PROOF. Consider any path $\omega = \langle \bar{l}, \mathbf{0} \rangle \xrightarrow{a_0, \mu_0} \langle l_1, \nu_1 \rangle \xrightarrow{a_1, \mu_1} \cdots \in \mathrm{Path}_{\bar{s}}$. We prove the lemma by induction on $i \in \mathbb{N}$. If $i = 0$, then by definition $\mathcal{D}_{[\omega]_\varepsilon}(i) = [\mathcal{D}_\omega(i)]_\varepsilon = 0$ as required. Now suppose that the lemma holds for some $i \in \mathbb{N}$. We have only two cases to consider.

- If $a_i \in \Sigma$, then $[\mathcal{D}_\omega(i+1)]_\varepsilon = [\mathcal{D}_\omega(i)]_\varepsilon$ and from Definition 5.7.8 we have $\mathcal{D}_{[\omega]_\varepsilon}(i+1) = \mathcal{D}_{[\omega]_\varepsilon}(i)$, and the result follows by induction.

- If $a_i \in \mathbb{R}$, then by Definition 5.7.8 we have:

$$\begin{aligned}\mathcal{D}_{[\omega]_\varepsilon}(i+1) &= \mathcal{D}_{[\omega]_\varepsilon}(i) + \big([\mathcal{D}_\omega(i+1)]_\varepsilon - [\mathcal{D}_\omega(i)]_\varepsilon\big) \\ &= [\mathcal{D}_\omega(i)]_\varepsilon + \big([\mathcal{D}_\omega(i+1)]_\varepsilon - [\mathcal{D}_\omega(i)]_\varepsilon\big) && \text{by induction} \\ &= [\mathcal{D}_\omega(i+1)]_\varepsilon && \text{as required.}\end{aligned}$$

Hence, the lemma holds by induction on $i \in \mathbb{N}$. □

This correspondence of the ε-digitisation of the paths of the dense-time semantics to those of the integer-time semantics allows us to prove that probabilistic reachability is digitisable.

Let $\mathcal{T} = (\text{Loc}, \bar{l}, \mathcal{X}, \Sigma, \text{inv}, \text{prob})$ be a closed, diagonal-free probabilistic timed automaton. For any set of locations $\text{Loc}' \subseteq \text{Loc}$, we denote the set of all states of $[\![\mathcal{T}]\!]_{\mathbb{T}}$ which correspond to these locations, that is, the set $\{\langle l, \nu\rangle \mid l \in \text{Loc}', \nu \in \mathbb{T}^{|\mathcal{X}|} \wedge \nu \triangleleft \text{inv}(l)\}$, by $F_{\mathbb{T}}^{\text{Loc}'}$.

THEOREM 5.7.10. [**94**] *For any closed, diagonal-free probabilistic timed automaton $\mathcal{T}$ and set of locations* $\text{Loc}' \subseteq \text{Loc}$*:*

$$\begin{aligned}\text{MaxProbReach}_{[\![\mathcal{T}]\!]_{\mathbb{R}}}(F_{\mathbb{R}}^{\text{Loc}'}) &= \text{MaxProbReach}_{[\![\mathcal{T}]\!]_{\mathbb{N}}}(F_{\mathbb{N}}^{\text{Loc}'}), \\ \text{MinProbReach}_{[\![\mathcal{T}]\!]_{\mathbb{R}}}(F_{\mathbb{R}}^{\text{Loc}'}) &= \text{MinProbReach}_{[\![\mathcal{T}]\!]_{\mathbb{N}}}(F_{\mathbb{N}}^{\text{Loc}'}).\end{aligned}$$

The proof of Theorem 5.7.10 proceeds by extending the notion of ε-digitisation from paths to adversaries and demonstrating that, for each adversary of the dense-time semantics, there exists an adversary of the integer-time semantics (and vice-versa) such that the probability measure of a set of paths of the adversary is preserved under ε-digitisation.

A similar digitisability result also holds for expected cost calculation. We restrict our attention to *non-negative* cost functions[1] such that $\mathbf{c}(t+t') = \mathbf{c}(t) + \mathbf{c}(t')$ for all $t, t' \in \mathbb{R}$, which is trivially satisfied by the cost function corresponding to expected time reachability.

THEOREM 5.7.11. [**93**] *For any closed, diagonal-free probabilistic timed automaton $\mathcal{T}$, set of locations* $\text{Loc}' \subseteq \text{Loc}$ *and non-negative cost function* $\mathbf{c}\colon \text{Act} \cup \mathbb{R} \to \mathbb{R}$ *such that $\mathbf{c}(t+t') = \mathbf{c}(t) + \mathbf{c}(t')$ for all $t, t' \in \mathbb{R}$:*

$$\begin{aligned}\text{MaxExpReach}_{[\![\mathcal{T}]\!]_{\mathbb{R}}}(F_{\mathbb{R}}^{\text{Loc}'}) &= \text{MaxExpReach}_{[\![\mathcal{T}]\!]_{\mathbb{N}}}(F_{\mathbb{N}}^{\text{Loc}'}), \\ \text{MinExpReach}_{[\![\mathcal{T}]\!]_{\mathbb{R}}}(F_{\mathbb{R}}^{\text{Loc}'}) &= \text{MinExpReach}_{[\![\mathcal{T}]\!]_{\mathbb{N}}}(F_{\mathbb{N}}^{\text{Loc}'}).\end{aligned}$$

The proof of Theorem 5.7.11 relies on the following lemma which gives bounds on the cumulative cost of reaching the target states in the dense-semantics.

LEMMA 5.7.12. *Let $\mathcal{M} = (\mathsf{PS}, L, \mathbf{c})$ be a labelled timed probabilistic system, where $\mathsf{PS} = (S, \bar{s}, \text{Act}, \mathbb{R}, \text{Steps})$, $\mathbf{c}\colon \text{Act} \cup \mathbb{T} \to \mathbb{R}$ is a cost function such that $\mathbf{c}(t+t') = \mathbf{c}(t) + \mathbf{c}(t')$ for all $t, t' \in \mathbb{R}$. For any path $\omega \in \text{Path}_{\bar{s}}$ and set of locations* $\text{Loc}' \subseteq \text{Loc}$*, we have:*

$$\text{cost}(F_{\mathbb{N}}^{\text{Loc}'})([\omega]_1) \leq \text{cost}(F_{\mathbb{R}}^{\text{Loc}'})(\omega) \leq \text{cost}(F_{\mathbb{N}}^{\text{Loc}'})([\omega]_0).$$

[1]The results presented here will also hold for non-positive cost functions.

Complete proofs of Theorems 5.7.10 and 5.7.11 can be found in [**93, 94**], respectively.

Both classes of properties (probabilistic and expected cost reachability) can be extended to include reachability of a state satisfying a formula which is a conjunction of atomic propositions and clock constraints. This can be achieved through a modification of the underlying probabilistic timed automaton which preserves the property of being diagonal-free and closed.

5.7.3. Integer-time semantics and model checking. The advantage of the digital clocks approach is that it avoids the need to build a Markov decision process over the regions or zones. From the construction given in Definitions 5.4.4 and 5.7.1, the integer-time semantic model will have only finitely many states. However, to ensure finite branching we must restrict the delays in the integer-time semantic models from $\mathbb{N}$ to some finite set. For example, we can restrict delays to have duration 1 only (or to any finite range $1, \ldots, N$) since any time transition of duration $n \in \mathbb{N}$ can be modelled by a sequence of time transitions of duration 1. By restricting to divergent adversaries we can additionally omit delays of duration 0.

Thus, we obtain a finite-state Markov decision process, and since both probabilistic and expected reachability can be expressed in PCTL with the help of the probabilistic and expectation operators, the verification can be performed by means of the model checking algorithms given in Section 3.4. In both cases, model checking reduces to solving a linear optimisation problem which admits an iterative solution method. We also note that algorithms for checking for the existence of divergent adversaries are given in [**45**].

5.8. Case Study: IEEE 1394 FireWire Root Contention

This case study concerns the "Tree Identify" protocol of the IEEE 1394 high performance serial bus (called "FireWire"), an efficient and widely-used multimedia protocol. It has a scalable architecture and is "hot-pluggable", meaning that devices can easily be added or removed from the network at any time.

The "Tree Identify" protocol of IEEE 1394 is a randomised leader election algorithm designed for connected, acyclic network topologies. It executes after a bus reset in the network, i.e. when a node is added or removed, and its purpose is to decide upon a leader (root) to act as the manager of the bus for subsequent phases of IEEE 1394. To carry out this process, the nodes communicate "be my parent" messages to each other, starting from the leaf nodes, in order to establish a tree topology. If two nodes contend the leadership (root contention), the contenders resolve the situation by exchanging a number of additional messages, subject to *time delays* and *electronic coin tosses*. Probabilistic timed automata, therefore, provide an ideal formalism for modelling the protocol.

We have analysed the following probabilistic aspects of the protocol: termination with probability 1; expected time and number of rounds required to resolve root contention; and the probability of resolving contention within certain time bounds. This has been done using both dense-time semantics [**47**] and integer-time semantics [**94**]. In this section, we describe the latter approach.

The model we consider is a probabilistic extension of the timed automata model analysed in [**120**]. It comprises two nodes, $\mathtt{Node}_1$ and $\mathtt{Node}_2$, and two wires, $\mathtt{Wire}_1$

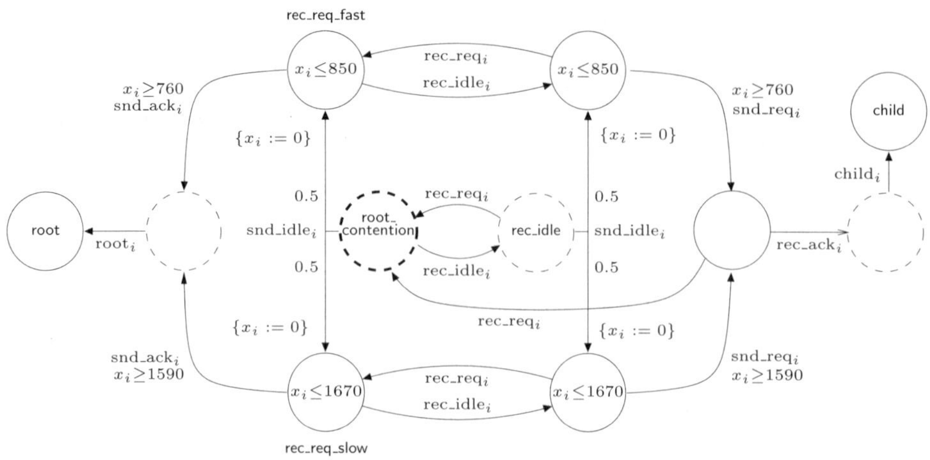

FIGURE 9. The probabilistic timed automaton $\texttt{Node}_i$.

and $\texttt{Wire}_2$, which is sufficient in order to analyse root contention. The full probabilistic timed automaton is built via parallel composition, as described in Definition 5.3.1: $\texttt{Impl} = \texttt{Node}_1 \| \texttt{Wire}_1 \| \texttt{Wire}_2 \| \texttt{Node}_2$,

The probabilistic timed automaton $\texttt{Node}_i$ is shown in Fig. 9. For simplicity, we have omitted invariant conditions or guards of the form true and some location names. Urgent locations, i.e. those in which time cannot pass, are drawn with dashed lines. The node starts in the urgent location root_contention (root contention has been detected). It immediately tosses a coin in order to determine whether it should wait for a short or a long time, i.e. move to location rec_req_fast or rec_req_slow, respectively. This coin toss corresponds to the probabilistic edge labelled by snd_idle$_i$. The event snd_idle$_i$ models transmission of an idle signal from $\texttt{Node}_i$ across its wire to the other node.

In the two locations rec_req_fast and rec_req_slow, the passage of time may mean that the clock x_i can reach a value which enables the left-pointing edges. In this case, the node sends an acknowledgement (event snd_ack$_i$) and then declares itself to be leader (event root$_i$), moving to location root. If, on the other hand, the $\texttt{Node}_i$ receives an idle signal from the other contending node (event rec_idle$_i$) before sending an acknowledgement, it is forced to move to the right via the edge labelled with rec_idle$_i$. In this case, after a certain amount of time has elapsed, it can send a request to the other node, asking it to be its parent (event snd_req$_i$). If $\texttt{Node}_i$ subsequently detects a parent request from the other node (event rec_req$_i$), it returns to the location root_contention, and restarts the whole process. If, though, the $\texttt{Node}_i$ detects an acknowledgement from the other node (event rec_ack$_i$), it proceeds to declare itself as the child by sending a child$_i$ event.

The communication medium between the nodes comprises two wires modelled as two-place buffers, along which signals are driven continuously. These are represented by the probabilistic automata $\texttt{Wire}_1$ and $\texttt{Wire}_2$. Figure 10 illustrates the general case, $\texttt{Wire}_i$. This model is the classical (non-probabilistic) timed automaton taken from [**120**], with the interpretation that all transitions are made with

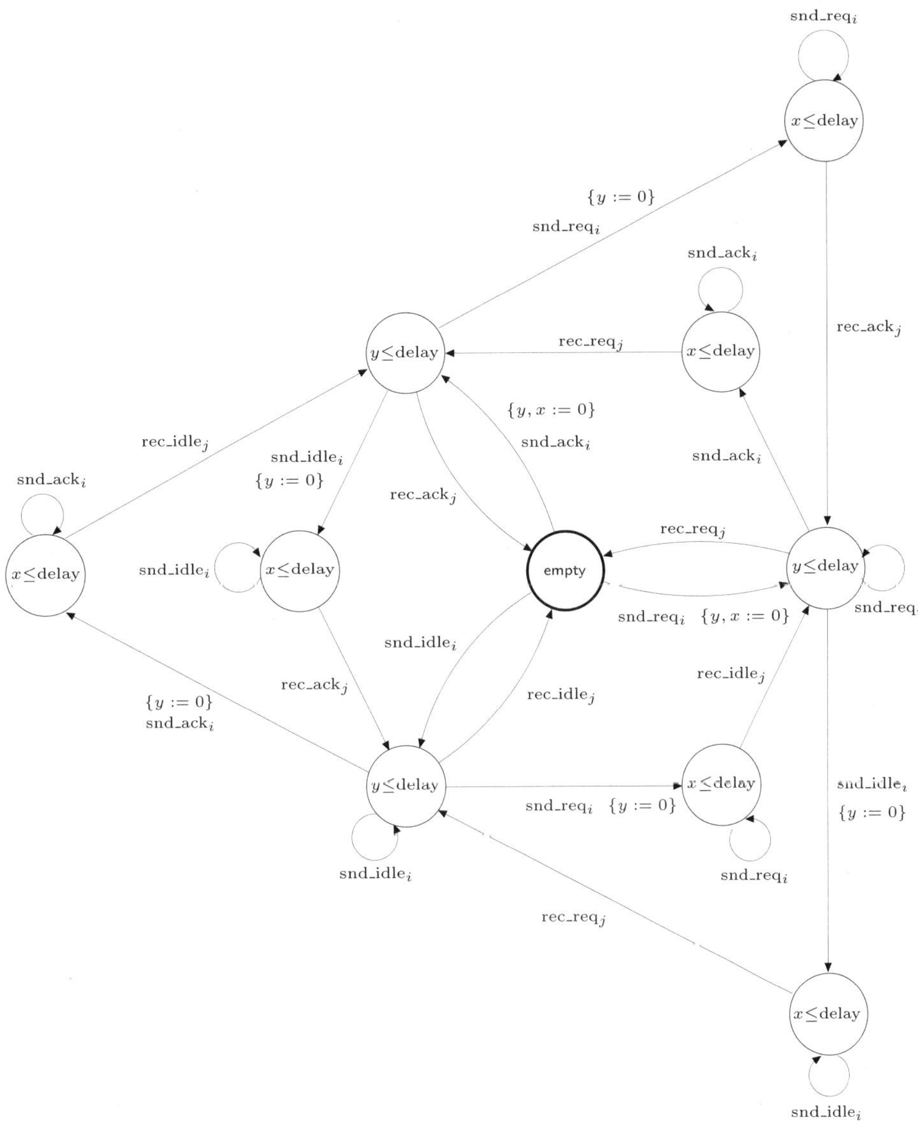

FIGURE 10. The probabilistic timed automaton $\texttt{Wire}_i$.

probability 1. In our analysis, we will consider two cases for the maximum transmission delay along the wires: 360 nanoseconds (ns) and 30ns. This models the distance between the two nodes, i.e. the length of the connecting wires. A delay of 360ns represents the assumption that the nodes are separated by a distance close to the maximum required for the correctness of the protocol (from the analysis of [**120**]). A delay of 30ns corresponds more closely to the maximum separation of nodes specified in the actual IEEE standard. In fact, in the second case, the correct value is 22.725ns so our figure of 30ns is an over-approximation. This is

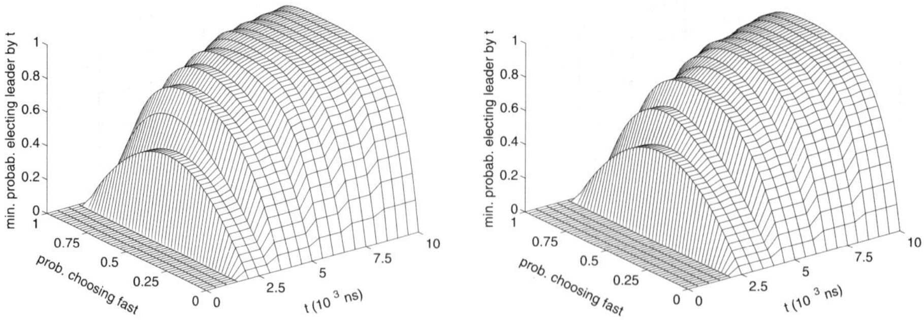

FIGURE 11. Minimum probability to elect a leader by time t.

for efficiency reasons: it allows us to use a time granularity of 10ns when we consider probabilistic model checking using digital clocks. In the following paragraphs, we will refer to the two cases (360ns and 30ns) as "long wire" and "short wire", respectively.

An interesting factor to consider in the analysis of the FireWire protocol is the advantage obtained by using a biased coin for the coin tosses which decide between short and long time delays. This is based on a conjecture from [**123**]. We will consider the effect of a range of biases. Note that, since the leader election algorithm is symmetric, we only consider the case where both nodes use the same bias (or indeed both use unbiased coins).

The properties of the root contention protocol we are interested in are:

- the minimum probability that a leader is eventually elected;
- the minimum probability that a leader is elected *within* time t;
- the maximum expected time to elect a leader;
- the maximum expected number of rounds until a leader is elected.

For the first property, our analysis showed that the minimum probability of eventually electing a leader is 1 for both wire lengths and for any biased coin which has a non-zero probability of flipping both heads and tails.

To compute time-bounded probabilities (i.e. the second property), we used an extended version of the model, where the original probabilistic timed automaton was put in parallel with a simple automaton which counts global time and prevents the protocol performing any transitions once the deadline has passed. Figure 11 plots the results obtained in this case, for both the short and long wire length and for a range of coin biases (i.e. the probability of choosing "fast") and time bounds t.

For properties involving expected time and number of rounds, we used two different cost functions. The first, for expected time, assigns cost 0 to all events and sets the cost of a time transition equal to the elapsed time. The second, for expected number of rounds, assigns a cost of 1 to each event which corresponds to the start of a new round, and 0 to all other events and time transitions. Figure 12 shows results for the maximum expected time to elect a leader using both wire lengths and a range of coin biases. The top graphs show a close-up view of where the minimum values occur. Figure 13 plots results for the maximum expected

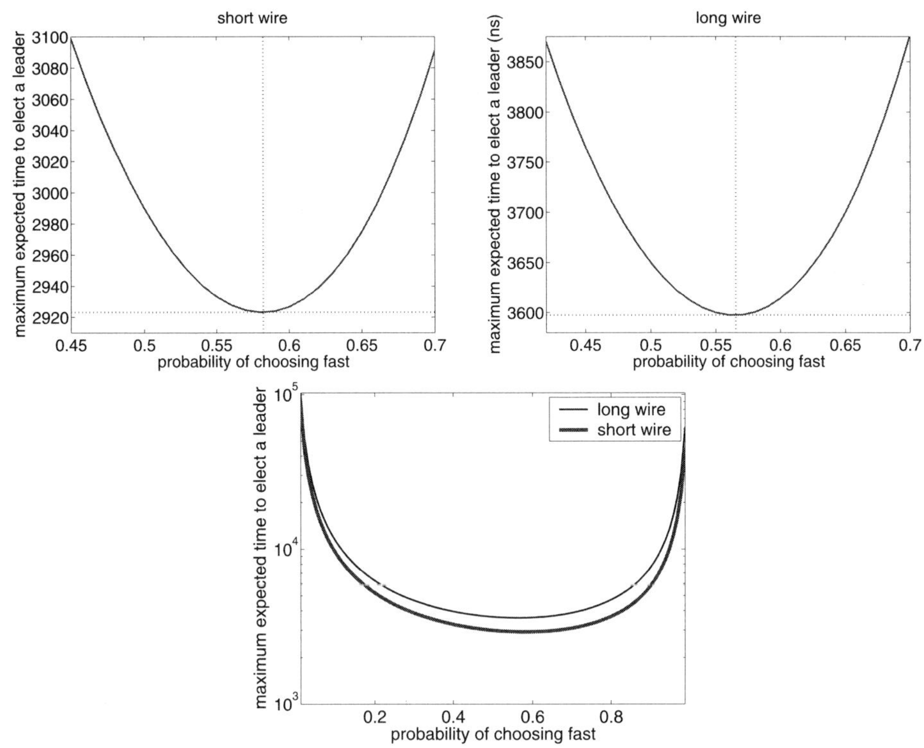

FIGURE 12. Maximum expected time to elect a leader.

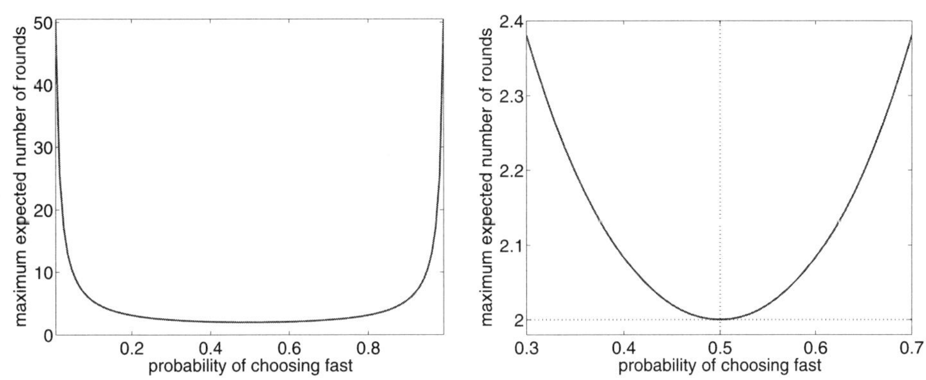

FIGURE 13. Maximum expected number of rounds to elect a leader.

number of rounds to elect a leader. Note that this is independent of the wire length used. Again, a close-up view of the minimum value is provided.

The statistics presented in Figures 11 and 12 demonstrate that the (timing) performance of the root contention protocol *can* be improved by using a biased coin which has a higher probability of flipping "fast". The intuition behind this result is that, although using such a biased coin decreases the likelihood of the nodes flipping different values, when nodes flip the same values there is a greater chance (i.e. when both flip "fast") that less time passes before they flip again. There is a compromise here; as the coin becomes more biased towards "fast" the probability of the nodes actually flipping different values (which is required for a leader to be elected) decreases even though the delay between coin flips will on average increase. This decrease in probability is demonstrated in Fig. 13, where the expected number of rounds, or alternatively the expected number of coin flips by one node, increases for any biased coin.

Furthermore, the results in Fig. 12 demonstrate that, for a shorter wire length, there is a greater advantage when using a biased coin. For the short wire length, the minimum occurs for a coin which flips "fast" with probability 0.58 while, for the long wire length, the minimum is near 0.56. This is because, for the short wire, there is a greater saving in time when both nodes flip "fast" than for a longer wire length, since the time required when both nodes flip "fast" is a constant delay given in the protocol plus a delay which is dependent on the wire length.

The expected number of rounds (see Fig. 13), on the other hand, obtains its minimum for a fair coin. This is unsurprising since it depends purely on the the nodes flipping different values and, if the nodes use a biased coin, the probability of them flipping the same value increases.

Finally, the results also demonstrate that the performance of the protocol is improved when a shorter wire is employed (the probability to elect a leader by time deadline t increases and the maximum expected time to elect a leader decreases). This is also to be expected since using a shorter wire means that the delay between the nodes sending messages decreases, and hence the nodes can reach agreement in less time. For further details about this case study, see [**94**] or the PRISM website [**111**]. Probabilistic timed automata have also been use to model and analyse random back-off schemes of physical layer protocols, as in e.g. the IEEE 802.11 Wireless LAN MAC protocol, see [**92**].

5.9. Discussion

Probabilistic timed automata and the region graph construction have also been considered in [**18, 60**]. In [**98**], model checking algorithms based on *forwards reachability* [**48, 126, 129**] over the zone graph are introduced; unfortunately, this approach can only be applied to probabilistic reachability properties and yields approximate probabilities. An approach based on *backwards reachability* [**67**], which is more efficient than the region graph approach and results in exact minimum/maximum probabilities, is given in [**91, 95**].

Probabilistic timed automata introduced in this chapter only admit integer clock resets. Models enhanced with *continuously distributed random delays* have been considered in [**2, 3**] and [**97**], but are generally much more complex. The model of [**2, 3**] does not permit any nondeterministic choice and only deals with qualitative probabilistic properties. In [**97**], probabilistic timed automata are extended with continuously distributed clock resets and the standard region graph construction is modified by subdividing the unit intervals in order to *approximate* the probability to within an error bound. This approach improves on the techniques of [**2, 3**] in that it allows the verification of *quantitative* probability bounds, but its complexity is prohibitive.

CHAPTER 6

Implementation

6.1. Introduction

In this final chapter, we consider some of the practical details regarding implementation of the probabilistic model checking techniques described in the preceding chapters. In recent years, such implementations have become available in a number of tools: $\mathsf{E} \vdash \mathsf{MC}^2$ [**68**], which supports model checking of CSL for CTMCs and PCTL for DTMCs, RAPTURE [**77**], which allows verification of a subset of PCTL for MDPs, and PRISM [**87**], which supports model checking for all three of these models. At present, there exists no direct tool support for model checking of probabilistic timed automata. Fortunately, as we saw in Chapter 5, model checking for PTAs reduces, via the region graph construction or digitisation, to MDP model checking. For the latter, the PRISM or RAPTURE tools can be used directly, and otherwise the timed automata model checker KRONOS [**48**] can be used to generate an MDP over zones [**47**] that can then be analysed with PRISM. Hence, in this chapter we focus on verification for DTMCs, CTMCs and MDPs.

The most fundamental issue in terms of a practical implementation of probabilistic model checking is the data structures which are used to store and manipulate the models. The main reason for this is a phenomenon known as the *state space explosion problem*, i.e. the fact that even the simplest real-world examples can result in models with enormous numbers of states. Since probabilistic models are generally represented by some form of matrix, the key problem becomes the storage and manipulation of very large matrices. These matrices do, however, usually contain relatively few entries since, despite there being a large number of states, each one will typically have a fairly small number of transitions. In Section 6.2, we discuss the *sparse matrix* data structure, which allows compact storage of matrices by only explicitly storing their non-zero elements. We also describe the implementation of matrix-vector multiplication which is one of the most fundamental operations required for probabilistic model checking.

In the remainder of the chapter, we focus on *symbolic* implementations of model checking, by which we mean those based on data structures called binary decision diagrams (BDDs) and their variants. The tools PRISM and RAPTURE both support this style of implementation. Symbolic approaches aim to generate an even more compact and efficient model representation by exploiting high-level structure and regularity. In particular, we will concentrate on a data structure called multi-terminal binary decision diagrams (MTBDDs). In Section 6.3 we describe the basic details of MTBDDs, in Section 6.4 we discuss how they can be used to efficiently represent DTMCs, CTMCs and MDPs and in Section 6.5 we give details about implementing the model checking algorithms themselves using the data structure. The latter is based on the actual implementation of the PRISM tool.

$$\begin{pmatrix} \cdot & 0.5 & \cdot & 0.5 \\ \cdot & \cdot & 1 & \cdot \\ 0.3 & \cdot & \cdot & 0.7 \\ 1 & \cdot & \cdot & \cdot \end{pmatrix}$$

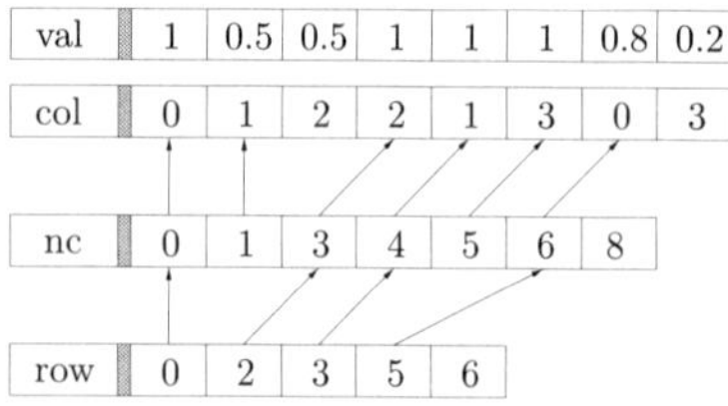

FIGURE 1. A 4×4 matrix and its sparse matrix representation.

6.2. Sparse Matrices

Two of the probabilistic models we consider in this book, DTMCs and CTMCs, are described by square, real-valued matrices. In practice, these matrices are often very large, but contain a relatively small number of non-zero entries. The transition matrix of a model with 10^6 states, for example, might have as few as 10 non-zero entries in each row. *Sparse matrices* are a well known data structure designed exactly for storing such matrices. The basic idea is that only the non-zero entries of a matrix are explicitly stored.

There are several different types of sparse storage scheme. The best one to use depends on what operations will be performed on the matrix. The scheme we will describe here is a common variant, often referred to as a *row-major* sparse matrix. Matrix entries are stored very compactly and are quick and easy to access. The main trade-off is that it is not efficient to make modifications to the matrix. The scheme is well suited to many typical applications of sparse matrices, such as the iterative solution methods required for probabilistic model checking. In particular, matrix-vector multiplication, the key operation for such methods, can be performed very efficiently using this sparse matrix scheme.

Assume that we wish to store a real-valued matrix $\mathbf{M}$. Since we intend to store matrices which represent probabilistic models, we will assume that the matrix is indexed over some state space S. For clarity of presentation, we will assume that this state space is $S = \{0, 1, 2, \ldots, |S| - 1\}$. We denote the number of non-zero entries in $\mathbf{M}$ by nnz.

The sparse matrix data structure stores information about the row index, column index and value of each matrix entry in three separate arrays, row, col and val. The arrays val and col each contain nnz elements and store the actual value and column index of each matrix entry, respectively, ordered by row. The third array, row, stores indices into the other two arrays: the ith element of row points to the start of the entries of row i in val and col.

EXAMPLE 6.2.1. Figure 1 shows an example of a matrix ($|S| = 4, \text{nnz} = 6$) and its corresponding sparse storage. For clarity, we show all zero entries of the matrix as dots. Both the matrix and the three arrays which store it are indexed from zero. To determine, for example, the entries in row 0 of the matrix, we read the values of row[0] and row[1]. Since these are 0 and 2, respectively, we can deduce that row 0 contains two entries and that they are stored in positions 0 and 1 of the other two arrays. Hence, row 0 consists of entries $(0, 1) = 0.5$ and $(0, 3) = 0.5$. Note that row contains $|S| + 1$ elements so that the number of entries in the final row can be deduced.

```
SparseMVMult(row, col, val, v)
for (i := 0 ... |S|−1)
    res[i] := 0
    l := row[i]
    h := row[i + 1]−1
    for (j := l ... h)
        res[i] := res[i] + val[j] × v[col[j]]
    endfor
endfor
return res
```

Algorithm 6. Multiplication of a sparse matrix with a vector.

The value of an arbitrary entry (r, c) in the matrix can be determined by looking up the rth and $(r+1)$th entries in row and then checking the column indices col[row[r]] to col[row[$r+1$] $-$ 1] for the value c. If the value c is not present, then $(r, c) = 0$. If it is present, then (r, c) is non-zero and its value can be found by looking up the value at the same position in val. In fact, for our purposes, we never require access to a single, random entry, but to all the entries at once, usually to perform a matrix-vector multiplication. The algorithm to perform such a multiplication, $\mathbf{M} \cdot \underline{v}$, is given in Algorithm 6. The matrix $\mathbf{M}$ is stored in the arrays row, col and val, as described above. The vector $\underline{v}$ and the vector representing the result $\mathbf{M} \cdot \underline{v}$ are stored in arrays v and res, respectively.

Let us assume that matrix entries are stored as 8 byte doubles and indices as 4 byte integers. This puts an upper limit on nnz of 2^{32} ($\approx 4.3 \times 10^9$). Given the storage of a typical workstation, this is ample. Using these figures, the amount of memory required to store a sparse matrix is $12\text{nnz} + 4(|S| + 1)$ bytes.

Note that it is also possible to order the entries in the sparse matrix by column, reversing the roles of the row and col arrays. This is known as a *column-major* scheme. In terms of storage, row-major and column-major schemes are identical. The principal difference in practice is that the former is better suited to matrix-vector multiplication ($\mathbf{M} \cdot \underline{v}$) and the latter is better suited to vector-matrix multiplication ($\underline{v} \cdot \mathbf{M}$). Which one is suitable in our case may depend on the particular model checking algorithm being implemented.

6.2.1. An extension for MDPs. The sparse matrix data structure can be extended in a fairly simple way to store MDPs. As discussed in Chapter 3, we can think of an MDP $(S, \bar{s}, \text{Act}, \text{Steps})$ as being represented by a non-square matrix **Steps**, but with each state described by several rows, rather than a single row as for DTMCs and CTMCs. In the sparse storage scheme for MDPs, we again store the value and column index of every matrix entry in two arrays val and col. This time, however, we require two levels of indexing to associate each entry with both a state of the MDP and a nondeterministic choice in that state. We name these arrays row and nc, respectively.

Example 6.2.2. The idea is illustrated most clearly through a simple example. Figure 2 gives the matrix representation, **Steps**, for a 4 state MDP and the sparse matrix based data structure which stores it. Again, we only show non-zero entries in the matrix. To determine the entries corresponding to state 0, we first read the

$$\left(\begin{array}{cccc} 1 & \cdot & \cdot & \cdot \\ \cdot & 0.5 & 0.5 & \cdot \\ \hline \cdot & \cdot & 1 & \cdot \\ \hline \cdot & 1 & \cdot & \cdot \\ \cdot & \cdot & \cdot & 1 \\ \hline 0.8 & \cdot & \cdot & 0.2 \end{array}\right)$$

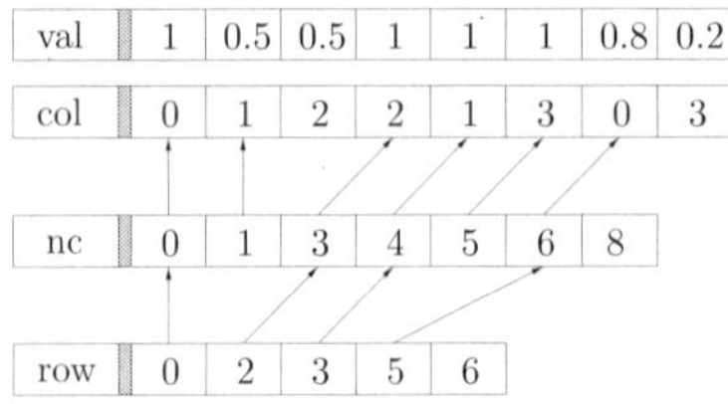

FIGURE 2. A 4 state MDP and its sparse storage.

```
SparseMVMultMax(row, nc, col, val, v)
for (i := 0 ... |S|-1)
    l_i := row[i]
    h_i := row[i+1]-1
    res[i] := -1
    for (j := l_i ... h_i)
        l_j := nc[j]
        h_j := nc[j+1]-1
        sum := 0
        for (k := l_j ... h_j)
            sum := sum + val[k] × v[col[k]]
        endfor
        if (sum > res[i]) then res[i] := sum
    endfor
endfor
return res
```

ALGORITHM 7. Combined multiplication and maximum operation for an MDP.

values of row[0] and row[1]. This reveals that state 0 contains two nondeterministic choices and that these have index 0 and 1 in the array nc. From nc[0] and nc[1], we see that the first nondeterministic choice comprises a single entry $(0,1) = 1$. Likewise, from nc[1] and nc[2], we can deduce that the second choice comprises two entries: $(0,1) = 0.5$ and $(0,2) = 0.5$.

From the algorithms for PCTL model checking of MDPs in Section 3.4, we see that the key operation required is a combined matrix-vector multiplication and maximum or minimum operation. This can be implemented on the sparse data structure in a very similar way to ordinary matrix-vector multiplication. The algorithm for the "maximum" case is given in Algorithm 7. It takes an MDP, stored in arrays row, nc, col and val, a vector stored in an array v and returns the result in an array res. The corresponding algorithm to determine the minimum is almost identical. Note that, since the PCTL model checking algorithms do not make reference to the action labels in the MDP, we have omitted them from the storage scheme presented here.

Again storing matrix entries as 8 byte doubles and indices as 4 byte integers, the amount of memory required to represent an MDP with this scheme can be computed as $4(|S|+1) + 4(\text{nnc}+1) + 12\text{nnz}$ bytes, where $|S|$ is the number of states, nnz is the number of non-zero entries and nnc is the total number of nondeterministic choices over all states, i.e. $\text{nnc} = \sum_{s \in S} |\,\text{Steps}(s)|$.

6.3. Multi-Terminal Binary Decision Diagrams

We now introduce the *multi-terminal binary decision diagram* (MTBDD) data structure. MTBDDs are actually an extension of *binary decision diagrams* (BDDs). BDDs were originally created by Lee [**100**] and Akers [**1**], popularised by Bryant [**31**], and applied to model checking by McMillan, Clarke and others [**34, 103**]. A BDD is a rooted, directed acyclic graph which represents a Boolean function of the form $f\colon \mathbb{B}^n \to \mathbb{B}$.

MTBDDs were first proposed in [**40**] and then developed independently in [**10, 39**]. In [**10**], they were christened algebraic decision diagrams (ADDs) but the two data structures are identical. MTBDDs extend BDDs by representing functions which can take values from an arbitrary set D, not just $\mathbb{B}$, i.e. functions of the form $f\colon \mathbb{B}^n \to D$. In the majority of cases, D is taken to be $\mathbb{R}$ and this is the policy we adopt here. Note that BDDs are in fact a special case of MTBDDs.

Let $\{\mathsf{x}_1, \ldots, \mathsf{x}_n\}$ be a set of distinct, Boolean variables which are totally ordered as follows: $\mathsf{x}_1 < \cdots < \mathsf{x}_n$. An MTBDD M over $\underline{\mathsf{x}} = (\mathsf{x}_1, \ldots, \mathsf{x}_n)$ is a rooted, directed acyclic graph. The vertices of the graph are known as *nodes*. Each node of the MTBDD is classed as either *non-terminal* or *terminal*. A non-terminal node m is labelled with a variable $\text{var}(\mathsf{m}) \in \underline{\mathsf{x}}$ and has exactly two children, denoted $\text{then}(\mathsf{m})$ and $\text{else}(\mathsf{m})$. A terminal node m is labelled by a real number $\text{val}(\mathsf{m})$ and has no children. Terminal and non-terminal nodes are often referred to simply as *terminals* and *non-terminals*, respectively.

The ordering $<$ over the Boolean variables is imposed upon the nodes of the MTBDD. For two non-terminals, m_1 and m_2, if $\text{var}(\mathsf{m}_1) < \text{var}(\mathsf{m}_2)$, then $\mathsf{m}_1 < \mathsf{m}_2$. If m_1 is a non terminal and m_2 is a terminal, then $\mathsf{m}_1 < \mathsf{m}_2$. We require that, for every non-terminal m in an MTBDD, $\mathsf{m} < \text{else}(\mathsf{m})$ and $\mathsf{m} < \text{then}(\mathsf{m})$.

Figure 3(a) shows an example of an MTBDD. The nodes are arranged in horizontal levels, one per Boolean variable. The variable $\text{var}(\mathsf{m})$ for a node m can be found at the left end of the level which contains it. The two children of a node m are connected to it by edges, a solid line for $\text{then}(\mathsf{m})$ and a dashed line for $\text{else}(\mathsf{m})$. Terminals are drawn as squares, instead of circles, and are labelled with their value $\text{val}(\mathsf{m})$. For clarity, we omit the terminal with value 0 and any edges which lead directly to it.

An MTBDD M over variables $\underline{\mathsf{x}} = (\mathsf{x}_1, \ldots, \mathsf{x}_n)$ represents a function mapping Boolean vectors to real numbers, $f_{\mathsf{M}}(\mathsf{x}_1, \ldots, \mathsf{x}_n)\colon \mathbb{B}^n \to \mathbb{R}$. The value of $f_{\mathsf{M}}(\mathsf{x}_1, \ldots, \mathsf{x}_n)$ is determined by tracing a path in M from the root node to a terminal, for each non-terminal m, taking the edge to $\text{then}(\mathsf{m})$ if $\text{var}(\mathsf{m})$ is 1 or $\text{else}(\mathsf{m})$ if $\text{var}(\mathsf{m})$ is 0. The function represented by the MTBDD in Fig. 3(a) is shown in Fig. 3(b). For example, the value of $f_{\mathsf{M}}(0,1,0)$ can be read as 9. We also use the notation $f_{\mathsf{M}}[\mathsf{x}_1 = 0, \mathsf{x}_2 = 1, \mathsf{x}_3 = 0]$ or $f_{\mathsf{M}}[\underline{\mathsf{x}} = (0,1,0)]$. This is often convenient when we wish to express the value of an MTBDD without explicit reference to the ordering of its Boolean variables.

We define a *minterm* of an MTBDD as a valuation of its Boolean variables which results in a non-zero value. The MTBDD in Fig. 3, for example, has 5

(a)

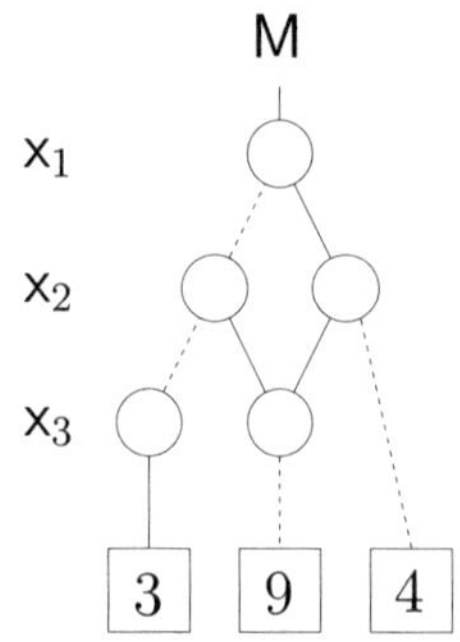

(b)

x_1	x_2	x_3	f_M
0	0	1	3
0	1	0	9
1	1	0	9
1	0	0	4
1	0	1	4
otherwise			0

FIGURE 3. An MTBDD M and the function it represents.

minterms, each corresponding to a row in the table. Note that there is not necessarily a one-to-one mapping between paths in the MTBDD and minterms. For example, the path to the 4 terminal in Fig. 3(a) corresponds to two different minterms.

For an MTBDD M, over variables $(x_1, \ldots, x_n)$, we define its *cofactor* $\mathsf{M}|_{x_i=b}$ (where $b = 0$ or 1) as the MTBDD over variables $(x_1, \ldots, x_{i-1}, x_{i+1}, \ldots, x_n)$ representing the function $f_{\mathsf{M}}(x_1, \ldots, x_{i-1}, b, x_{i+1}, \ldots, x_n)$. Conversely, the function f_{M} for an MTBDD M, with top variable x_1, can be evaluated recursively using its cofactors:

$$f_{\mathsf{M}} = x_1 \cdot f_{\mathsf{M}|_{x_1=1}} + (1 - x_1) \cdot f_{\mathsf{M}|_{x_1=0}}.$$

Note that, because of the recursive nature of the data structure, each node of an MTBDD is itself an MTBDD. Hence, we often consider an MTBDD and the node which represents it (i.e. its root node) to be interchangeable.

The reason that MTBDDs can often provide compact storage is because they are stored in a reduced form. Firstly, if nodes m and m' are identical (i.e. $\mathrm{var}(\mathsf{m}) = \mathrm{var}(\mathsf{m}')$, $\mathrm{then}(\mathsf{m}) = \mathrm{then}(\mathsf{m}')$ and $\mathrm{else}(\mathsf{m}) = \mathrm{else}(\mathsf{m}')$), only one copy of the node is stored. We refer to this as *sharing* of nodes. Secondly, if a node m satisfies $\mathrm{then}(\mathsf{m}) = \mathrm{else}(\mathsf{m})$, it is removed and any incoming edges from levels above are redirected to its unique child. We will refer to this as a *skipped level.* Figure 4 illustrates the way that the MTBDD in Fig. 3 has been reduced:

(a) shows the full binary tree representing the same function;
(b) shared nodes have been merged; and
(c) nodes with identical children have also been removed, introducing skipped levels.

Here, unless stated otherwise, we assume that all MTBDDs are fully reduced in this way. Under this assumption, and for a fixed ordering of Boolean variables, the data structure can be shown to be *canonical,* meaning that there is a one-to-one correspondence between MTBDDs and the functions they represent. As we will see later, this canonicity property has crucial implications for performance.

Another important consideration from a practical point of view is *variable ordering.* The size of an MTBDD (i.e. number of nodes) representing a given function is extremely sensitive to the ordering of its Boolean variables. This has a direct effect on both the storage requirements for the data structure and the time needed to perform operations on it. We will discuss this issue further in Section 6.4.3.

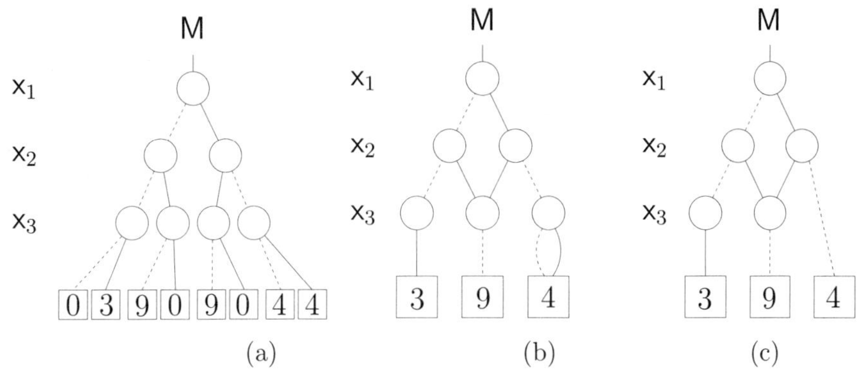

FIGURE 4. Reducing an MTBDD.

The size of an MTBDD is also affected by the number of terminals it contains, or equivalently, the number of distinct values taken by the function which it represents. Compact MTBDDs can only be obtained by maximising sharing within the nodes of the graph. A high number of terminals reduces the capacity for sharing and so increases the number of nodes in the MTBDD.

6.3.1. Operations. Described below are a selection of the most common operations which can be performed on MTBDDs. Some of these are also applicable to BDDs. We treat BDDs simply as a special case of MTBDDs. In the following, we assume that M, M_1 and M_2 are MTBDDs over the set of variables $\underline{\mathsf{x}} = (\mathsf{x}_1, \ldots, \mathsf{x}_n)$.

- $\textsc{Const}(c)$, where $c \in \mathbb{R}$, creates a new MTBDD with the constant value c, i.e. a single terminal m, with $\mathrm{val}(\mathsf{m}) = c$.
- $\textsc{Apply}(op, \mathsf{M}_1, \mathsf{M}_2)$, where op is a binary operation over the reals (e.g. $+$, $-$, $\times$, $\div$, min, max, etc.), returns the MTBDD representing the function $f_{\mathsf{M}_1}\ op\ f_{\mathsf{M}_2}$. If M_1 and M_2 are BDDs, op can also be a Boolean operation ($\wedge$, $\vee$, $\rightarrow$, etc.). For clarity, $\textsc{Apply}$ operations are often expressed in infix notation, for example:
  - $\mathsf{M}_1 \times \mathsf{M}_2 = \textsc{Apply}(\times, \mathsf{M}_1, \mathsf{M}_2)$,
  - $\mathsf{M}_1 \wedge \mathsf{M}_2 = \textsc{Apply}(\wedge, \mathsf{M}_1, \mathsf{M}_2)$.
- $\textsc{Not}(\mathsf{M})$, where M is a BDD, returns the BDD representing the function $\neg f_{\mathsf{M}}$. As above, $\textsc{Not}(\mathsf{M})$ can be abbreviated to $\neg\,\mathsf{M}$.
- $\textsc{Abs}(\mathsf{M})$ returns the MTBDD representing the function $|f_{\mathsf{M}}|$, giving the absolute value of the original one.
- $\textsc{Threshold}(\mathsf{M}, \bowtie, c)$, where $\bowtie$ is a relational operator ($<$, $>$, $\leq$, $\geq$, etc.) and $c \in \mathbb{R}$, returns the BDD M' with $f_{\mathsf{M}'}$ equal to 1 if $f_{\mathsf{M}} \bowtie c$ and 0 otherwise.
- $\textsc{FindMin}(\mathsf{M})$ returns the real constant equal to the minimum value of f_{M}.
- $\textsc{FindMax}(\mathsf{M})$ returns the real constant equal to the maximum value of f_{M}.
- $\textsc{Abstract}(op, \underline{\mathsf{x}}, \mathsf{M})$, where op is a commutative and associative binary operation over the reals, returns the result of *abstracting* all the variables in $\underline{\mathsf{x}}$ from M by applying op over all possible values taken by the variables. For example, $\textsc{Abstract}(+, (\mathsf{x}_1), \mathsf{M})$ would give the MTBDD representing the function $f_{\mathsf{M}|_{\mathsf{x}_1=0}} + f_{\mathsf{M}|_{\mathsf{x}_1=1}}$ and $\textsc{Abstract}(\times, (\mathsf{x}_1, \mathsf{x}_2), \mathsf{M})$ would give the

MTBDD representing the function $f_{\mathsf{M}|_{\mathsf{x}_1=0,\mathsf{x}_2=0}} \times f_{\mathsf{M}|_{\mathsf{x}_1=0,\mathsf{x}_2=1}} \times f_{\mathsf{M}|_{\mathsf{x}_1=1,\mathsf{x}_2=0}} \times f_{\mathsf{M}|_{\mathsf{x}_1=1,\mathsf{x}_2=1}}$. In the latter, $\mathsf{M}|_{\mathsf{x}_1=b_1,\mathsf{x}_2=b_2}$ is equivalent to $(\mathsf{M}|_{\mathsf{x}_1=b_1})|_{\mathsf{x}_2=b_2}$. When M is a BDD, we also have:
  - ThereExists($\underline{\mathsf{x}}, \mathsf{M}$) $\equiv$ Abstract($\vee, \underline{\mathsf{x}}, \mathsf{M}$),
  - ForAll($\underline{\mathsf{x}}, \mathsf{M}$) $\equiv$ Abstract($\wedge, \underline{\mathsf{x}}, \mathsf{M}$)
- ReplaceVars($\mathsf{M}, \underline{\mathsf{x}}, \underline{\mathsf{y}}$), where $\underline{\mathsf{y}} = (\mathsf{y}_1, \ldots, \mathsf{y}_n)$, returns the MTBDD M' over variables $\underline{\mathsf{y}}$ with $f_{\mathsf{M}'}(b_1, \ldots, b_n) = f_{\mathsf{M}}(b_1, \ldots, b_n)$ for all $(b_1, \ldots, b_n) \in \mathbb{B}^n$.

6.3.2. Vectors and matrices. From the point of view of using MTBDDs to implement probabilistic model checking, one of the most interesting applications of MTBDDs is the representation of vectors and matrices. This was investigated in the original work which introduced the data structure [**10, 39**]. Consider a real-valued vector $\underline{v}$ of length 2^n. We can think of $\underline{v}$ as a mapping from indices to reals, i.e. $\underline{v}\colon \{0, \ldots, 2^n - 1\} \to \mathbb{R}$. Given an encoding of the 2^n indices into n Boolean variables, i.e. a bijection $\mathrm{enc}\colon \{0, \ldots, 2^n - 1\} \to \mathbb{B}^n$, we can represent $\underline{v}$ as an MTBDD v over variables $\underline{\mathsf{x}} = (\mathsf{x}_1, \ldots, \mathsf{x}_n)$. We say that v represents $\underline{v}$ if and only if $f_{\mathsf{v}}[\underline{\mathsf{x}} = \mathrm{enc}(i)] = \underline{v}(i)$ for $0 \le i \le 2^n - 1$.

This idea extends easily to matrices. We can think of a 2^n by 2^n matrix, $\mathbf{M}$, as a mapping from $\{0, \ldots, 2^n - 1\} \times \{0, \ldots, 2^n 1\}$ to $\mathbb{R}$. Again, assuming an encoding $\mathrm{enc}\colon \{0, \ldots, 2^n - 1\} \to \mathbb{B}^n$, we can represent $\mathbf{M}$ by an MTBDD M over $2n$ variables, n of which encode row indices and n of which encode column indices. Using *row variables* $\underline{\mathsf{x}} = (\mathsf{x}_1, \ldots, \mathsf{x}_n)$ and *column variables* $\underline{\mathsf{y}} = (\mathsf{y}_1, \ldots, \mathsf{y}_n)$, we say that M represents $\mathbf{M}$ if and only if $f_{\mathsf{M}}[\underline{\mathsf{x}} = \mathrm{enc}(i), \underline{\mathsf{y}} = \mathrm{enc}(j)] = \mathbf{M}(i, j)$ for $0 \le i, j \le 2^n - 1$. Note that we can also encode vectors and matrices with sizes which are not powers of two simply by padding them with extra rows and/or columns, all set to 0.

Example 6.3.1. Figure 5 shows an example of a matrix $\mathbf{M}$ and an MTBDD M representing it. The accompanying table illustrates how the matrix entries can be read off the MTBDD. We are representing a 4×4 matrix so we use two row variables $(\mathsf{x}_1, \mathsf{x}_2)$ and two column variables $(\mathsf{y}_1, \mathsf{y}_2)$. We use the standard binary encoding for integers, i.e. $0 \to 00$, $1 \to 01$, $2 \to 10$ and $3 \to 11$. Each minterm of the MTBDD represents a non-zero entry in the matrix. Consider, for example, the minterm $f_{\mathsf{M}}(1, 1, 0, 1) = 5$. In this case, $\mathsf{x}_1 = 1$ and $\mathsf{x}_2 = 0$, so the row index is 2. Similarly, $\mathsf{y}_1 = 1$ and $\mathsf{y}_2 = 1$, giving a column index of 3. This corresponds to the matrix entry $(2, 3) = 5$.

Observe that, in Fig. 5, row and column variables are ordered alternately. This is a well known heuristic to minimise MTBDD size. We will discuss this issue again in Section 6.4.3. Another advantage of the ordering is that it allows a convenient division of the matrix into its component submatrices. Figure 6 illustrates this. If the matrix $\mathbf{M}$ is represented by MTBDD M, the top and bottom halves of $\mathbf{M}$ are represented by the cofactor MTBDDs $\mathsf{M}|_{\mathsf{x}_1=0}$ and $\mathsf{M}|_{\mathsf{x}_1=1}$, respectively. Since x_1 is the top variable in the ordering, these cofactors can be found simply by following the else and then edges. By repeating this process on the next variable in the ordering, y_1, the four quadrants of $\mathbf{M}$ can be obtained. As well as providing convenient access to submatrices, this can be beneficial for implementing matrix operations which can be expressed recursively.

A good example of this is matrix-matrix multiplication. An algorithm to compute the product of two matrices represented as MTBDDs was first presented in

$$\mathbf{M} = \begin{pmatrix} 0 & 8 & 0 & 5 \\ 2 & 0 & 0 & 5 \\ 0 & 0 & 0 & 5 \\ 0 & 0 & 2 & 0 \end{pmatrix}$$

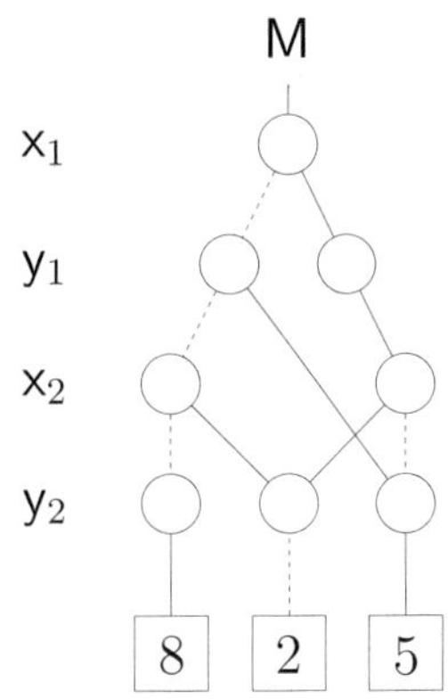

x_1	y_1	x_2	y_2	f_{M}	Entry in $\mathbf{M}$
0	0	0	1	8	$(0,1) = 8$
0	0	1	0	2	$(1,0) = 2$
0	1	0	1	5	$(0,3) = 5$
0	1	1	1	5	$(1,3) = 5$
1	1	1	0	2	$(3,2) = 2$
1	1	0	1	5	$(2,3) = 5$

FIGURE 5. A matrix $\mathbf{M}$ and an MTBDD M representing it.

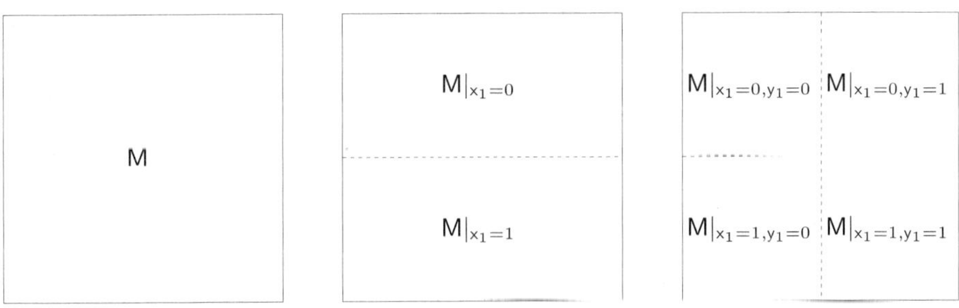

FIGURE 6. Submatrix access via cofactors.

[**40**]. It was shown that by taking all the products of matrix entries first, and summing them afterwards, the multiplication could be performed with an APPLY and an ABSTRACT operation. In [**39**] an improved, recursive algorithm was proposed, based on the decomposition of a matrix into quadrants, as shown in Fig. 6, and the fact that matrix-matrix multiplication can be computed in terms of these quadrants:

$$\begin{pmatrix} \mathbf{A}_1 & \mathbf{A}_2 \\ \mathbf{A}_3 & \mathbf{A}_4 \end{pmatrix} = \begin{pmatrix} \mathbf{B}_1 & \mathbf{B}_2 \\ \mathbf{B}_3 & \mathbf{B}_4 \end{pmatrix} \cdot \begin{pmatrix} \mathbf{C}_1 & \mathbf{C}_2 \\ \mathbf{C}_3 & \mathbf{C}_4 \end{pmatrix} \iff \begin{array}{l} \mathbf{A}_1 = \mathbf{B}_1 \cdot \mathbf{C}_1 + \mathbf{B}_2 \cdot \mathbf{C}_3 \\ \mathbf{A}_2 = \mathbf{B}_1 \cdot \mathbf{C}_2 + \mathbf{B}_2 \cdot \mathbf{C}_4 \\ \mathbf{A}_3 = \mathbf{B}_3 \cdot \mathbf{C}_1 + \mathbf{B}_4 \cdot \mathbf{C}_3 \\ \mathbf{A}_4 = \mathbf{B}_3 \cdot \mathbf{C}_2 + \mathbf{B}_4 \cdot \mathbf{C}_4 \end{array}$$

In [**10**], the algorithm was improved further, observing that it was not always necessary to split the matrix at every level of recursion. The latter work also provided a comparison, in terms of empirical results on some benchmark matrices, of all three methods and concluded that, in general, their method performed better.

Note that, as discussed previously, one of the principal operations required for probabilistic model checking is matrix-vector multiplication. Fortunately, the three algorithms can easily be adapted to this case. For an MTBDD M over variables $\underline{\mathsf{x}}$ and $\underline{\mathsf{y}}$ representing matrix $\mathbf{M}$, and an MTBDD v over $\underline{\mathsf{x}}$ representing column vector $\underline{v}$, we will denote by $\textsc{MVMult}(\mathsf{M}, \mathsf{v})$ the function which returns the MTBDD over $\underline{\mathsf{x}}$, representing the column vector $\mathbf{M} \cdot \underline{v}$. It is also trivial to adapt this algorithm to perform multiplication of a vector by a matrix.

Several other useful operations on vectors and matrices can be implemented with basic MTBDD functions. For example, any pointwise operation such as addition or scalar multiplication can be performed with $\textsc{Apply}$ and vectors and matrices can be transposed using the $\textsc{ReplaceVars}$ function. One very commonly required matrix is the identity matrix. Hence, we introduce the notation $\textsc{Identity}(\underline{\mathsf{x}}, \underline{\mathsf{y}})$ to denote the MTBDD (in fact, BDD) representing the identity matrix with row variables $\underline{\mathsf{x}}$ and column variables $\underline{\mathsf{y}}$.

6.3.3. Implementation issues. We now discuss a few important topics regarding the practical implementation of BDDs and MTBDDs. The details given here are based on the CUDD (Colorado University Decision Diagram) package of Fabio Somenzi [**121**] which, unlike the majority of BDD software, also supports MTBDDs. Most of the issues we discuss here are applicable to all decision diagram packages. Papers which discuss these in more detail can be found in, e.g. [**30, 128**].

We first examine how the data structures are stored. We have seen above that MTBDDs are kept in a reduced form where no identical nodes are duplicated. In practice, this is taken one step further. All of the MTBDDs in use at any one time are maintained in one large data structure, effectively a very large MTBDD with multiple root nodes. This means that nodes are *never* duplicated, even across different MTBDDs. At any point during an MTBDD operation where a new node needs to be created, it is first verified whether or not such a node already exists and, if so, it is reused.

To ensure that such checks can be performed rapidly, the nodes are stored in a set of hash tables, one for each level (i.e. MTBDD variable). This structure is usually known as the *unique table*. Terminal MTBDD nodes are also stored in this fashion. However, since their values are stored as double precision floating points numbers, in practice two terminals are considered to be identical if the difference between their values is less than some small threshold ε. In the CUDD package, the default value for ε is 10^{-12}. In PRISM, this is lowered to 10^{-15}, almost equal to the precision of a double, to minimise round-off error in numerical computations.

In normal operation, MTBDDs are constantly being created and destroyed. Hence, it is necessary to keep track of which nodes are currently needed and which are not. This issue is complicated by the fact that the nodes are stored in one shared structure and a given node may be contained in several MTBDDs. The solution is to keep a *reference count* on each node, denoting how many MTBDDs in current use contain it. If the reference count reaches zero, it is known as a *dead node*. Since removing a dead node from the unique table potentially involves restructuring the whole data structure, dead nodes are only removed periodically, typically when their number exceeds some predefined threshold. This process is know as *garbage collection.*

The total amount of memory required to store an MTBDD is directly proportional to its number of nodes. In the CUDD package, each non-terminal node comprises: one integer storing the level (i.e. which variable it is labelled with); two pointers, one for the then child and one for the else child; one integer storing the reference count; and one pointer used to maintain the node in a linked list as part of a hash table. Assuming that both integers and pointers are stored in 4 bytes, this gives a total of 20 bytes per node. Terminal nodes store slightly different information but this can still be fitted into the same amount of space.

The existence of the unique table is fundamental to the efficient operation of an MTBDD package. Clearly, storing the MTBDDs in this way minimises the amount of storage space required. More importantly, though, it also has ramifications for the speed at which MTBDD operations can be performed. Checking whether two MTBDDs are identical is easy: it reduces to verifying that the root nodes are stored in the same place in the unique table, i.e. a comparison of pointers. Because of the canonicity property of MTBDDs, checking whether two MTBDDs represent the same function is equally trivial.

The true value of this comes in the implementation of another fundamental aspect of MTBDD packages: the *computed table.* This is essentially a cache for storing the results of operations on MTBDDs. Before any operation is performed, the cache is checked to see if it has been executed previously. If so, the result can simply be reused. If not, the result is computed, stored in the cache and then returned. Typically, many operations *are* repeated several times, particularly since operations are usually performed recursively and there is often a great deal of node sharing in MTBDDs. Lookups in this cache can only be implemented efficiently if it is quick to compare the equality of two MTBDDs.

In theory, any single operation is only ever performed once. Practically, of course, there is usually an upper limit on the amount of memory allocated to the cache so this is not true. It is still fair to say, though, that the computed table can result in a dramatic improvement in performance.

6.4. Model Representation with MTBDDs

We now move on to consider the problem of using MTBDDs to represent probabilistic models. In the next section, we explain how to do this for DTMCs, CTMCs and MDPs. In the subsequent two chapters, we consider two issues which can have a dramatic effect on the efficiency of this process: firstly, the *encoding* used and, secondly, the *variable ordering* chosen.

6.4.1. Representing DTMCs, CTMCs and MDPs. As we have seen in earlier chapters, DTMCs and CTMCs are both represented by square, real-valued matrices. From their inception, MTBDDs [**10, 39, 40**] have been used to represent matrices, a process we described above in Section 6.3.2. The basic idea is that a matrix can be thought of as a function mapping pairs of indices to real numbers. Given an encoding of these indices into Boolean variables, we can instead view the matrix as a function mapping Boolean variables to real numbers, which is exactly what an MTBDD represents.

In (6.3.1) earlier, matrix indices were simply integers and we encoded them using their standard binary representation. The transition probability matrix **P** of a DTMC or the transition rate matrix $\mathbf{R}$ of a CTMC, however, is indexed by the states of some state space S. Hence, what we actually need is an encoding of S into MTBDD variables. Clearly, one approach would be to enumerate the elements of S, assigning each one a unique integer, and then proceed as before. As we will discuss in Section 6.4.2, by taking a more structured approach to the encoding, the efficiency of this process can be dramatically improved.

Representing MDPs with MTBDDs is more complex than DTMCs or CTMCs since the nondeterminism must also be encoded. An MDP $(S, \bar{s}, \mathrm{Act}, \mathrm{Steps})$ is not described by a transition matrix, but by the function Steps mapping each state of S to a set of nondeterministic choices, each of which corresponds to a probability distribution over S. Assuming, however, that the maximum number of nondeterministic choices in any state is m, we can reinterpret Steps as a function of the form $S \times \{1, \ldots, m\} \times S \to [0, 1]$. Assuming that we have an encoding into Boolean variables, not just of the state space S, but also of the set $\{1, \ldots, m\}$, we can consider Steps as a function mapping Boolean variables to real numbers and hence represent it as an MTBDD.

We will use variables $\underline{\mathsf{x}} = (\mathsf{x}_1, \ldots, \mathsf{x}_n)$ and $\underline{\mathsf{y}} = (\mathsf{y}_1, \ldots, \mathsf{y}_n)$ to range over source and destination states (i.e. rows and columns in the DTMC or CTMC setting) and variables $\underline{\mathsf{z}} = (\mathsf{z}_1, \ldots, \mathsf{z}_k)$ to encode $\{1, \ldots, m\}$. We will refer to $\underline{\mathsf{x}}$ and $\underline{\mathsf{y}}$ as row and column variables, as before, and $\underline{\mathsf{z}}$ as *nondeterministic variables*. The idea of encoding nondeterministic choice in an MDP as a third index to represent it as an MTBDD can be traced back to [**11**] but the application or practicality of it was not considered until [**51, 109**]. The latter [**109**] contains significantly more details about the process of representing MDPs as MTBDDs. In a non-model checking context, details of similar representations for MDPs can be found in, for example, [**55, 72**].

We have seen in Chapter 3 and above, in Section 6.2.1, that when implementing model checking of MDPs it is often useful to consider Steps as non-square matrix **Steps**, where each row corresponds to a single nondeterministic choice. This allows certain parts of the model checking algorithm to be conveniently expressed as matrix-vector multiplication. Similarly, we can think about the MTBDD representation described above as storing a matrix with $m \cdot |S|$ rows, encoded by variables $\underline{\mathsf{x}}$ and $\underline{\mathsf{z}}$, and $|S|$ columns, encoded by variables $\underline{\mathsf{y}}$. We will return to this idea when we consider an MTBDD-based implementation of model checking in Section 6.5.

6.4.2. Encoding heuristics. The use of MTBDDs to encode probabilistic models has been proposed on a number of occasions, for example [**10, 13, 61**]. The issue of developing an *efficient* encoding, however, was first considered by Hermanns et al. in [**70**]. One of the main contributions of the paper is a set of "rules of thumb" for deriving compact MTBDD encodings of CTMCs from descriptions in high-level formalisms such as process algebras and queueing networks. This extends previous work [**53, 54**] which considers the efficient encoding of non-probabilistic process algebra descriptions into BDDs.

The key observation of Hermanns et al. is that one should try to preserve *structure* and *regularity* from the high-level description of the CTMC in its MTBDD encoding. For example, in the process-algebraic setting, a system is typically described as the parallel composition of several sequential components. They show that it is more efficient to first obtain a separate encoding for each of these components, and only then combine them into a global encoding. Regularity in the high-level description can be translated into regularity in the low-level MTBDD representation. This results in an increase in the number of shared nodes and, subsequently, a decrease in the size of the data structure.

Similar findings are presented in [**51, 69, 81, 109, 119**], which consider the problem of deriving an efficient MTBDD representation for probabilistic models described in the stochastic process algebras and the input description language of the PRISM tool, respectively. In the latter case, it is found that maintaining a close correlation between variables in the high-level model description language and Boolean variables at the MTBDD level provides a compact, structured representation. [**109**] also considers the issue of encoding nondeterminism in MDPs. In general, the conclusion is that the best way to generate an efficient MTBDD representation of a probabilistic model is to perform a direct translation from the high-level description.

6.4.3. Variable ordering heuristics. Another important factor to consider when developing an efficient MTBDD model representation is the selection of an ordering for the Boolean variables of the MTBDD. It is well known that this can have a dramatic effect on the size of the data structure. For BDDs, in particular, this issue has received an enormous amount of attention in the literature. Because MTBDDs and BDDs are so similar, many of the principles involved translate naturally. One key result, from [**27, 124**], is that the problem of determining the optimal variable ordering for a BDD is NP-hard. Because of this, the traditional approach to finding a good ordering is to rely on *heuristics*.

We have already seen one example of such an ordering heuristic. When building the MTBDD to represent a matrix in (6.3.1), we interleaved the Boolean variables encoding the row and column indices. The ordering was shown to provide a convenient decomposition of the matrix into its constituent submatrices. More importantly, though, this heuristic can dramatically reduce MTBDD size. The idea was first presented in [**54**]. We can illustrate its usefulness on a simple example.

In Fig. 7, we show two MTBDDs (in fact BDDs) representing the 8×8 identity matrix. Both use variables $\mathsf{x}_1, \mathsf{x}_2, \mathsf{x}_3$ to encode row indices and $\mathsf{y}_1, \mathsf{y}_2, \mathsf{y}_3$ to encode column indices. The first MTBDD (a) uses the interleaved variable ordering $\mathsf{x}_1 < \mathsf{y}_1 < \mathsf{x}_2 < \mathsf{y}_2 < \mathsf{x}_3 < \mathsf{y}_3$ and the second (b) uses $\mathsf{x}_1 < \mathsf{x}_2 < \mathsf{x}_3 < \mathsf{y}_1 < \mathsf{y}_2 < \mathsf{y}_3$. Even on this small example, the difference is clear. In fact, for the general case, the size of the MTBDD for the $2^n \times 2^n$ identity matrix is $\mathcal{O}(n)$ or $\mathcal{O}(2^n)$ for the interleaved and non-interleaved orderings respectively.

The effect of the variable ordering on this example can be explained as follows. Note that, in terms of MTBDD variables, the actual function being represented by the MTBDDs in Fig. 7 is $(\mathsf{x}_1 = \mathsf{y}_1) \wedge (\mathsf{x}_2 = \mathsf{y}_2) \wedge (\mathsf{x}_3 = \mathsf{y}_3)$. Consider what happens when we traverse the MTBDDs from top to bottom, trying to determine the value of the function for some assignment of the variables. Effectively, each node encodes the values of all the variables in levels above it. For example, in the first MTBDD, after two levels, we have established whether or not $\mathsf{x}_1 = \mathsf{y}_1$. If so, we will be positioned at the single node on the x_2 level. If not, we will have already moved to the zero constant node. In either case, from this point on, the values of x_1 and y_1 are effectively irrelevant, since in the function being represented, x_1 and y_1 relate only to each other. In the second MTBDD, however, there is a gap in the ordering between x_1 and y_1. After the first level, we "know" the value of x_1 but cannot "use" it until the fourth level. In the meantime, we must consider all

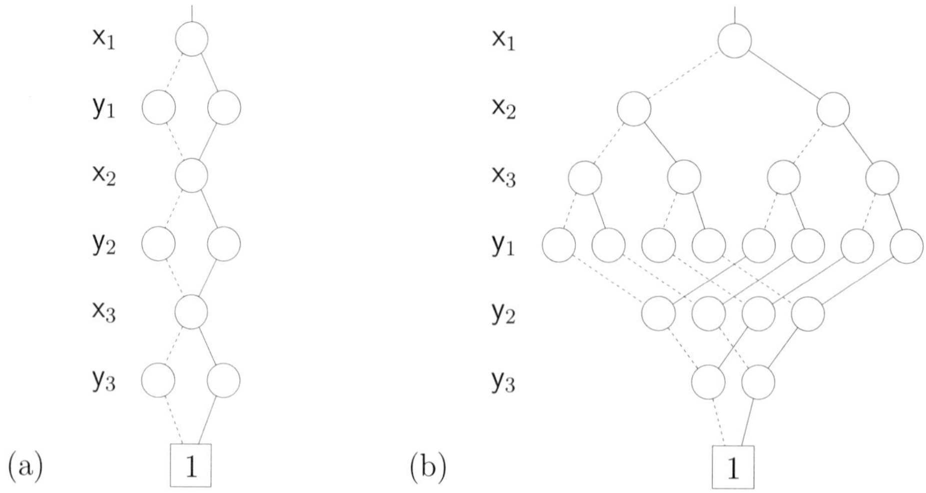

FIGURE 7. Alternative BDD variable orderings for the 8×8 identity matrix: (a) interleaved (b) non-interleaved.

possible values of x_2 and x_3, causing a blow-up in the number of nodes required. As a general heuristic, it is always advisable to place Boolean variables which are related to each other close together in the variable ordering.

The effectiveness of the interleaving heuristic is also known to extend to the case of representing transition matrices as MTBDDs. The reasoning is similar. Every path through the MTBDD represents an element of the matrix which, in turn, represents a transition of the model being represented. In many cases, especially if the model is constructed as an asynchronous parallel composition of processes, in each such transition only a few pairs of x_i, y_i variables will have different values. Hence, the majority of pairs can be seen as closely related and benefit from being position together in the ordering.

These heuristics or variants of them can be applied in several other situations. For instance, we can reuse the same ideas at a higher level. In a structured representation of a model, as we have advocated above, different parts of the model will be encoded separately and hence with different Boolean variables. Since the variables corresponding to each component are inevitably related, they should be grouped together, see e.g. [**69, 109**]. Additional optimisations can be made in the case of representing MTBDDs since these use additional variables to encode nondeterminism. We might opt, for example, to place all these variables at the start or at the end of the ordering. In fact, empirical results presented in [**109**] show that a more efficient approach is to allocate nondeterministic variables to each different component of the model and group the variables with the other Boolean variables for that component.

6.5. MTBDD-Based Model Checking

We now discuss how, using the MTBDD-based model representation schemes described in the previous section, the probabilistic model checking algorithms of

ϕ	$\text{SAT}(\phi)$
true	$\text{CONST}(1)$
a	$\text{ENCODE}(a)$
$\neg\phi$	$\neg\text{SAT}(\phi)$
$\phi_1 \wedge \phi_2$	$\text{SAT}(\phi_1) \wedge \text{SAT}(\phi_2)$

ALGORITHM 8. Model checking for non-probabilistic PCTL and CSL operators.

Chapters 2, 3 and 4 can be implemented. The interested reader can find further details in a number of places including [**11–13, 16, 51, 63, 69, 78, 109, 119**].

The model checking algorithms take as input a probabilistic model (either a DTMC, an MDP or a CTMC) and a property expressed in the corresponding probabilistic temporal logic (either PCTL or CSL) and return the set of states which satisfy the property, usually via computation of a vector of probabilities or expected costs. We have seen in the preceding section how to represent DTMCs, MDPs and CTMCs as MTBDDs. We also saw in Section 6.3.2 how real-valued vectors can be stored. As a special case of the latter, we can also represent sets of states using MTBDDs (in fact BDDs) by storing the $0-1$ vector representing the set's *characteristic function.* For a subset S' of the set of states S, the characteristic function $\chi_{S'} : S \to \{0,1\}$ maps states to 0 or 1, returning 1 if the state is in the set S' and 0 otherwise.

The model checking algorithms for PCTL and CSL proceed by traversing the parse tree for the logical formula and recursively evaluating the set of states which satisfy each subformula. Hence, the implementation of each operator can be considered separately. For the non-probabilistic operators $(\text{true}, a, \neg, \wedge)$, common to PCTL and CSL, model checking is relatively simple. The symbolic implementation is illustrated in Fig. 8. We denote by $\text{SAT}(\phi)$ the function which computes the BDD for the set of states satisfying the formula ϕ. Note that, since the model checking algorithm is recursive, we can assume that any subformulas have already been model checked. For example, when computing $\text{SAT}(\neg\phi)$, we can assume that the result of $\text{SAT}(\phi)$ is known. For an atomic proposition a, we denote by $\text{ENCODE}(a)$ the function which returns the BDD representing the set $L(a)$. The work required for this will depend on exactly how the information about atomic propositions is represented; however, the process is typically quite straightforward.

The majority of the work in implementing an MTBDD-based version of probabilistic model checking comes in developing the algorithms for the $\mathcal{P}_{\bowtie p}[\psi]$, $\mathcal{E}_{\bowtie c}[\phi]$ and $\mathcal{S}_{\bowtie p}[\phi]$ operators of PCTL and CSL, where computation of probabilities and expected costs is required. The basic overall procedure is the same in each case. We first generate an MTBDD, probs or costs, representing the appropriate vector of probabilities or expected costs, respectively. These values are then compared to the bounds in the logical operators to generate the BDD storing the result:

$$\text{SAT}(\mathcal{P}_{\bowtie p}[\psi]) = \text{THRESHOLD}(\mathsf{probs}, \bowtie, p),$$
$$\text{SAT}(\mathcal{E}_{\bowtie c}[\phi]) = \text{THRESHOLD}(\mathsf{costs}, \bowtie, c),$$
$$\text{SAT}(\mathcal{S}_{\bowtie p}[\phi]) = \text{THRESHOLD}(\mathsf{probs}, \bowtie, p).$$

In the next two sections, we discuss how the arrays probs and costs are actually calculated. We divide our discussion into two parts, the first covering the implementation of graph-based algorithms, such as precomputation algorithms, and the

```
PROB0(phi_1, phi_2)
1. sol := phi_2
2. done := false
3. while (done = false)
4.     sol' := sol ∨ (phi_1 ∧ THEREEXISTS(y, T ∧ REPLACEVARS(sol, x, y)))
5.     if (sol' = sol) then done := true
6.     sol := sol'
7. endwhile
8. return ¬sol
```

ALGORITHM 9. The MTBDD version of the PROB0 algorithm.

second focusing on numerical computation. Rather than attempting to present a comprehensive coverage, we instead highlight a number of important issues and present some illustrative examples. Full details of the symbolic implementation of the majority of the PCTL and CSL operators can be found in [**109**].

6.5.1. Graph-based computation. The probabilistic model checking algorithms in the earlier chapters of this book make use of a number of graph-based algorithms, i.e. those that work solely on the underlying graph structure of the probabilistic model. The most prominent examples are the precomputation algorithms which constitute the first step in several parts of the PCTL and CSL model checking algorithms. These determine the states for which the relevant probability or expected cost is exactly 0 or 1. Other examples are the algorithms needed to determine the set of bottom strongly connected components in a CTMC and the process of computing the set of reachable states of a model, usually performed when the model is first constructed.

In this section, we will focus on the precomputation algorithm PROB0, described in Chapter 3 (see Algorithm 1 on page 109), which illustrates many of the issues which arise in implementing all the graph-based algorithms required. PROB0 is used when performing model checking of a PCTL "until" formula $\mathcal{P}_{\bowtie p}[\phi_1 \mathcal{U} \phi_2]$ for a DTMC. It computes the set of states s for which $p_s(\phi_1 \mathcal{U} \phi_2) = 0$. Let us assume that the sets $\mathrm{Sat}(\phi_1)$ and $\mathrm{Sat}(\phi_2)$ have already been computed and are represented by phi_1 and phi_2, BDDs over the variables $\underline{\mathsf{x}}$, and that the transition probability matrix $\mathbf{P}$ of the DTMC is represented by the MTBDD P with row variables $\underline{\mathsf{x}}$ and column variables $\underline{\mathsf{y}}$. The result of the algorithm will be a BDD over variables $\underline{\mathsf{x}}$ representing the set of states s for which $p_s(\phi_1 \mathcal{U} \phi_2) = 0$.

The algorithm only considers the existence of a transition between two states, not the actual probability of making that transition. Hence, it is more efficient to use a BDD T, representing the transition relation of the underlying graph than to use the P itself. This can be computed as:

$$\mathsf{T} := \textsc{Threshold}(\mathsf{P}, >, 0).$$

Algorithm 9 gives the MTBDD algorithm for PROB0.

Like all the precomputation algorithms, PROB0 uses a fixpoint computation to generate its result. The key aspects of the implementation of the fixpoint are as follows. The set of states being computed is stored in a BDD sol over variables $\underline{\mathsf{x}}$. First, sol is initialised to some value (line 1). The algorithm then repeatedly computes a new set sol' based on sol (line 4) and compares it to the previous set

```
PctlUntil(phi1, phi2)
1. sno := Prob0(phi1, phi2)
2. syes := Prob1(phi1, phi2, sno)
3. s? := ¬ (sno ∨ syes)
4. P' := s? × P
5. A := Identity(x, y) − P'
6. b := syes
7. probs := SolveJacobi(A, b, b)
8. return probs
```

ALGORITHM 10. The PCTLUNTIL algorithm.

(line 5). The iterative process is stopped when the set does not change, i.e. the BDDs sol and sol′ are identical. The computation is guaranteed to terminate since there are finitely many states. The set generated by the fixpoint is then given by sol. Note that, in this case, the required result is actually the states *not* accumulated by the fixpoint, hence we negate sol before returning it.

As for the actual computation performed in each iteration, the original algorithm translates easily into BDDs. Note the close correspondence between line 4 of Algorithm 1 on page 109 and line 4 of Algorithm 9 above:

$$R' := R \cup \{s \in \mathrm{Sat}(\phi_1) \mid \exists s' \in R \,.\, \mathbf{P}(s,s') > 0\},$$

$$\mathsf{sol}' := \mathsf{sol} \vee \big(\mathsf{phi}_1 \wedge \textsc{ThereExists}(\underline{\mathsf{y}}, \mathsf{T} \wedge \textsc{ReplaceVars}(\mathsf{sol}, \underline{\mathsf{x}}, \underline{\mathsf{y}})\big).$$

Furthermore, this consists entirely of simple BDD operations which can be performed efficiently. The other important part of the algorithm, checking if the BDDs sol and sol′ are identical, can be done in constant time due to the canonicity property of (MT)BDDs and the efficient data structures used to store them, as discussed in Section 6.3.

6.5.2. Numerical computation. We now move on to discuss MTBDD-based implementation of numerical computation. As in the previous section, there is a considerable amount of overlap between the various algorithms required for PCTL and CSL model checking. Hence, we limit our coverage to a few examples which illustrate all of the relevant points. We cover two cases: the PCTL $\mathcal{P}_{\bowtie p}[\phi_1 \mathcal{U} \phi_2]$ operator for both DTMCs and MDPs.

6.5.2.1. *The* PCTL $\mathcal{P}_{\bowtie p}[\phi_1 \mathcal{U} \phi_2]$ *operator for* DTMCs. Model checking for this operator over DTMCs was discussed in Section 2.4. The MTBDD version is given in Fig. 10. It is essentially the same as the one proposed in [**13**]. The probabilities are computed by the MTBDD algorithm PCTLUNTIL. It takes as input two BDDs, phi_1 and phi_2, representing the sets of states $\mathrm{Sat}(\phi_1)$ and $\mathrm{Sat}(\phi_2)$, respectively, and returns an MTBDD representing the vector of probabilities $p_s(\phi_1 \mathcal{U} \phi_2)$ for each state s. It also uses the MTBDD P representing the transition probability matrix of the DTMC.

The first step (lines 1–3) determines the sets S^{no}, S^{yes} and $S^{?}$. This uses the precomputation algorithms PROB0 and PROB1, the implementation of which was discussed in the previous section. Secondly (lines 4–6), the linear equation system $\mathbf{A} \cdot \underline{x} = \underline{b}$ is constructed, as described in Section 2.4. The matrix $\mathbf{A}$ and vector $\underline{b}$ are represented by MTBDDs A and b respectively. We first build the matrix $\mathbf{P}'$

```
SolveJacobi(A, b, init)
1.  d := Abstract(max, y, A × Identity(x, y))
2.  A' := A × Const(−1) × ¬Identity(x, y)
3.  sol := init
4.  done := false
5.  while (done = false)
6.      sol' := MVMult(A', sol)
7.      sol' := sol' + b
8.      sol' := sol' ÷ d
9.      if (MaxDiff(sol, sol') < ε) then
10.         done := true
11.     endif
12.     sol := sol'
13. endwhile
14. return sol
```

ALGORITHM 11. The SOLVEJACOBI algorithm.

(represented by P') which is equal to the matrix $\mathbf{P}$ but with the rows corresponding to states not in $S^?$ set to zero. This is done by a pointwise multiplication, using the APPLY operator, of P and $\mathsf{s}_?$, the latter representing $S^?$. The matrix $\mathbf{A} = \mathbf{I} - \mathbf{P}'$ is then constructed using the IDENTITY and APPLY functions. In line 7, the solution of the linear equation system $\mathbf{A} \cdot \underline{x} = \underline{b}$ is computed, using the Jacobi iterative method. This constitutes the bulk of the work and is contained in a separate algorithm, SOLVEJACOBI.

We use iterative methods for solving linear equation systems, rather than alternative, direct methods such as Gaussian elimination or L/U decomposition. This is because we are aiming to study large probabilistic models, which will produce very large linear equation systems. Direct methods usually require modifications to the matrix $\mathbf{A}$, which are costly both in terms of space and time.

This argument applies regardless of the data structure being used. In the symbolic case, however, it is particularly relevant. Work by [**10**] showed that MTBDDs are very poorly suited to methods such as Gaussian elimination. As discussed in Section 6.4, the effectiveness of MTBDDs relies heavily on them being used to store regular, structured information. Modifications to the matrix $\mathbf{A}$, such as those made by Gaussian elimination, inevitably lead to a significant loss in regularity and a consequent blow-up in the size of the MTBDD. Furthermore, the operations required to perform these modifications work on individual elements, rows and columns of the matrix. These are particularly difficult to implement on inherently recursive data structures such as MTBDDs. The iterative methods, on the other hand, do not modify the matrix throughout the computation and can be implemented with matrix-vector multiplication, for which efficient MTBDD algorithms exist.

The problem of implementing iterative solution methods using MTBDDs was first considered in [**61**], which implemented steady-state probability calculation using the Power method. [**70**] extended this work, also presenting MTBDD algorithms for the Jacobi and Gauss–Seidel methods. In Fig. 11, we give the function SOLVEJACOBI, the MTBDD implementation of the Jacobi method.

The algorithm can be compared to the description of the Jacobi method we give in Appendix B. Note that there are two alternative formulations presented there: one expressed in terms of operations on individual matrix elements; and another in terms of matrix-vector multiplication. We select the latter because, as described above, it is far more efficient to implement in MTBDDs. The first three lines of Fig. 11 set up the MTBDDs A', d and sol which will be used in the main iterative loop. In terms of the description of the Jacobi method in Appendix B, A' corresponds to the matrix $\mathbf{L}+\mathbf{U}$ and d stores the diagonal values from the matrix $\mathbf{D}$. The MTBDD sol represents the solution vector.

The main part of SOLVEJACOBI is the loop in lines 4–13. Each iteration computes the next approximation to the solution vector, sol', from the previous one, sol. This is done with one matrix-vector multiplication and a pointwise addition and division on a vector. Each iteration also contains a convergence check which compares sol and sol' to determine whether or not the method should be terminated. Various stopping criteria can be used, as discussed in Appendix B. We check if the maximum relative difference between elements of the two vectors is below some threshold ε. We assume the presence of a function MAXDIFF($\mathsf{v}_1, \mathsf{v}_2$) which computes this difference between two vectors represented by MTBDDs v_1 and v_2. This could be done with basic MTBDD operations, e.g. FINDMAX(ABS(($\mathsf{sol}' - \mathsf{sol}) \div \mathsf{sol}'$)). In some MTBDD implementations, for example the CUDD package [**121**], there are also built-in operations which can be used to compute this directly from the MTBDDs. For a discussion of MTBDD-based versions of other iterative algorithms for solution of linear equation systems, see [**70, 109, 119**].

6.5.2.2. *The PCTL* $\mathcal{P}_{\bowtie p}[\phi_1 \mathcal{U} \phi_2]$ *operator for* MDPs. As our second example, we describe model checking of the same PCTL operator, but for MDPs. We consider the case where $\bowtie p$ defines an upper bound, and hence the algorithm computes the probabilities $p_s^{\max}(\phi_1 \mathcal{U} \phi_2)$. In this case, it makes no difference whether we consider all adversaries or fair adversaries only. The model checking algorithm was given in Section 3.4.3. The MTBDD version, PCTLUNTILMAX, can be seen in Fig. 12. This was first presented in [**51**].

As was the case in the previous section, there exist alternative, direct methods to perform this model checking problem. It can be reformulated as a linear programming (LP) problem, as detailed in Section 3.4.3, and solved with classic LP techniques such as the Simplex algorithm. The applicability of MTBDDs to the Simplex method was investigated in [**89**] and it was found that the situation is similar to that of Gaussian elimination: its reliance on access to individual elements, rows and columns of the matrix makes it impractical for symbolic implementation.

The overall structure of PCTLUNTILMAX is very similar to PCTLUNTIL from the previous section: firstly, we use precomputation algorithms to determine the sets S^{no}, S^{yes} and $S^{?}$; secondly, we make modifications to the MTBDD representing the model; and thirdly, we execute an iterative method using this modified MTBDD.

Note that an MDP is represented by a transition function Steps, not a transition matrix, as is the case for a DTMC. As described in Section 3.4.1 though, from an implementation point of view it is useful to think of Steps as a non-square matrix **Steps**, where there are several rows corresponding to each state of the MDP. The key component of each iteration of the MDP model checking algorithms then becomes a matrix-vector multiplication using this non-square matrix.

```
PctlUntilMax(phi_1, phi_2)
1.   s_no := Prob0A(phi_1, phi_2)
2.   s_yes := Prob1E(phi_1, phi_2)
3.   s_? := ¬ (s_no ∨ s_yes)
4.   Steps' := s_? × Steps
5.   probs := s_yes
6.   done := false
7.   while (done = false)
8.       probs' := MVMult(Steps', probs)
9.       probs' := Abstract(max, z, probs')
10.      probs' := probs' + yes
12.      if (MaxDiff(probs, probs') < ε) then
13.          done := true
14.      endif
15.      probs := probs'
16.  endwhile
16.  return probs
```

ALGORITHM 12. The PCTLUNTILMAX algorithm

We can use the same idea here. The MTBDD Steps, representing the MDP, is over row variables $\underline{x}$, column variables $\underline{y}$, and nondeterministic variables $\underline{z}$. If we instead treat Steps as representing a matrix with $\underline{x}$ and $\underline{z}$ for row variables and $\underline{y}$ for column variables, we can perform this operation using the MTBDD MVMULT function. This is done in line 8 of Fig. 12. The other important step, computing the maximum probability for each state, can be carried out easily using the ABSTRACT(max, ·, ·) operator. The convergence check is performed as in the previous PCTLUNTIL algorithm.

6.6. Discussion

More advanced MTBDD-based implementation techniques for model checking of probabilistic systems have also been developed. While the methods outlined in the previous section have successfully been used in the verification of probabilistic models with as many as 10^{10} states [**84, 90**], it has been found that the time and space requirements for MTBDD-based numerical computation can often make it infeasible. This is because, despite providing a compact, structured representation of the probabilistic model, MTBDDs are poor for storing large, unstructured vectors of probabilities. One successful approach to resolving this problem is found in [**88, 109**], which combines the advantages of symbolic (MTBDD-based) and explicit (sparse matrix based) techniques, representing the model as an MTBDD but solution vectors explicitly as arrays. In fact, a modified form of the symbolic data structure, known as *offset-labelled* MTBDDs is used for this purpose. This work is implemented in the PRISM model checker [**87**].

Another field of research which specialises in efficient analysis of structured, probabilistic models is *Kronecker* techniques. These are based on representing the transition matrix of a CTMC (or DTMC) compactly as a Kronecker algebraic expression of smaller matrices. Numerical solution is applied directly to this representation of the matrix. The original application of Kronecker algebra to CTMC storage can be traced back to [**110**], but a large number of developments and improvements have been made since then, see for example [**32, 36, 80**]. The Kronecker approach has also been specifically applied to the problem of CSL model checking [**33**]. Another line of work in this field which is very closely related to MTBDDs is the *matrix diagram* data structure [**35**], which implements Kronecker methods using decision diagram data structures.

Finally, we mention an alternative direction which attempts to reduce the memory burden of probabilistic model checking: [**82, 83**] discuss *out-of-core* approaches to CTMC analysis, the basic idea being that large data structures, such as those for storing solution vectors are kept on disk, rather than in RAM, and retrieved as necessary. While important progress has been made in all of these areas, more work is needed to improve the scalability of the techniques with the aim of making probabilistic model checking more amenable to real-world, industrial case studies.

APPENDIX A

Measure Theory and Probability

A *discrete probability distribution* over a countable set Q is a function $\mu\colon Q \to [0,1]$ such that $\sum_{q\in Q}\mu(q) = 1$. For a possibly uncountable set Q', let $\mathsf{Dist}(Q')$ be the set of such distributions over countable subsets of Q'. For some element $q \in Q$, let the *point distribution* $\mu_q \in \mathsf{Dist}(Q)$ be the distribution which assigns probability 1 to q. In the following, we assume some familiarity with probability and measure theory, see for example [**24**].

DEFINITION A.0.1. Let Ω be an arbitrary non-empty set and $\mathcal{F}$ a family of subsets of Ω. We say that $\mathcal{F}$ is a *field* on Ω if:

(1) the empty set $\emptyset$ is in $\mathcal{F}$;
(2) whenever A is an element of $\mathcal{F}$, the complement of A, denoted $\bar{A}$, is in $\mathcal{F}$;
(3) whenever A and B are elements of $\mathcal{F}$, then $A \cup B$ is in $\mathcal{F}$.

A field of subsets $\mathcal{F}$ is called a *σ-field* if it is field which is closed under countable union: whenever $A_i \in \mathcal{F}$ for $i \in \mathbb{N}$, then $\bigcap_{i\in\mathbb{N}} A_i$ is also in $\mathcal{F}$.

The elements of a σ-field are called *measurable sets*, and $(\Omega, \mathcal{F})$ is called a *measurable space*. A σ-field *generated* by a family of sets $\mathcal{A}$, denoted $\sigma(\mathcal{A})$, is the smallest σ-field that contains $\mathcal{A}$ which exists by the following proposition.

PROPOSITION A.0.2. *For any non-empty set Ω and $\mathcal{A}$ a family of subsets of Ω, there exists a unique smallest σ-field containing $\mathcal{A}$.*

DEFINITION A.0.3 (Probability Measure and Space). Let $(\Omega, \mathcal{F})$ be a measurable space. A function $\mu\colon \mathcal{F} \to [0,1]$ is a *probability measure* on $(\Omega, \mathcal{F})$ and $(\Omega, \mathcal{F}, \mu)$ a *probability space*, if μ satisfies the following properties:

(1) $\mu(\Omega) = 1$;
(2) if $A_1, A_2, \ldots$ is a disjoint sequence of elements of $\mathcal{F}$, then $\mu(\cup_i A_i) = \sum_i \mu(A_i)$.

The measure μ is also referred to as a *probability distribution*. The set Ω is called the *sample space*, and the elements of $\mathcal{F}$ *events*.

In order to go from a notion of size defined on a family of subsets $\mathcal{A}$ to an actual measure on the σ-field generated by $\mathcal{A}$, we need an extension theorem. The following [**24**] is a typical example.

DEFINITION A.0.4. A family $\mathcal{F}$ of subsets of Ω is called a *semi-ring* if

(1) the empty set $\emptyset$ is in $\mathcal{F}$;
(2) whenever A and B are elements of $\mathcal{F}$, then $A \cap B$ is also in $\mathcal{F}$;
(3) if $A \subseteq B$ are in $\mathcal{F}$, then there are *finitely* many *pairwise disjoint* subsets $C_1, \ldots, C_k \in \mathcal{F}$ such that $B - A = \cup_{i=1}^{k} C_i$.

This is not the form of the definition most commonly used in algebra because of the strange last condition but it is precisely the property that holds for "hyperrectangles" in $\mathbb{R}^n$ and, more importantly here, for the *cylinder sets* defined in earlier chapters of this book.

THEOREM A.0.5. *Suppose that $\mathcal{F}$ is a semi-ring on X and $\mu\colon \mathcal{F} \to [0, \infty]$ satisfies*

(1) $\mu(\emptyset) = 0$,
(2) *μ is finitely additive and*
(3) *μ is countably subadditive.*

Then μ extends to a measure on the σ-field generated by $\mathcal{F}$.

The proof of this theorem may be found in a standard text on probability and measure, for example [**24**]. It is straightforward to check that the "measures" we defined on cylinder sets in earlier chapters satisfy the hypotheses of the above theorem. Hence, these can be extended to the measures used for the interpretation of the logics PCTL, CSL and PTCTL.

APPENDIX B

Iterative Solution Methods

As we saw in Chapters 2 and 4, model checking for several operators of PCTL and CSL requires the solution of a system of linear equations. We can assume that this system is of the traditional form $\mathbf{A} \cdot \underline{x} = \underline{b}$, where $\mathbf{A}$ is a real-valued matrix, $\underline{b}$ is a real-valued vector, and $\underline{x}$ is a vector containing the solution to be determined. We assume also that matrices and vectors are indexed over a state space $S = \{0, 1, \ldots, |S| - 1\}$, i.e. matrices are of size $|S| \times |S|$ and vectors are of size $|S|$. A linear equation system of the form $\underline{x} \cdot \mathbf{A} = \underline{b}$ can be solved by transposing it to $\mathbf{A}^T \cdot \underline{x}^T = \underline{b}^T$.

Solving a system of linear equations is, of course, a well known and much studied problem. Typically, methods for their solution fall into two distinct classes: *direct* methods and *iterative* methods. Direct methods compute the exact solution (within the bounds of numerical error) in a fixed number of steps. Examples include Gaussian elimination and L/U decomposition. Iterative methods compute successive approximations to the solution, terminating when the sequence of solutions has converged to within some pre specified accuracy. Examples include the Jacobi and Gauss–Seidel methods.

As with many applications, probabilistic model checking typically produces systems where $\mathbf{A}$ is large and sparse. Direct methods are not well suited to solving such systems because of a phenomenon known as *fill-in*. Modifications made to $\mathbf{A}$ as the computation progresses usually increase the number of non-zero elements. For large systems, the resulting growth in required storage space can make the computation infeasible. Fortunately, this is not the case for iterative methods. With the exception of some possible initialisation steps, the matrix is not modified at all during computation, and hence there is no increase in memory consumption.

In this appendix, we describe four of the most common iterative methods for solving systems of linear equations: Jacobi, Gauss Seidel, JOR and SOR. The general procedure for all four methods is as follows. Starting with some initial estimate, each iteration produces an increasingly accurate approximation to the solution of the linear equation system. The approximation computed in the kth iteration is denoted $\underline{x}^{(k)}$. We refer to this as the *solution vector* or *iteration vector*. Using the same notation, we denote the initial estimate as $\underline{x}^{(0)}$.

Each estimate $\underline{x}^{(k)}$ is computed from the previous one $\underline{x}^{(k-1)}$. The iterative process is stopped when the solution vector is judged to have converged sufficiently. The criteria for convergence can vary. One approach is to stop when the maximum difference between elements of consecutive vectors is less than some threshold ε:

$$\max_i |\underline{x}^{(k)}(i) - \underline{x}^{(k-1)}(i)| < \varepsilon.$$

One potential problem with this, though, is that if the solution vector contains very small values (i.e. less than ε), the process may be terminated prematurely. A better approach is to check the *relative* difference between elements:

$$\max_i \left(\frac{|\underline{x}^{(k)}(i) - \underline{x}^{(k-1)}(i)|}{|\underline{x}^{(k)}(i)|} \right) < \varepsilon$$

B.1. The Jacobi Method

The Jacobi method is based on the observation that the ith equation of the linear equation system $\mathbf{A} \cdot \underline{x} = \underline{b}$:

$$\sum_{j=0}^{|S|-1} \mathbf{A}(i,j) \cdot \underline{x}(j) = \underline{b}(i)$$

can be rearranged as:

$$\underline{x}(i) = \left(\underline{b}(i) - \sum_{j \neq i} \mathbf{A}(i,j) \cdot \underline{x}(j) \right) \Big/ \mathbf{A}(i,i)$$

On the basis of this, in the Jacobi method, the ith element of the kth iteration vector is computed from the elements of the $(k-1)$th vector as:

$$\underline{x}^{(k)}(i) := \left(\underline{b}(i) - \sum_{j \neq i} \mathbf{A}(i,j) \cdot \underline{x}^{(k-1)}(j) \right) \Big/ \mathbf{A}(i,i).$$

Note that for the systems which arise from PCTL or CSL model checking, the diagonal elements $\mathbf{A}(i,i)$ will always be non-zero. For our purposes, it is also useful to express a single iteration of the Jacobi method in terms of matrices and vectors, rather than individual elements:

$$\underline{x}^{(k)} := \mathbf{D}^{-1} \cdot \left((\mathbf{L} + \mathbf{U}) \cdot \underline{x}^{(k-1)} + \underline{b} \right)$$

where $\mathbf{D} - (\mathbf{L} + \mathbf{U})$ is a partitioning of the matrix $\mathbf{A}$ into its diagonal, lower-triangular, and upper-triangular elements respectively, i.e. $\mathbf{D}$ contains all the diagonal entries of $\mathbf{A}$ and $\mathbf{L} + \mathbf{U}$ contains all the non-diagonal entries, negated. In this setting, the main operation required to perform a single iteration is a matrix-vector multiplication. Note also that the Jacobi method requires two vectors to be stored: one for the previous iteration's vector, and one for the current one.

B.2. The Gauss–Seidel Method

The Jacobi method can be improved by observing that the kth update to the ith vector element can actually use the new value $\underline{x}^{(k)}(j)$ rather than the old value $\underline{x}^{(k-1)}(j)$ for $j < i$. This gives rise to the Gauss–Seidel method:

$$\underline{x}^{(k)}(i) := \left(\underline{b}(i) - \sum_{j<i} \mathbf{A}(i,j) \cdot \underline{x}^{(k)}(j) - \sum_{j>i} \mathbf{A}(i,j) \cdot \underline{x}^{(k-1)}(j) \right) \Big/ \mathbf{A}(i,i).$$

This usually converges much faster than the Jacobi method. Furthermore, it only requires a single vector to be stored: after each new vector entry is computed, its old value is no longer required and can be overwritten.

As with the Jacobi method, we can show that one iteration of the Gauss–Seidel method can be performed using matrix-vector multiplication:

$$\underline{x}^{(k)} := (\mathbf{D} - \mathbf{L})^{-1} \cdot (\mathbf{U} \cdot \underline{x}^{(k-1)} + \underline{b})$$

where $\mathbf{D}$, $\mathbf{L}$ and $\mathbf{U}$ are as for the Jacobi method above. In this case, though, it may not be a sensible option since the computation of the inverse of $(\mathbf{D} - \mathbf{L})$ may require significant extra work or result in fill-in. Furthermore, with this formulation, we again need to store two vectors rather than one.

B.3. Over-Relaxation Methods

Both the Jacobi and Gauss–Seidel methods can be improved with a technique called *over-relaxation.* In an iteration, the new value of each vector element is first computed as described above, and then subjected to a weighted average between itself and the corresponding vector element from the previous iteration. The weights for this average are determined according to a parameter ω. For example, using the Jacobi method as a basis, we get:

$$\underline{x}^{(k)}(i) := (1-\omega)\cdot\underline{x}^{(k-1)}(i) + \omega\cdot\left(\underline{b}(i) - \sum_{j\neq i}\mathbf{A}(i,j)\cdot\underline{x}^{(k-1)}(j)\right)\Big/\mathbf{A}(i,i).$$

This is known as the JOR (Jacobi Over-Relaxation) method. The same technique can be applied to the Gauss–Seidel method, in which case it is referred to as the SOR (Successive Over-Relaxation) method. The idea is that the methods converge faster for a well chosen value of ω. It is known that JOR and SOR will only converge for $\omega \in (0,2)$. Unfortunately, it is usually either impossible or impractical to compute the optimum value of ω. Typically, one must rely on heuristics to select a good value. The interested reader is referred to, for example [**122**], for more details. Note also that, technically, for $\omega < 1$, this technique should be referred to as *under-relaxation*, but for convenience it is often still known as over-relaxation.

Bibliography

1. S. Akers, *Binary decision diagrams*, IEEE Trans. Comput. **C-27** (1978), no. 6, 509–516.
2. R. Alur, C. Courcoubetis, and D. Dill, *Model-checking for probabilistic real-time systems*, Proc. International Colloquium on Automata, Languages and Programming (ICALP'91) (M. Richter, C. Smith, R Wiehagen, and T. Zeugmann, eds.), Lecture Notes in Comput. Sci., vol. 510, Springer, Berlin, 1991, pp. 115–136.
3. ———, *Verifying automata specifications of probabilistic real-time systems*, REX Workshop, Real-Time: Theory in Practice (J. de Bakker, C. Huizing, W. de Roever, and G. Rozenberg, eds.), Lecture Notes in Comput. Sci., vol. 600, Springer, Berlin, 1992, pp. 28–44.
4. ———, *Model-checking in dense real-time*, Inform. and Comput. **104** (1993), no. 1, 2–34.
5. R. Alur and D. Dill, *A theory of timed automata*, Theoret. Comput. Sci. **126** (1994), no. 2, 183–235.
6. E. Asarin, O. Maler, and A. Pnueli, *On discretization of delays in timed automata and digital circuits*, CONCUR'98 (R. de Simone and D. Sangiorgi, eds.), Lecture Notes in Comput. Sci., vol. 1466, Springer, Berlin, 1998, pp. 470–484.
7. J. Aspnes and M. Herlihy, *Fast randomized consensus using shared memory*, J. Algorithms **15** (1990), no. 1, 441–460.
8. A. Aziz, K. Sanwal, V. Singhal, and R. Brayton, *Verifying continuous time Markov chains*, 8th Internat. Conf. on Computer Aided Verification (CAV'96) (R. Alur and T. Henzinger, eds.), Lecture Notes in Comput. Sci., vol. 1102, Springer, Berlin, 1996, pp. 269–276.
9. A. Aziz, V. Singhal, F. Balarin, R. Brayton, and A. Sangiovanni–Vincentelli, *It usually works: The temporal logic of stochastic systems*, 7th Internat. Conf. on Computer Aided Verification (CAV'95) (P. Wolper, ed.), Lecture Notes in Comput. Sci., vol. 939, Springer, Berlin, 1995, pp. 155–165.
10. I. Bahar, E. Frohm, C. Gaona, G. Hachtel, E.Macii, A. Pardo, and F. Somenzi, *Algebraic decision diagrams and their applications*, IEEE/ACM Internat. Conf. on CAD, IEEE Comput. Soc. Press, Santa Clara, CA, 1993, pp. 188–191.
11. C. Baier, *On algorithmic verification methods for probabilistic systems*, Habilitation Thesis, Fakultät für Mathematik & Informatik, Universität Mannheim, 1998.
12. C. Baier and E. Clarke, *The algebraic mu-calculus and* MTBDD*s*, 5th Workshop on Logic, Language, Inform. and Comput. (WOLLIC'98) (R. de Queiroz and M. Finger, eds.), 1998, pp. 27–38.
13. C. Baier, E. Clarke, V. Hartonas-Garmhausen, M. Kwiatkowska, and M. Ryan, *Symbolic model checking for probabilistic processes*, 24th International Colloquium on Automata, Languages and Programming (ICALP'97) (P. Degano, R. Gorrieri, and A. Marchetti-Spaccamela, eds.), Lecture Notes in Comput. Sci., vol. 1256, Springer, Berlin, 1997, pp. 430–440.
14. C. Baier, B. Haverkort, H. Hermanns, and J.-P. Katoen, *Model checking continuous-time Markov chains by transient analysis*, 12th Internat. Conf. on Computer Aided Verification (CAV'00) (A. Emerson and A. Sistla, eds.), Lecture Notes in Comput. Sci., vol. 1855, Springer, Berlin, 2000, pp. 358–372.
15. ———, *On the logical characterisation of performability properties*, 27th International Colloquium on Automata, Languages and Programming (ICALP'00) (U. Montanari, J. Rolim, and E. Welzl, eds.), Lecture Notes in Comput. Sci., vol. 1853, Springer, Berlin, 2000, pp. 780–792.
16. C. Baier, J.-P. Katoen, and H. Hermanns, *Approximate symbolic model checking of continuous-time Markov chains*, 10th Internat. Conf. on Concurrency Theory (CONCUR'99) (J. Baeten and S. Mauw, eds.), Lecture Notes in Comput. Sci., vol. 1664, Springer, Berlin, 1999, pp. 146–161.

17. C. Baier and M. Kwiatkowska, *Model checking for a probabilistic branching time logic with fairness*, Distributed Computing **11** (1998), 125–155.
18. D. Beauquier, *On probabilistic timed automata*, Theoret. Comput. Sci. **292** (2003), no. 1, 65–84.
19. J. Beauquier, M. Gradinariu, and C. Johnen, *Memeory space requirements for self-stabilizing leader election protocols*, Proc. ACM Symposium on Principles of Distributed Computing (PODC'99), ACM Press, New York, 1990, pp. 199–208.
20. G. Behrmann, A. Fehnker, T. Hune, K. Larsen, P. Pettersson, and J. Romijn, *Efficient guiding towards cost-optimality in* UPPAAL, Proc. Tools and Algorithms for the Construction and Analysis of Systems (TACAS 2001) (T. Margaria and W. Yi, eds.), Lecture Notes in Comput. Sci., vol. 2031, Springer, Berlin, 2001, pp. 174–188.
21. L. Benini, A. Bogliolo, G. Paleologo, and G. De Micheli, *Policy optimization for dynamic power management*, IEEE Trans. Computer-Aided Design of Integrated Circuits and Systems **8** (2000), no. 3, 299–316.
22. D. Bertsekas and J. Tsitsiklis, *An analysis of stochastic shortest path problems*, Math. Oper. Res. **16** (1991), no. 3, 580–595.
23. D. Beyer, *Improvements in* BDD*-based reachability analysis of timed automata*, Proc. Symp. Formal Methods Europe (FME'01) (J. Oliveira and P. Zave, eds.), Lecture Notes in Comput. Sci., vol. 2021, Springer, Berlin, 2001, pp. 318–343.
24. P. Billingsley, *Probability and measure*, 3rd ed., John Wiley and Sons, New York, 1995.
25. A. Bianco and L. de Alfaro, *Model checking of probabilistic and nondeterministic systems*, Foundations of Software Technology and Theoretical Computer Science (P. Thiagarajan, ed.), Lecture Notes in Comput. Sci., vol. 1026, Springer, Berlin, 1995, pp. 499–513.
26. H. Bohnenkamp, P. van der Stok, H. Hermanns, and F. Vaandrager, *Cost-optimisation of the* IPv4 *zeroconf protocol*, Proc. Performance and Dependability Symposium (IPDS 2003), IEEE Press, Washington, 2003, pp. 531–540.
27. B. Bollig and I. Wegner, *Improving the variable ordering of* OBDDs *is* NP*-complete*, IEEE Trans. Comput. **45** (1996), no. 9, 993–1006.
28. D. Bosnacki, *Digitization of timed automata*, 4th International Workshop on Formal Methods for Industrial Critical Systems FMICS '99 (Italy, 1999), 1999, pp. 283-302.
29. A. Bouajjani, S. Tripakis, and S. Yovine, *On-the-fly symbolic model checking for real-time systems*, Proc. IEEE Real-Time Systems Symposium (RTSS'97), IEEE Press, Washington, 1997, pp. 25–35.
30. K. Brace, R. Rudell, and R. Bryant, *Efficient implementation of a* BDD *package*, 27th Design Automation Conference (DAC'90), IEEE Comput. Society Press, Los Alamitos, CA, 1990, pp. 40–45.
31. R. Bryant, *Graph-based algorithms for Boolean function manipulation*, IEEE Trans. Comput. **C-35** (1986), no. 8, 677–691.
32. P. Buchholz, G. Ciardo, S. Donatelli, and P. Kemper, *Complexity of Kronecker operations on sparse matrices with applications to the solution of Markov models*, ICASE Report 97–66; Inst. for Comput. Appl. in Sci. and Engr. (Hampton, 1997) (submitted for publication).
33. P. Buchholz, J.-P. Katoen, P. Kemper, and C. Tepper, *Model-checking large structured Markov chains*, J. Logic and Algebraic Programming **56** (2003), no. 1–2, 69–97.
34. J. Burch, E. Clarke, K. McMillan, D. Dill, and J. Hwang, *Symbolic model checking*: 10^{20} *states and beyond*, 5th Annual IEEE Symposium on Logic in Comput. Sci. (LICS'90), IEEE Comput. Society Press, Los Alamitos, CA, 1990, pp. 428–439.
35. G. Ciardo and A. Miner, *A data structure for the efficient Kronecker solution of GSPNs*, Proc. 8th International Workshop on Petri Nets and Performance Models (PNPM'99) (P. Buchholz and M. Silva, eds.), IEEE Press, Washington, 1999, pp. 22–31.
36. G. Ciardo and M. Tilgner, *On the use of Kronecker operators for the solution of generalized stocastic Petri nets*, ICASE Report 96-35, Institute for Comput. Appl. Sci. and Engr., 1996.
37. S. Cheshire, B. Adoba, and E. Guttman, *Dynamic configuration of* IPv4 *link-local addresses*, IETF Draft, 2001;
`www.ietf.org/internet-drafts/draft-ietf-zeroconf-ipv4-linklocal-07.txt`, 2002.
38. E. Clarke, E. Emerson, and A. Sistla, *Automatic verification of finite-state concurrent systems using temporal logics*, ACM Trans. Programming Languages and Systems **8** (1986), no. 2, 244–263.

39. E. Clarke, M. Fujita, P. McGeer, K. McMillan, J. Yang, and X. Zhao, *Multi-terminal binary decision diagrams: An efficient data structure for matrix representation*, Formal Methods in System Design: An International J. **10** (1997), nos. 2-3, 149–169.
40. E. Clarke, K. McMillan, X. Zhao, M. Fujita, and J. Yang, *Spectral transforms for large Boolean functions with applications to technology mapping*, Formal Methods in System Design **10** (1997), nos. 2-3, 54–60.
41. C. Courcoubetis and M. Yannakakis, *Verifying temporal properties of finite state probabilistic programs*, 29th Annual Symposium of Foundations of Computer Science (FOCS'88), IEEE Press, Washington, 1988, pp. 338–345.
42. ______, *Markov decision processes and regular events*, 17th International Colloquium on Automata, Languages and Programming (ICALP'90) (M. Paterson, ed.), Lecture Notes in Comput. Sci., vol. 443, Springer, Berlin, 1990, pp. 336–349.
43. ______, *The complexity of probabilistic verification*, J. ACM **42** (1995), no. 4, 857–907.
44. L. de Alfaro, *Formal verification of probabilistic systems*, Ph.D. Thesis, Stanford University, 1997.
45. ______, *Computing minimum and maximum reachability times in probabilistic systems*, 10th Internat. Conf. on Concurrency Theory (CONCUR'99) (J. Baeten and S. Mauw, eds.), Lecture Notes in Comput. Sci., vol. 1664, Springer, Berlin, 1999, pp. 66–81.
46. P. D'Argenio, B. Jeannet, H. Jensen, and K. Larsen, *Reachability analysis of probabilistic systems by successive refinements*, 1st Joint International Workshop on Process Algebra and Probabilistic Methods, Performance Modeling and Verification (PAPM/PROBMIV'01) (L. de Alfaro and S. Gilmore, eds.), Lecture Notes in Comput. Sci., vol. 2165, Springer, Berlin, 2001, pp. 39–56.
47. C. Daws, M. Kwiatkowska, and G. Norman, *Automatic verification of the* IEEE *1394 root contention protocol with* KRONOS *and* PRISM, 7th International Workshop on Formal Methods for Industrial Critical Systems (FMICS'02), Electronic Notes in Theoret. Comput. Sci. (R. Cleaveland and H. Garavel, eds.), vol. 66, no. 2, Elsevier, Amsterdam, 2002.
48. C. Daws, A. Olivero, S. Tripakis, and S. Yovine, *The tool* KRONOS, Hybrid Systems III (R. Alur, T. Henzinger, and E. Sontag, eds.), Lecture Notes in Comput. Sci., vol. 1066, Springer, Berlin, 1996, pp. 208–219.
49. C. Daws and S. Yovine, *Two examples of verification of multirate timed automata with* KRONOS, Proc. IEEE Real-Time Systems Symposium (RTSS'95), IEEE Press, Washington, 1995, pp. 66–75.
50. L. de Alfaro, T. Henzinger, and R. Jhala, *Compositional methods for probabilistic systems*, 12th Internat. Conf. on Concurrency Theory (CONCUR'01) (K. Larsen and M. Nielsen, eds.), Lecture Notes in Comput. Sci., vol. 2154, Springer, Berlin, 2001, pp. 351–365.
51. L. de Alfaro, M. Kwiatkowska, G. Norman, D. Parker, and R. Segala, *Symbolic model checking of concurrent probabilistic processes using* MTBDD*s and the Kronecker representation*, 6th Internat. Conf. on Tools and Algorithms for the Construction and Analysis of Systems (TACAS'00) (S. Graf and M. Schwartzbach, eds.), Lecture Notes in Comput. Sci., vol. 1785, Springer, Berlin, 2000, pp. 395–410.
52. J. Desharnais, A. Edalat, and P. Panangaden, *Bisimulation for labelled Markov processes*, Inform. and Comput. **179** (2002), no. 2, 163–193.
53. A. Dsouza and B. Bloom, *Generating BDDs for symbolic model checking in* CCS, 7th Internat. Conf. on Computer Aided Verification (CAV'95) (P. Wolper, ed.), Lecture Notes in Comput. Sci., vol. 939, Springer, Berlin, 1995, pp. 16–30.
54. R. Enders, T. Filkorn, and D. Taubner, *Generating* BDD*s for symbolic model checking in* CCS, 3rd International Workshop on Computer Aided Verification (CAV'91) (K. Larsen and A. Skou, eds.), Lecture Notes in Comput. Sci., vol. 575, Springer, Berlin, 1991, pp. 203–213.
55. Z. Feng and E. Hansen, *Symbolic heuristic search for factored Markov decision processes*, 18th National Conference on Artificial Intelligence (AAAI'02), AAAI Press, California, 2002.
56. M. Fischer, N. Lynch, and M. Paterson, *Impossibility of distributed commit with one faulty process*, J. ACM **32** (1985), no. 5, 374–382.
57. B. Fox and P. Glynn, *Computing Poisson probabilities*, Comm. ACM **31** (1988), no. 4, 440–445.
58. R. German, *Performance analysis of communication systems: Modeling with non-markovian stochastic petri nets*, John Wiley and Sons, 2000.

59. A. Gollu, A. Puri, and P. Varaiya, *Discretization of timed automata*, IEEE Conference on decision and control, IEEE Press, Washington, 1994, pp. 957–958.
60. H. Gregersen and H. Jensen, *Formal design of reliable real time systems*, Masters Thesis, Aalborg University, 1995.
61. G. Hachtel, E. Macii, A. Pardo, and F. Somenzi, *Probabilistic analysis of large finite state machines*, 31st Design Automation Conference (DAC'94), ACM Press, New York, 1994, pp. 270–275.
62. H. Hansson and B. Jonsson, *A logic for reasoning about time and reliability*, Formal Aspects of Computing **6** (1994), no. 4, 512–535.
63. V. Hartonas-Garmhausen, *Probabilistic symbolic model checking with engineering models and applications*, Ph.D. Thesis, Carnegie Mellon University, 1998.
64. B. Haverkort, L. Cloth, H. Hermanns, J.-P. Katoen, and C. Baier, *Model checking performability properties*, Internat. Conf. on Dependable Systems and Networks (DSN'02), IEEE Computer Society, Washington, 2002, pp. 103–112.
65. T. Henzinger, *The temporal specification and verification of real-time systems*, Ph.D. Thesis, Stanford University, 1991.
66. T. Henzinger, Z. Manna, and A. Puneli, *What good are digital clocks?*, Proc. Automata, Languages and Programming (ICALP'92) (W. Kuich, ed.), Lecture Notes in Comput. Sci., vol. 623, Springer, Berlin, 1992, pp. 545–558.
67. T. Henzinger, X. Nicollin, J. Sifakis, and S. Yovine, *Symbolic model checking for real-time systems*, Inform. and Comput. **111** (1994), no. 2, 193–244.
68. H. Hermanns, J.-P. Katoen, J. Meyer-Kayser, and M. Siegle, *A Markov chain model checker*, 6th Internat. Conf. on Tools and Algorithms for the Construction and Analysis of Systems (TACAS'00) (S. Graf and M. Schwartzbach, eds.), Lecture Notes in Comput. Sci., vol. 1785, Springer, Berlin, 2000, pp. 347–362.
69. H. Hermanns, M. Kwiatkowska, G. Norman, D. Parker, and M. Siegle, *On the use of* MTBDD*s for performability analysis and verification of stochastic systems*, J. of Logic and Algebraic Programming: Special Issue on Probabilistic Techniques for the Design and Analysis of Systems **56** (2003), no. 1-2, 23–67.
70. H. Hermanns, J. Meyer-Kayser, and M. Siegle, *Multi terminal binary decision diagrams to represent and analyse continuous time Markov chains*, 3rd International Workshop on Numerical Solution of Markov Chains (NSMC'99) (B. Plateau, W. Stewart, and M. Silva, eds.), Prensas Universitarias de Zaragoza, Prensas Universitarias de Zaragoza, Zaragoza, 1999, pp. 188–207.
71. J. Hillston, *A compositional approach to performance modelling*, Ph.D. Thesis, University of Edinburgh, 1994.
72. J. Hoey, R. St-Aubin, A. Hu, and C. Boutilier, SPUDD: *Stochastic planning using decision diagrams*, 15th Annual Conference on Uncertainty in Artificial Intelligence (UAI'99) (K. Laskey and H. Pade, eds.), Morgan Kaufmann, San Francisco, pp. 279–288.
73. Michael Huth and Marta Kwiatkowska, *Quantitative analysis and model checking*, LICS'97, IEEE Press, Washington, 1997, pp. 111–122.
74. IEEE *standard for a high performance serial bus*, IEEE Computer Soc., Standard 1394–1995, 1996.
75. A. Israeli and M. Jalfon, *Token management schemes and random walks yield self-stabilizing mutual exclusion*, Proc. ACM Symposium on Principles of Distributed Computing (PODC'90), ACM Press, New York, 1990, pp. 119–131.
76. A. Itai and M. Rodeh, *Symmetry breaking in distributed networks*, Inform. and Comput. **88** (1990), no. 1, 60–87.
77. B. Jeannet, P. D'Argenio, and K. Larsen, *Rapture: A tool for verifying Markov decision processes*, Tools Day, 13th Internat. Conf. on Concurrency Theory (CONCUR'02), 2002.
78. J.-P. Katoen, M. Kwiatkowska, G. Norman, and D. Parker, *Faster and symbolic* CTMC *model checking*, 1st Joint International Workshop on Process Algebra and Probabilistic Methods, Performance Modeling and Verification (PAPM/PROBMIV'01) (L. de Alfaro and S. Gilmore, eds.), Lecture Notes in Comput. Sci., vol. 2165, Springer, Berlin, 2001, pp. 23–38.
79. J. Kemeny, J. Snell, and A. Knapp, *Denumerable Markov chains*, 2nd ed., Springer-Verlag, PUBL ADDRESS, 1976.
80. P. Kemper, *Numerical analysis of superposed* GSPN*s*, IEEE Trans. on Software Engr. **22** (1996), no. 9, 615–628.

81. M. Kuntz and M. Siegle, *Deriving symbolic representations from stochastic process algebras*, 2nd Joint International Workshop on Process Algebra and Probabilistic Methods, Performance Modeling and Verification (PAPM/PROBMIV'02) (H. Hermanns and R. Segala, eds.), Lecture Notes in Comput. Sci., vol. 2399, Springer, Berlin, 2002, pp. 188–206.
82. M. Kwiatkowska and R. Mehmood, *Out-of-core solution of large linear systems of equations arising from stochastic modelling*, 2nd Joint International Workshop on Process Algebra and Probabilistic Methods, Performance Modeling and Verification (PAPM/PROBMIV'02) (H. Hermanns and R. Segala, eds.), Lecture Notes in Comput. Sci., vol. 2399, Springer, Berlin, 2002, pp. 135–151.
83. M. Kwiatkowska, R. Mehmood, G. Norman, and D. Parker, *A symbolic out-of-core solution method for Markov models*, Proc. Workshop on Parallel and Distributed Model Checking (PDMC'02), Electronic Notes in Theoretical Computer Science, vol. 68.4, Elsevier, Amsterdam, 2002.
84. M. Kwiatkowska and G. Norman, *Verifying randomized byzantine agreement*, Proc. Formal Techniques for Networked and Distributed Systems (FORTE'02) (D. Peled and M. Vardi, eds.), Lecture Notes in Comput. Sci., vol. 2529, Springer, Berlin, 2002, pp. 194–209.
85. M. Kwiatkowska, G. Norman, and A. Pacheco, *Model checking* CSL *until formulae with random time bounds*, 2nd Joint International Workshop on Process Algebra and Probabilistic Methods, Performance Modeling and Verification (PAPM/PROBMIV'02) (H. Hermanns and R. Segala, eds.), Lecture Notes in Comput. Sci., vol. 2399, Springer, Berlin, 2002, pp. 152–168.
86. ———, *Model checking expected time and expected reward formulae with random time bounds*, Proc. 2nd Euro-Japanese Workshop on Stochastic Risk Modelling for Finance, Insurance, Production and Reliability (T. Dohi, N. Limnios, and S. Osak, eds.), 2002, pp. 282–291.
87. M. Kwiatkowska, G. Norman, and D. Parker, PRISM: *Probabilistic symbolic model checker*, Proc. TOOLS'02, Lecture Notes in Comput. Sci., vol. 2324, Springer, Berlin, 2002, pp. 200–204.
88. ———, *Probabilistic symbolic model checking with* PRISM: *A hybrid approach*, 8th Internat. Conf. on Tools and Algorithms for the Construction and Analysis of Systems (TACAS'02) (J.-P. Katoen and P. Stevens, eds.), Lecture Notes in Comput. Sci., vol. 2280, Springer, Berlin, 2002, pp. 52–66.
89. M. Kwiatkowska, G. Norman, D. Parker, and R. Segala, *Symbolic model checking of concurrent probabilistic systems using* MTBDD*s and simplex*, Tech. Report CSR-99-1, University of Birmingham, 1999.
90. M. Kwiatkowska, G. Norman, and R. Segala, *Automated verification of a randomized distributed consensus protocol using Cadence* SMV *and* PRISM, 13th Internat. Conf. on Computer Aided Verification (CAV'01) (G. Berry, H. Comon, and A. Finkel, eds.), Lecture Notes in Comput. Sci., vol. 2102, Springer, Berlin, 2001, pp. 194–206.
91. M. Kwiatkowska, G. Norman, and J. Sproston, *Symbolic computation of maximal probabilistic reachability*, 13th Internat. Conf. on Concurrency Theory (CONCUR'01) (K. Larsen and M. Nielsen, eds.), Lecture Notes in Comput. Sci., vol. 2154, Springer, Berlin, 2001, pp. 169–183.
92. ———, *Probabilistic model checking of the IEEE 802.11 wireless local area network protocol*, Proc. Joint International Workshop on Process Algebra and Probabilistic Methods, Performance Modeling and Verification (PAPM/PROBMIV'02) (H. Hermanns and R. Segala, eds.), Lecture Notes in Comput. Sci., vol. 2399, Springer, Berlin, 2002, pp. 169–187.
93. ———, *Performance analysis of probabilistic timed automata using digital clocks*, Proc. Formal Modeling and Analysis of Timed Systems (FORMATS 2003) (K. Larsen and P. Niebert, eds.), Lecture Notes in Comput. Sci., Springer, Berlin, 2003 (to appear).
94. ———, *Probabilistic model checking of deadline properties in the* IEEE 1394 *FireWire root contention protocol*, Special Issue of Formal Aspects of Computing **14** (2003), 295–318.
95. ———, *Symbolic computation of minimal probabilistic reachability*, Tech. Report CSR-03-3, University of Birmingham, 2003.
96. M. Kwiatkowska, G. Norman, R. Segala, and J. Sproston, *Automatic verification of real-time systems with discrete probability distributions*, ARTS'99, (J.-P. Katoen, ed.), Lecture Notes in Comput. Sci., vol. 1601, Springer, Berlin, 1999.

97. ———, *Verifying quantitative properties of continuous probabilistic timed automata*, CONCUR 2000–Concurrency Theory (C. Palamidessi, ed.), Lecture Notes in Comput. Sci., vol. 1877, Springer, Berlin, 2000, pp. 123–137.
98. ———, *Automatic verification of real-time systems with discrete probability distributions*, Theoret. Comput. Sci. **282** (2002), 101–150.
99. K. Larsen and A. Skou, *Bisimulation through probabilistic testing*, Inform. and Comput. **94** (1991), 1–28.
100. C. Lee, *Representation of switching circuits by binary-decision programs*, Bell System Technical J. **38** (1959), 985–999.
101. N. Lynch, R. Segala, and F. Vaandrager, *Compositionality for probabilistic automata*, Proc. CONCUR'03 (R. Amadio and D. Lugiez, eds.), Lecture Notes in Comput. Sci., vol. 2761, Springer, Berlin, 2003, pp. 57–71.
102. A. McIver and C. Morgan, *Games, probability and the quantitative mu-calculus qmu*, Logic for Programming, Artificial Intelligence and Reasoning (M. Baaz and A. Voronkov, eds.), vol. 2514, Springer, Berlin, 2002, pp. 292–310.
103. K. McMillan, *Symbolic model checking*, Kluwer Academic Publishers, 1993.
104. G. Norman, D. Parker, M. Kwiatkowska, S. Shukla, and R. Gupta, *Formal analysis and validation of continuous time Markov chain based system level power management strategies*, 7th Annual IEEE International Workshop on High Level Design Validation and Test (HLDVT'02) (W. Rosenstiel, ed.), IEEE Computer Soc., Washington, 2002, pp. 45–50.
105. ———, *Using probabilistic model checking for dynamic power management*, Proc. 3rd Workshop on Automated Verification of Critical Systems (AVoCS'03) (M. Leuschel, S. Gruner, and S. Lo Presti, eds.), Technical Report DSSE-TR-2003-2, University of Southampton, 2003, pp. 202–215.
106. J. Ouaknine and J. Worrell, *Revisiting digitization, robustness, and decidability for timed automata*, Proc of LICS 03 (Canada, 2003), IEEE Computer Soc. Press, Washington, 2003.
107. ———, *Universality and language inclusion for open and closed timed automata*, Proc. HSCC '03: Hybrid Systems: Computation and Control (O. Maler and A. Pnueli, eds.), Lecture Notes in Comput. Sci., vol. 2623, Springer, Berlin, 2003, pp. 375–388.
108. G. Paleologo, L. Benini, A. Bogliolo, and G. Micheli, *Policy optimization for dynamic power management*, Design Automation Conference, IEEE Computer Sci., Washington, 1988, pp. 182–187.
109. D. Parker, *Implementation of symbolic model checking for probabilistic systems*, Ph.D. Thesis, University of Birmingham, 2002.
110. B. Plateau, *On the stochastic structure of parallelism and synchronisation models for distributed algorithms*, 1985 ACM SIGMETRICS Conference on Measurement and Modeling of Computer Systems, Performance Evaluation Review, vol. 13, no. 2, ACM Press, New York, 1985, pp. 147–153.
111. PRISM *website*, http://www.cs.bham.ac.uk/~dxp/prism/.
112. M. Puterman, *Markov decision processes: Discrete stochastic dynamic programming*, John Wiley and Sons Publ., New York, 1994.
113. Q. Qiu and M. Pedram, *Dynamic power management based on continuous-time Markov decision processes*, Design Automation Conference, ACM Press, New York, 1999, pp. 555–561.
114. Q. Qiu, Q. Wu, and M. Pedram, *Stochastic modeling of a power-managed system: Construction and optimization*, IEEE Trans. Comput. Aided Design **20** (2001), no. 9, 1200–1217.
115. M. Rabin, *N-process mutual exclusion with bounded waiting by $4\log_2 N$-valued shared variable*, J. Comput. System Sci. **25** (1982), no. 1, 66–75.
116. A. Schrijver, *Theory of linear and integer programming*, John Wiley and Sons Publ., New York, 1986.
117. R. Segala and N. Lynch, *Probabilistic simulations for probabilistic processes*, 5th Internat. Conf. on Concurrency Theory (CONCUR'94) (B. Jonsson and J. Parrow, eds.), Lecture Notes in Comput. Sci., vol. 836, Springer, Berlin, 1994, pp. 481–496.
118. ———, *Probabilistic simulations for probabilistic processes*, Nordic J. Comput. **2** (1995), no. 2, 250–273.
119. M. Siegle, *Behaviour analysis of communication systems: Compositional modelling, compact representation and analysis of performability properties*, Habilitation Thesis, Universität Erlangen-Nürnberg, 2002.

120. D. P. L. Simons and M. I. A. Stoelinga, *Mechanical verification of the* IEEE 1394a *root contention protocol using* UPPAAL 2k, Internat. J. Software Tools for Technology Transfer (STTT) **3** (2001), no. 4, 469–485.
121. F. Somenzi, CUDD: *Colorado University decision diagram package*, Public software, Colorado Univeristy, Boulder; http://vlsi.colorado.edu/~fabio/, 1997.
122. W. Stewart, *Introduction to the numerical solution of Markov chains*, Princeton University Press, NJ, 1994.
123. M. Stoelinga, *Alea jacta est: Verification of probabilistic, real-time and parametric systems*, Ph.D. Thesis, University of Nijmegen, 2002.
124. S. Tani, K. Hamaguchi, and S. Yajima, *The complexity of the optimal variable ordering problems of shared binary decision diagrams*, 4th International Symposium on Algorithms and Computation (ISAAC'93) (K. Ng, P. Raghavan, N. Balasubramanian, and F. Chin, eds.), Lecture Notes in Comput. Sci., vol. 762, Springer, Berlin, 1993, pp. 389–398.
125. R. Tarjan, *Depth-first search and linear graph algorithms*, SIAM J. Comput. **1** (1972), 146–160.
126. S. Tripakis, *The formal analysis of timed systems in practice*, Ph.D. Thesis, Université Joseph Fourier, 1998.
127. M. Vardi, *Automatic verification of probabilistic concurrent finite state programs*, 26th Symposium on Foundations of Computer Science (FOCS'85), IEEE Comp. Soc. Press, Los Alamitos, CA, 1985, pp. 327–338.
128. B. Yang, R. Bryant, D. O'Hallaron, A. Biere, O. Coudert, G. Janssen, R. Ranjan, and F. Somenzi, *A performance study of* BDD-*based model checking*, 2nd Internat. Conf. on Formal Methods in Computer-Aided Design (FMCAD'98) (G. Gopalakrishnan and P. Windley, eds.), Lecture Notes in Comput. Sci., vol. 1522, Springer, Berlin, 1998, pp. 255–289.
129. W. Yi, P. Pettersson, and M. Daniels, *Automatic verification of real-time communicating systems by constraint-solving*, 7th Internat. Conf. on Formal Description Techniques (Switzerland, 1994) (D. Hogrefe and S. Leue, eds.), IFIP Conference Proc., vol. 6, Chapman & Hall, London, 1994, pp. 223–238.